17th International Conference on the Physics of Highly Charged Ions 2014

Journal of Physics: Conference Series Volume 583

San Carlos de Bariloche, Argentina
31 August – 5 September 2014

ISBN: 978-1-63439-914-2
ISSN: 1742-6588

Printed from e-media with permission by:

Curran Associates, Inc.
57 Morehouse Lane
Red Hook, NY 12571

Printed by Curran Associates, Inc. (2015)

Institute of Physics
Dirac House, Temple Back
Bristol BS1 6BE UK

Phone: 44 1 17 929 7481
Fax: 44 1 17 920 0979

techtracking@iop.org

Additional copies of this publication are available from:

Curran Associates, Inc.
57 Morehouse Lane
Red Hook, NY 12571 USA
Phone: 845-758-0400
Fax: 845-758-2634
Email: curran@proceedings.com
Web: www.proceedings.com

TABLE OF CONTENTS

The 17th edition of the International Conference on the Physics of Highly Charged Ions (HCI 2014) was held in San Carlos de Bariloche, in the southern region of Argentina known as Patagonia, from August 31 to September 5, 2014. This meeting corresponds to a series of HCI conferences, which has been held every other year since 1982 in cities in Europe, USA, Japan and China. This was the first time that the conference took place in Latin America. This edition was organized by a Local Committee made up of physicists mainly from the cities of Bariloche and Rosario and also from Buenos Aires and Bahía Blanca, all sites where research on Atomic Collisions is developed. The conference was attended by delegates coming from 18 countries, more that 23% of whom were women.

The field of highly charged ions has seen in recent years a promising evolution originating from bold progress in theory and significant advances in experimental techniques. The HCI conferences aim at bringing together experimentalists and theoreticians from as wide a range of fields as, for instance, Fundamental Aspects, Structure and Spectroscopy, Collisions with Electrons, Ions, Atoms and Molecules, Interaction with Clusters, Surfaces and Solids, Interactions with Photons and Plasmas, Strong Field Processes, and Production, Experimental Developments and Applications.

The Scientific Programme, selected by an International Advisory Board, included 5 Review Lectures, 11 Progress Reports, 1 Local Report and 24 Special Reports. In addition, the results of 132 contributed works were presented as poster communications and a Public Lecture on 'The wonders of the Southern Skies' was delivered by an Argentinean expert. Thus, a wide range of subjects comprising a balanced mix of topics was covered throughout the course of the conference.

The HCI 2014 was a resounding success for the international and local communities, from both the scientific and social aspects, considering that the attendees and accompanying persons had the opportunity to enjoy a comfortable and friendly ambiance as well as the beautiful landscapes of the region including big lakes and high mountains. During the conference vivid and interesting scientific discussions were maintained by the participants in a relaxed environment.

The Local Organizing Committee would like to recognize the generous contributions made by various institutions: International Pure and Applied Physics

(IUPAP), Comisión Nacional de Energía Atómica (CNEA), Consejo Nacional de Investigaciones Científicas y Técnicas (CONICET), Investigación Aplicada S. E. (INVAP), Patagonia Norte Division of CONICET, Instituto Balseiro and NEC Corporation.

The Local Chairs would like to express their gratitude to all the other members of the Local Organizing Committee who collaborated with a great effort and in countless ways to the success of the conference. In particular, we thank Omar Fojón for the outstanding collaboration received, and Juan Manuel Monti, who worked hard for the production of the Book of Abstracts and other related activities. We also would like to thank Renata Della Picca, Juana Gervasoni and Silvina Seguí, for their generous work in printing the book of abstracts, bags, banners, etc. and for their continuous help at the Registration Desk. We extend our gratitude to the younger members of the Local Organizing Committee, Marcos Feole, Francisco Navarrete and Juan Martín Randazzo for their help with the poster logistics; and to Luis Rodríguez for ensuring that all the presentations were ready and on schedule. We thank María Silvia Gravielle, our colleague and liaison in Buenos Aires, for helping delegates to get their visas on time; Sebastián Otranto and Jocelyn Hanssen, for their early work collecting and sorting the talk's proposals and Carlos Stia, for helping with the revision of the style of the manuscripts. We also thank M. Galassi, P Focke and D. Fregenal. We recognize Sergio Suárez, for getting funding from private sources and Roberto Garibotti, for his help with the edition of the book of abstracts. We thank Lynn van Brook for her work as conference secretary, and Silvana Peralta and Natalia Mastrángelo, for the web development and graphic design.

Finally, we express our acknowledgements to the scientists that participated during the refereeing process of the present Proceedings.

Raúl Barrachina

Flavio Colavecchia

Roberto Rivarola

Local Chairs, HCI 2014

December 2014

Journal of Physics: Conference Series **583** (2015) 011002 doi:10.1088/1742-6596/583/1/011002

Local Organizing Committee

Chairs

- R. D. Rivarola (IFIR/Rosario)
- R. Barrachina (CAB/Bariloche)
- F. D. Colavecchia(CAB/Bariloche)

Members

- R. Della Picca (Bariloche)
- O. Fojón (Rosario)
- M. Feole (Bariloche)
- P. Focke (Bariloche)
- D. Fregenal (Bariloche)
- M. Galassi (Rosario)
- C. R. Garibotti (Bariloche)
- J. Gervasoni (Bariloche)
- M. S. Gravielle (Buenos Aires)
- J. Hanssen (Rosario)
- S. López (Buenos Aires)
- J. M. Monti (Rosario)
- F. Navarrete (Bariloche)
- S. Otranto (Bahía Blanca)
- J. M. Randazzo (Bariloche)
- L. Rodriguez (Bariloche)
- S. Seguí (Bariloche)
- S. Suárez (Bariloche)
- C. Stia (Rosario)

Journal of Physics: Conference Series **583** (2015) 011003 doi:10.1088/1742-6596/583/1/011003

International Advisory Board

Journal of Physics: Conference Series **583** (2015) 011004 doi:10.1088/1742-6596/583/1/011004

Peer review statement

All papers published in this volume of *Journal of Physics: Conference Series* have been peer reviewed through processes administered by the proceedings Editors. Reviews were conducted by expert referees to the professional and scientific standards expected of a proceedings journal published by IOP Publishing.

Journal of Physics: Conference Series **583** (2015) 012001 doi:10.1088/1742-6596/583/1/012001

Nuclear recoil correction to the g factor of boron-like argon

A A Shchepetnov[1,2], D A Glazov[1,2,3], A V Volotka[1,3], V M Shabaev[1], I I Tupitsyn[1] and G Plunien[3]

[1] Department of Physics, St. Petersburg State University, Oulianovskaya 1, Petrodvorets, 198504 St. Petersburg, Russia

[2] State Scientific Centre "Institute for Theoretical and Experimental Physics" of National Research Centre "Kurchatov Institute", B. Cheremushkinskaya st. 25, 117218 Moscow, Russia

[3] Institut für Theoretische Physik, Technische Universität Dresden, Mommsenstraße 13, 01062 Dresden, Germany

E-mail: d.glazov@spbu.ru

Abstract. The nuclear recoil effect to the g factor of boron-like ions is investigated. The one-photon-exchange correction to the nuclear recoil effect is calculated in the non-relativistic approximation for the nuclear recoil operator and in the Breit approximation for the interelectronic-interaction operator. The screening potential is employed to estimate the higher-order contributions. The updated g-factor values are presented for the ground $^2P_{1/2}$ and first excited $^2P_{3/2}$ states of B-like argon $^{40}\mathrm{Ar}^{13+}$, which are presently being measured by the ARTEMIS group at GSI.

1. Introduction

During the last 15 years, the g-factor measurements in low-Z ions have reached an accuracy of 10^{-10} [1, 2, 3, 4, 5] and motivated corresponding theoretical investigations. In particular, the most accurate value of the electron mass was obtained in these studies [6]. The case of Li-like silicon manifests presently the most accurate verification of the many-electron QED effects in magnetic field [5, 7]. Experimental and theoretical investigations of the g factor of heavy few-electron ions will provide stringent tests of bound-state QED in strong nuclear field. Moreover, they will serve for an independent determination of the fine structure constant, provided simultaneous investigations of H-like and B-like heavy ions of the same isotope will be performed [8, 9].

First measurement of the g factor of a B-like highly charged ion sensitive to the QED effects was performed in Ref. [10]. The ARTEMIS project presently implemented at GSI will use the laser-microwave double-resonance spectroscopy to measure with ppb accuracy the Zeeman splittings of both ground state and first excited state in B-like argon [11]. Corresponding theoretical predictions for the g factor and the non-linear effects in magnetic field have been reported in Ref. [12]. In this contribution we report on the recent progress for the nuclear recoil effect evaluated with more rigorous consideration of the screening correction. Namely, the contribution of the one-photon-exchange diagrams for the nuclear recoil effect has been calculated in the non-relativistic approximation. Total results for the g factor of $^2P_{1/2}$ and $^2P_{3/2}$

states of B-like argon presented here also include more accurate values of the interelectronic-interaction correction of the order $1/Z^2$ and higher.

The relativistic units ($\hbar = m = c = 1$) and the Heaviside charge unit ($\alpha = e^2/(4\pi), e < 0$) are used in the paper. Electron-to-nucleus mass ratio is written as m/M for clarity.

2. Nuclear recoil effect

The theory of the nuclear recoil effect for the atomic g factor to the leading orders in the parameter αZ was developed in a number of papers, see, e.g. Refs. [13, 14, 15, 16, 17, 18] and references therein. The rigorous QED theory valid to all orders in αZ and to first order in electron-to-nucleus mass ratio m/M was developed in Ref. [19]. In Ref. [20] the corresponding numerical results were presented for 1s state. Since the contributions of the second and higher orders in m/M are negligible at the present level of accuracy, we do not consider them in the present paper.

First, we introduce the one-electron and the many-electron parts of the nuclear recoil correction, and present the former as the sum of the low-order and the higher-order terms:

$$\Delta g_{\rm rec} = \Delta g_{\rm rec}^{\rm 1-el,L} + \Delta g_{\rm rec}^{\rm 1-el,H} + \Delta g_{\rm rec}^{\rm m-el}\,. \tag{1}$$

The lower-order term is given by the expression [19]

$$\begin{aligned}\Delta g_{\rm rec}^{\rm 1-el,L} &= \frac{1}{M_J}\frac{m}{M}\Bigg\{ -\langle a|\Big[\mathbf{r}\times(\mathbf{p}-\mathbf{D}(0))\Big]_z|a\rangle \\ &+ \sum_n^{\varepsilon_n\neq\varepsilon_a} \frac{\langle a|\left[\mathbf{r}\times\boldsymbol{\alpha}\right]_z|n\rangle}{\varepsilon_a-\varepsilon_n}\langle n|\Big[\mathbf{p}^2-\mathbf{p}\cdot\mathbf{D}(0)-\mathbf{D}(0)\cdot\mathbf{p}\Big]|a\rangle\Bigg\}\,. \end{aligned}\tag{2}$$

Here $|a\rangle$ is the one-electron reference state, M_J is the z-projection of the total angular momentum, while the z-axis is directed along the external magnetic field, $\boldsymbol{\alpha}$ stands for the vector of Dirac matrices, $\mathbf{p} = -i\boldsymbol{\nabla}$ is the momentum operator, and

$$\mathbf{D}(\omega) = \alpha Z\left(\boldsymbol{\alpha}\frac{\exp(i\omega r)}{r} + \boldsymbol{\nabla}(\boldsymbol{\alpha}\cdot\boldsymbol{\nabla})\frac{\exp(i\omega r)-1}{\omega^2 r}\right)\,. \tag{3}$$

While $\Delta g_{\rm rec}^{\rm 1-el,L}$ gives the one-electron nuclear recoil correction complete in orders $m/M(\alpha Z)^0$ and $m/M(\alpha Z)^2$, the higher-order term $\Delta g_{\rm rec}^{\rm 1-el,H}$ contains contributions of the order $m/M(\alpha Z)^3$ and higher. Numerical evaluation of this term to all orders in αZ was done in [20] for 1s state only. In this work, we estimate the uncertainty due to unknown value of $\Delta g_{\rm rec}^{\rm 1-el,H}$ as $(\alpha Z)^3\Delta g_{\rm rec}^{\rm 1-el,L}$.

The many-electron part of the nuclear recoil correction for an atom with one electron over closed shells can be found from Eqs. (73) and (92) of Ref. [19] employing the formalism of redefined vacuum [21]. In this way, we derive the following expression:

$$\begin{aligned}\Delta g_{\rm rec}^{\rm m-el} &= \frac{2}{M_J}\frac{m}{M}\sum_c\Bigg\{\Big[\langle a|\mathbf{r}|c\rangle\times\langle c|[\mathbf{p}-\mathbf{D}(\Delta)]|a\rangle\Big]_z \\ &- \sum_n^{\varepsilon_n\neq\varepsilon_a}\frac{\langle a|\left[\mathbf{r}\times\boldsymbol{\alpha}\right]_z|n\rangle}{\varepsilon_a-\varepsilon_n}\langle n|\Big[\mathbf{p}-\mathbf{D}(\Delta)\Big]|c\rangle\cdot\langle c|\Big[\mathbf{p}-\mathbf{D}(\Delta)\Big]|a\rangle \\ &- \sum_n^{\varepsilon_n\neq\varepsilon_c}\frac{\langle c|\left[\mathbf{r}\times\boldsymbol{\alpha}\right]_z|n\rangle}{\varepsilon_c-\varepsilon_n}\langle n|\Big[\mathbf{p}-\mathbf{D}(\Delta)\Big]|a\rangle\cdot\langle a|\Big[\mathbf{p}-\mathbf{D}(\Delta)\Big]|c\rangle \\ &+\langle a|\mathbf{D}'(\Delta)|c\rangle\cdot\langle c|\Big[\mathbf{p}-\mathbf{D}(\Delta)\Big]|a\rangle\Big(\langle a|\left[\mathbf{r}\times\boldsymbol{\alpha}\right]_z|a\rangle-\langle c|\left[\mathbf{r}\times\boldsymbol{\alpha}\right]_z|c\rangle\Big)\Bigg\}\,. \end{aligned}\tag{4}$$

Here the summation over $|c\rangle$ runs over all closed-shell electrons, $\mathbf{D}'(\omega) = \partial\mathbf{D}(\omega)/\partial\omega$, and $\Delta = \varepsilon_a - \varepsilon_c$.

We calculate the contributions $\Delta g_{\rm rec}^{\rm 1-el,L}$ and $\Delta g_{\rm rec}^{\rm m-el}$ according to Eqs. (2) and (4) for $^2P_{1/2}$ and $^2P_{3/2}$ states. The numerical computation is performed employing the standard algebra for angular coefficients and the dual kinetic balance approach [22] to construct the finite basis set of the radial functions. Apart from the Coulomb potential (with account for the finite nuclear size), we use the core-Hartree and Kohn-Sham screening potentials. The explicit expressions for this potentials can be found e.g. in [23], while the examples of their numerical implementations and applications in various atomic structure calculations can be found, e.g. in Refs. [7, 24, 25, 26, 27]. The results for B-like argon are presented in the first and second lines of Table 1.

The next step in our consideration is to take into account the interelectronic-interaction beyond the screening-potential approximation, namely, to calculate the first-order correction to the nuclear recoil effect within the perturbation theory. In the non-relativistic limit Eqs. (2) and (4) yield together the well-known expression [13]

$$\Delta g_{\rm rec}^{\rm non-rel} = \frac{1}{M_J} \langle A|W_{\rm rec}|A\rangle \,, \tag{5}$$

$$W_{\rm rec} = \frac{m}{M} \sum_{j,k} [\mathbf{r}_j \times \mathbf{p}_k]_z \,. \tag{6}$$

Here $|A\rangle$ is the reference-state many-electron wave function in the non-interacting-electrons approximation, i.e. the Slater determinant for $(1s)^2\,(2s)^2\,2p_j$ configuration. We employ the approximation given by Eqs. (5), (6) to evaluate the first-order interelectronic-interaction correction to the nuclear recoil effect. The interaction operator is taken in the Breit approximation,

$$H_{\rm int} = \alpha \sum_{j<k} \left(\frac{1 - \boldsymbol{\alpha}_j \cdot \boldsymbol{\alpha}_k}{r_{jk}} - \frac{1}{2} [(\boldsymbol{\alpha}_j \cdot \boldsymbol{\nabla}_j), [(\boldsymbol{\alpha}_k \cdot \boldsymbol{\nabla}_k), r_{jk}]] \right) . \tag{7}$$

The general expression for this contribution is

$$\Delta g_{\rm rec}^{(1)} = \frac{2}{M_J} \sum_{N}^{E_N \neq E_A} \frac{\langle A|W_{\rm rec}|N\rangle\langle N|H_{\rm int}|A\rangle}{E_A - E_N} \,, \tag{8}$$

where the summation runs over the complete spectrum of the many-electron states $|N\rangle$, constructed as the Slater determinants from the one-electron solutions of the Dirac equation. Substitution of the two-electron operators $W_{\rm rec}$ and $H_{\rm int}$ leads to the rather lengthy formulae, which are not presented here, therefore. The structure of these formulae can be easily understood from the corresponding diagrams depicted in Fig. 1. The first diagram corresponds to the "one-electron" part of $W_{\rm rec}$, i.e. to the case of $j = k$ in Eq. (6). Other diagrams correspond to the "two-electron" part of $W_{\rm rec}$, i.e. to the case of $j \neq k$. We calculate $\Delta g_{\rm rec}^{(1)}$ numerically for the pure Coulomb potential, as well as for the core-Hartree and Kohn-Sham screening potentials. When the screening potential is included in zeroth-order in the Dirac equation, the corresponding counter-term shall be taken into account for the first-order correction $\Delta g_{\rm rec}^{(1)}$. It can be written as the following replacement in Eq. (8),

$$H_{\rm int} \to H_{\rm int} - \sum_j V_{\rm scr}(r_j) \,. \tag{9}$$

The diagrams for the counter-term are shown in Fig. 2.

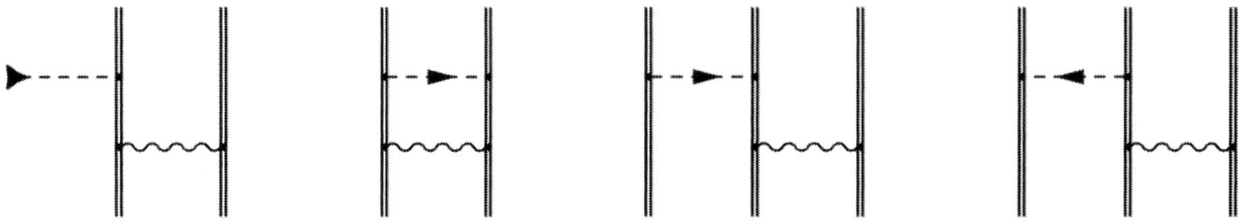

Figure 1. Diagrams corresponding to $\Delta g^{(1)}_{\rm rec}$ contribution to the g factor, given by Eq. (8): the first-order interelectronic-interaction correction to the nuclear recoil effect. The dashed lines with the triangle and with the arrow correspond to the one-electron and many-electron parts of the non-relativistic recoil operator $W_{\rm rec}$ (6).

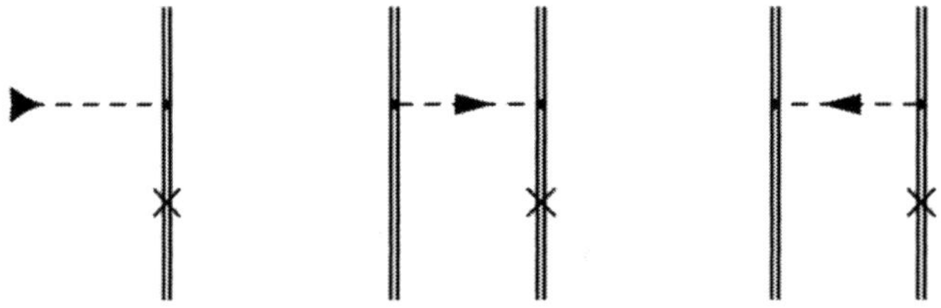

Figure 2. Additional counter-term diagrams for the screening potential.

Table 1. Individual contributions to the nuclear recoil correction to the g factor of B-like argon $^{40}\mathrm{Ar}^{13+}$ ($m/M = 13.7308 \cdot 10^{-6}$) for $^2P_{1/2}$ and $^2P_{3/2}$ states. The units are 10^{-6}.

	$^2P_{1/2}$			$^2P_{3/2}$		
	Coulomb	core-Hartree	Kohn-Sham	Coulomb	core-Hartree	Kohn-Sham
$\Delta g^{\rm 1-el,L}_{\rm rec}$	−18.208	−18.223	−18.221	−9.099	−9.109	−9.108
$\Delta g^{\rm m-el}_{\rm rec}$	7.548	8.218	8.332	3.698	4.064	4.119
$\Delta g^{(1)}_{\rm rec}$	1.434	0.929	0.790	0.704	0.459	0.387
sum	−9.226	−9.076	−9.099	−4.697	−4.586	−4.602
$\Delta g_{\rm rec}$			−9.10(13)			−4.60(9)

In Table 1 the terms $\Delta g^{\rm 1-el,L}_{\rm rec}$, $\Delta g^{\rm m-el}_{\rm rec}$, and $\Delta g^{(1)}_{\rm rec}$ are presented for the case of B-like argon $^{40}\mathrm{Ar}^{13+}$ for Coulomb, core-Hartree and Kohn-Sham potentials. The Kohn-Sham value is taken for the final result, while the uncertainty is estimated as the difference between the values for Coulomb and Kohn-Sham potentials. This rather conservative estimation, i.e. 100% of the contribution of higher orders, is based on the observation, that the effect of the screening potential for $\Delta g^{\rm 1-el,L}_{\rm rec}$ and $\Delta g^{\rm m-el}_{\rm rec}$ constitutes only 50% of the total interelectronic-interaction correction obtained. Another minor uncertainty is related to $\Delta g^{\rm 1-el,H}_{\rm rec}$, its estimation $((\alpha Z)^3 \Delta g^{\rm 1-el,L}_{\rm rec})$ amounts to 0.041 and 0.021 for $^2P_{1/2}$ and $^2P_{3/2}$ states, respectively. In comparison to the previously published values [12], the obtained accuracy of the nuclear recoil effect is 2 times better for $^2P_{1/2}$ state and 4 times better for $^2P_{3/2}$ state.

3. *g* factor of boron-like argon

Table 2 represents the individual contributions to the g factors of B-like argon for the ground $[(1s)^2\,(2s)^2\,2p]\,^2P_{1/2}$ and first excited $[(1s)^2\,(2s)^2\,2p]\,^2P_{3/2}$ states. As compared to the previous

Table 2. Individual contributions to the g factor of boron-like argon for $^2P_{1/2}$ and $^2P_{3/2}$ states.

		$^2P_{1/2}$	$^2P_{3/2}$
Dirac value		0.663 775 447	1.331 030 389
Finite nuclear size		0.000 000 000	0.000 000 000
One-photon exchange	$\sim 1/Z$	0.000 657 525	0.000 481 188
Many-photon exchange	$\sim 1/Z^{2+}$	−0.000 007 5 (4)	−0.000 003 (3)
One-loop QED	$\sim \alpha$	−0.000 769 9 (5)	0.000 779 6 (8)
Higher-order QED	$\sim \alpha^{2+}$	0.000 001 2 (1)	−0.000 001 2 (1)
Nuclear recoil		−0.000 009 1 (2)	−0.000 004 6 (1)
Total		0.663 647 7 (7)	1.332 282 (3)

compilation [12], two terms are improved: the interelectronic interaction of the second and higher orders ($1/Z^{2+}$) and the nuclear recoil effect. Evaluation of the latter has been presented in the previous section. The higher-order interelectronic-interaction correction was calculated in [12] for the ground state within the Breit approximation employing the large-scale configuration interaction method with the Dirac-Fock and Dirac-Fock-Sturm basis functions (CI-DFS). The 10% uncertainty was ascribed to it due to the moderate basis employed in the calculations and relatively poor convergence of the result with respect to the basis size. Recently we have performed calculations with larger basis and found justification for 2 times smaller uncertainty of the result. It is supported by the independent calculation of the $1/Z^2$-term performed within the perturbation theory. The corresponding calculation for the $^2P_{3/2}$ state yields the new value for this contribution, while the all-order CI-DFS result is still in demand. The details on this calculation will be published elsewhere. In total, we have an improvement by a factor of 1.5 for the g factor of the $^2P_{1/2}$ state.

Acknowledgments

The work was supported in part by DFG (Grant No. VO 1707/1-2), by GSI, by RFBR (Grants No. 14-02-31316 and 13-02-00630), by SPbSU (Grants No. 11.38.269.2014, and No. 11.38.261.2014), and by the Helmholtz-Rosatom grant provided via FAIR–Russia Research Center.

References

[1] Häffner H *et al* 2000 *Phys. Rev. Lett.* **85** 5308
[2] Verdú J L *et al* 2004 *Phys. Rev. Lett.* **92** 093002
[3] Sturm S *et al* 2011 *Phys. Rev. Lett.* **107** 023002
[4] Sturm S *et al* 2013 *Phys. Rev. A* **87** 030501(R)
[5] Wagner A *et al* 2013 *Phys. Rev. Lett.* **110** 033003
[6] Sturm S *et al* 2014 *Nature* **506** 467
[7] Volotka A V *et al* 2014 *Phys. Rev. Lett.* **112** 253004
[8] Shabaev V M *et al* 2006 *Phys. Rev. Lett.* **96** 253002
[9] Volotka A V and Plunien G 2014 *Phys. Rev. Lett.* **113** 023002
[10] Soria Orts R *et al* 2007 *Phys. Rev. A* **76** 052501
[11] von Lindenfels D *et al* 2013 *Phys. Rev. A* **87** 023412
[12] Glazov D A *et al* 2013 *Phys. Scr.* **T156** 014014
[13] Phillips M 1949 *Phys. Rev.* **76** 1803
[14] Faustov R N 1970 *Phys. Lett. B* **33** 422; 1970 *Nuovo Cimento A* **69** 37
[15] Grotch H 1970 *Phys. Rev. A* **2** 1605
[16] Grotch H and Hegstrom R A 1971 *Phys. Rev. A* **4** 59
[17] Yelkhovsky A 2001 Recoil correction to the magnetic moment of a bound electron *Preprint* hep-ph/0108091

[18] Pachucki K 2008 *Phys. Rev. A* **78** 012504
[19] Shabaev V M 2001 *Phys. Rev. A* **64** 052104
[20] Shabaev V M and Yerokhin V A 2002 *Phys. Rev. Lett.* **88** 091801
[21] Shabaev V M 2002 *Phys. Rep.* **356** 119
[22] Shabaev V M *et al* 2004 *Phys. Rev. Lett.* **93** 130405
[23] Cowan R 1981 *The Theory of Atomic Spectra* (University of California Press, Berkeley, CA)
[24] Sapirstein J and Cheng K T 2001 *Phys. Rev. A* **63** 032506; 2002 *Phys. Rev. A* **66** 042501; 2003 *Phys. Rev. A* **67** 022512
[25] Glazov D A *et al* 2006 *Phys. Lett. A* **357** 330
[26] Yerokhin V A, Artemyev A N and Shabaev V M, 2007 *Phys. Rev. A* **75** 062501
[27] Artemyev A N *et al* 2007 *Phys. Rev. Lett.* **98**, 173004

Journal of Physics: Conference Series **583** (2015) 012002 doi:10.1088/1742-6596/583/1/012002

Laser spectroscopy of the ground-state hyperfine structure in H-like and Li-like bismuth

J Vollbrecht[1], Z Andelkovic[2], A Dax[3], W Geithner[2], C Geppert[4], C Gorges[4], M Hammen[5], V Hannen[1], S Kaufmann[4], K König[4], Y Litvinov[2], M Lochmann[4], B Maass[4], J Meisner[6], T Murboeck[4], W Nörtershäuser[4], R Sánchez[2], S Schmidt[4,7], M Schmidt[6], M Steck[2], T Stöhlker[2,8,9], R C Thompson[10], J Ullmann[4,2,8] and C Weinheimer[1]

[1] Institut für Kernphysik, Westfälische Wilhelms Universität, 48149 Münster, Germany
[2] GSI Helmholtzzentrum für Schwerionenforschung, 64291 Darmstadt, Germany
[3] Paul Scherrer Institut, 5232 Villigen PSI, Switzerland
[4] Institut für Kernphysik, Technische Universität Darmstadt , 64289 Darmstadt, Germany
[5] Helmholtz Institut Mainz, 55099 Mainz, Germany
[6] PTB Braunschweig, Elektrische Energiemesstechnik, 38116 Braunschweig, Germany
[7] Institut für Kernchemie, Johannes Gutenberg Universität Mainz, 55128 Mainz, Germany
[8] Helmholtz-Institut Jena, 07743 Jena, Germany
[9] IOQ, Friedrich-Schiller-Universität Jena, 07743 Jena, Germany
[10] Department of Physics, Imperial College, London SW7 2AZ, United Kingdom

E-mail: jonas.vollbrecht@wwu.de

Abstract. The LIBELLE experiment performed at the experimental storage ring (ESR) at the GSI Helmholtz Center in Darmstadt aims for the determination of the ground state hyperfine (HFS) transitions and lifetimes in hydrogen-like ($^{209}Bi^{82+}$) and lithium-like ($^{209}Bi^{80+}$) bismuth. The study of HFS transitions in highly charged ions enables precision tests of QED in extreme electric and magnetic fields otherwise not attainable in laboratory experiments. While the HFS transition in H-like bismuth was already observed in earlier experiments at the ESR, the LIBELLE experiment succeeded for the first time to measure the HFS transition in Li-like bismuth in a laser spectroscopy experiment.

1. Introduction

Highly charged ions provide a testing ground for QED calculations in extreme electric (up to 10^{16} V/cm) and magnetic (up to 10^4 T) fields that cannot be created in the laboratory with conventional methods (like lasers and superconducting magnets). This approach has been used since the 1990s with various isotopes in laser spectroscopy as well as in x-ray emission spectroscopy experiments (see references in [1]).

To put the results in a theoretical context precise QED calculations have to be performed, where a major issue is the large uncertainty of nuclear structure corrections. Particularly the uncertainty of the Bohr-Weisskopf effect, which arises due to the spatially smeared out magnetic moment distribution in the nucleus, is comparable in size to the total contribution of QED corrections and hinders a direct test of QED. To tackle this problem, Shabaev et al. [2] proposed to use a new approach by introducing the so-called specific difference $\Delta' E$ between the HFS splittings in H-like ($\Delta E^{(1s)}$) and Li-like ($\Delta E^{(2s)}$) configurations of the same isotope

$$\Delta' E = \Delta E^{(2s)} - \xi \Delta E^{(1s)} . \tag{1}$$

The parameter ξ that acts to cancel the contribution of the Bohr-Weisskopf effect in the specific difference is largely model independent and can be calculated to high accuracy [3, 4]. A suitable candidate for an experimental determination of the specific difference is the bismuth isotope ^{209}Bi. Here the ground state HFS transitions in both, the H-like and in the Li-like system, are accessible to high precision laser spectroscopy measurements.
While the $\Delta E^{(1s)}$ splitting in $^{209}Bi^{82+}$has already been observed experimentally in earlier laser spectroscopy measurements at the Experimental Storage Ring (ESR) at the GSI Helmholtz Center in Darmstadt [5], searches for the HFS transition in $^{209}Bi^{80+}$have been unsuccessful for a long time. The main obstacles for this case are the long transition wavelength predicted to be at $\lambda_{\text{theor}} = 1555.3(3)$ nm [4] in the ions rest frame, which is outside the sensitivity of typical photo-multipliers used for fluorescence detection, and the long lifetime of the HFS state of $\tau_{\text{theor}} = 82.0(1.4)$ ms [6] leading to low signal rates in the experiment.
The LIBELLE collaboration overcame these difficulties using a newly designed detection system [7] and managed to detect the HFS transition in $^{209}Bi^{80+}$in 2011 for the first time in a laser spectroscopy experiment [8]. During the same experiment we also re-measured the hyperfine transition in $^{209}Bi^{82+}$and made a first measurement of the lifetime of the HFS state in $^{209}Bi^{80+}$. The accuracy of the experimental wavelength result was limited by uncertainties in the voltage calibration of the electron cooler present in the storage ring in order to cool the ion beam. The voltage of the device that ultimately determines the velocity of the cooled ions had an uncertainty of approximately $5 \cdot 10^{-4}$ limiting the overall precision of the measurement. For this reason a second beam-time with a in situ measurement of the cooler voltage, using a precision HV divider provided by the Physikalisch-Technische Bundesanstalt Braunschweig (PTB), and with an improved data acquisition system was performed in 2014. During this second beam-time both HFS transitions were re-measured with an expected accuracy $< 10^{-4}$ and lifetime data for the HFS states in $^{209}Bi^{82+}$and $^{209}Bi^{80+}$were taken.
These data will provide a first assessment of the QED contributions to the specific difference of the HFS transitions and pave the way for trap-assisted experiments like SPECTRAP [9] that will allow to study HFS transitions in highly charged ions at rest reaching relative accuracies in the 10^{-7} regime.

2. Experimental setup

For the LIBELLE experiment, either hydrogen-like or lithium-like bismuth ions were provided by the GSI accelerator system, injected into the ESR and stored at 400 MeV/u. This represents an ion speed of $\beta \approx 0.71$. This has the advantage, that the transition wavelengths are Doppler shifted into a region convenient for (anti)collinear laser excitation and fluorescence detection according to

$$\lambda_{\text{lab}}^{\uparrow\uparrow} = \lambda_0 \frac{1}{\gamma(1+\beta)} \quad ; \quad \lambda_{\text{lab}}^{\uparrow\downarrow} = \lambda_0 \frac{1}{\gamma(1-\beta)} \ . \tag{2}$$

Subsequently the ions are cooled in the electron cooler to a momentum spread of $\Delta p/p \approx 10^{-5}$. The acceleration voltage of the electron cooler finally also determines the velocity of the ions. The ions are compressed into two bunches of about 10 m length each by applying an RF voltage to one of the radio-frequency cavities installed in the ring. One bunch, referred to as signal bunch, is brought into overlap with the pulsed excitation laser inside the cooler straight section, while the other bunch is not illuminated by the laser and is used as reference for background correction (see figure 1).
Fluorescence photons emitted by the HFS resonance in H-like bismuth are reflected by a mirror system installed in the vacuum pipe towards two view-ports and detected by two solar blind photomultiplier. This mirror system, which has also been used in earlier studies of HFS states in highly charged ions at the ESR [10], collects mainly photons emitted at angles between 15° and 60° relative to the beam direction.

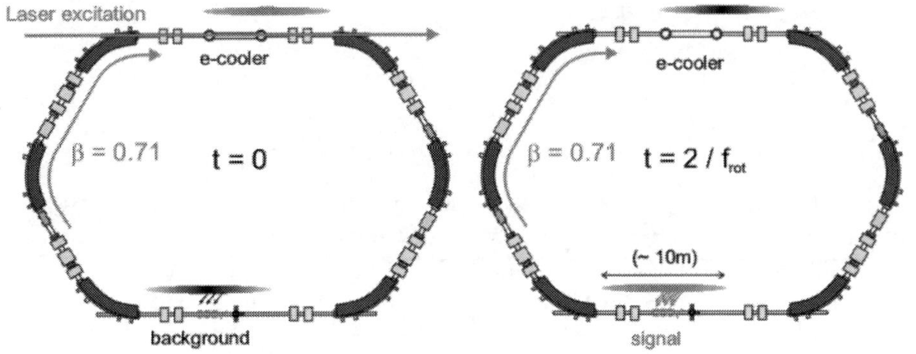

Figure 1. Schematic view of the experiment at the Experimental Storage Ring at GSI.

For the detection of the resonance in Li-like bismuth it is mandatory to collect the most forward emitted photons, as only these are Doppler shifted to a wavelength regime where the applied photomultiplier has the largest quantum efficiency. For this purpose a dedicated detection system was developed by the University of Münster [7]. It contains a movable parabolic mirror made of Oxygen-free high thermal conductivity (OFHC) copper with a central 3 cm slit for the ions to pass through. Fluorescence photons emitted at angles $\lesssim 20°$ relative to the ion beam are reflected by the mirror towards a highly selected low-noise photomultiplier tube model R1017 from Hamamatsu with a maximum quantum efficiency of 16% for the most forward emitted photons (see figure 2, left). The PMT is located in a Peltier cooler housing outside the vacuum beam-line.

Light at the Doppler-shifted transition wavelengths of about 590 nm and 640 nm for H-like and Li-like ions, respectively, was produced by a pulsed dye laser delivering a typical pulse energy of $\approx$ 100 mJ at $\leq$ 10 ns pulse length and 30 Hz repetition rate. Temporal overlap between laser pulse and ion bunch in the interaction zone inside the electron cooler was achieved by synchronizing the pump laser Q-switch signal with the bunch-generating RF voltage. By scanning the laser wavelength across a region around the predicted value for the HFS transition, this setup enabled the detection of the 2s hyperfine splitting in $^{209}Bi^{80+}$ (see figure 2, right).

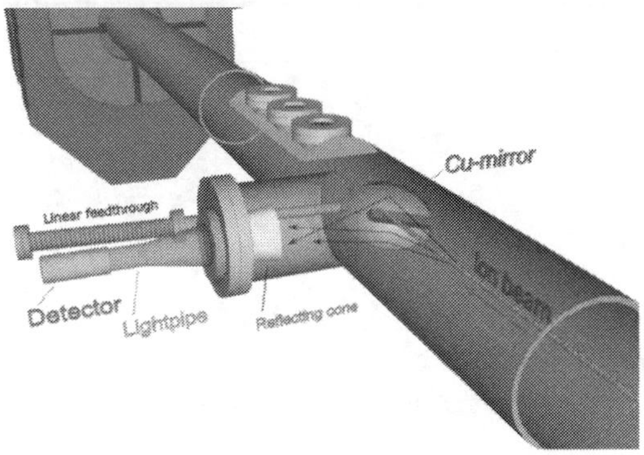

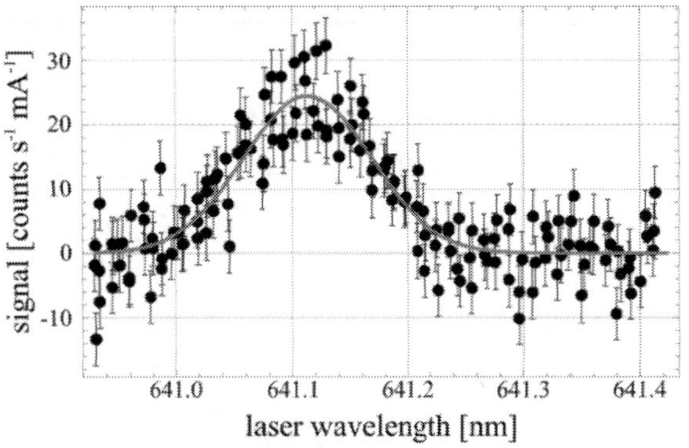

Figure 2. Left: Schematic view of the detection system for forward emitted fluorescence photons. Right: Wavelength scan over the resonance region of $^{209}Bi^{80+}$, i.e. detected photon rate normalized to the beam intensity versus wavelength in the laboratory system [11].

Journal of Physics: Conference Series **583** (2015) 012002 doi:10.1088/1742-6596/583/1/012002

3. Results of the 2011 beam-time

After data analysis of the data taken in 2011, the transition wavelength for H-like and Li-like bismuth in the laboratory frame were determined to be

$$\lambda_{\text{lab}}^{(82+)} = 591.183\,(26)\ \text{nm} \quad ; \quad \lambda_{\text{lab}}^{(80+)} = 641.112\,(24)\ \text{nm} \tag{3}$$

respectively [8]. In order to obtain the wavelength in the reference frame of the ions we used the voltage applied to the electron cooler (see [8]). However this voltage could not be measured directly and an extrapolation had to be performed. This introduces an additional systematic uncertainty. A thorough analysis with corrections extracted from previous work performed by Lochmann [11] and Jöhren [12] yielded the HFS transition wavelength in the rest frame of

$$\lambda_{\text{rest}}^{(82+)} = 243.76\,(5)(2)\ \text{nm} \quad ; \quad \lambda_{\text{rest}}^{(80+)} = 1554.66\,(33)(10)\ \text{nm}\,. \tag{4}$$

Here the first uncertainty arises from the voltage calibration while other uncorrelated effects are summed up in the second uncertainty contribution. For further details please see Ref. [8].

In this beam-time we were also able to measure the lifetime of the HFS state of Li-like bismuth. For this purpose a shutter was installed in front of the entrance window to the ESR blocking the pulsed laser beam in regular intervals. During the measurement the laser wavelength was fixed to the resonance wavelength. A cycle was started where, after the ions interacted with the laser, the shutter was closed to let the excited ions decay over a time window $[0,\, t_c]$ with $t_c \approx 1000$ ms. Subsequently the shutter was opened again for an interval $[t_c,\, 1500\ \text{ms}]$ to repopulate the upper HFS state.

To extract the lifetime, a fit model was constructed taking into account the processes of excitation, stimulated emission and spontaneous emission of fluorescence photons in order to be able to use the data both from the excitation phases as well as from the pure decay phases of the measurement. The fit model is given by

$$f(t) = \begin{cases} s \cdot N_2(0) \exp\left(-\frac{t}{\tau}\right) & \text{for } t \leq t_c \ \text{(laser off)} \\ s \cdot \left[\frac{a}{b} + \left(N_2(t_c) - \frac{a}{b}\right) \exp\left(-b\,(t - t_c)\right)\right] & \text{for } t > t_c \ \text{(laser on)} \end{cases} \tag{5}$$

where $a = \frac{g_2}{g_1} N \frac{\phi}{\tau}$, $b = \frac{1}{\tau}\left[\left(\frac{g_2}{g_1} + 1\right)\phi + 1\right]$ and $\phi = \frac{\lambda^3 u}{8\pi h}$. Here u corresponds to the spectral radiance and N to the total ion number, while $N_2(t)$ denotes the number of excited ions at a given time. g_1 and g_2 count the number of magnetic sub-states of the ground state ($F = 4$) and the excited state ($F = 5$) of the HFS transition, respectively, ϕ describes the time averaged laser power and s is a scaling factor taking into account experimental parameters like measurement time and observed solid angle.

The fit gives a preliminary result for the lifetime of the HFS state in $^{209}\text{Bi}^{80+}$in the lab frame of

$$\tau_{\text{lab}}^{(80+)} = 106.9(8.0)\ \text{ms}\,. \tag{6}$$

The given uncertainty is purely statistical. Separate fits of the decay and excitation phases resulted in lifetimes consistent to the fit over the combined data. As the lifetime measurement could use only 2-3 hours of the beam-time, there was no time to investigate systematic effects on the lifetime during these first measurements.

4. Preliminary results of the 2014 beam-time

To overcome the dominant uncertainty introduced by the electron cooler voltage a second beam-time took place in March 2014 with special emphasis on the voltage calibration. For this purpose a high-precision divider for voltages up to a maximum of 250 kV provided by PTB

Braunschweig allowed for the first time an in-situ measurement of the high voltage and improved the uncertainty to the $< 10^{-4}$ level. Data were recorded for the ground state HFS transitions in $^{209}Bi^{82+}$and $^{209}Bi^{80+}$. Data analysis regarding the transition wavelengths is currently ongoing. To improve the statistics of the lifetime measurements and collect data for both $^{209}Bi^{82+}$and $^{209}Bi^{80+}$, a data acquisition system with improved timing capability was set up. With this system it was possible to collect data for the lifetime of H-like bismuth whenever the laser was tuned on resonance (here the lifetime in the rest frame is < 400 μs and therefore much shorter than the interval between laser shots of 33 ms). Due to the long lifetime of the HFS state in Li-like bismuth, a shutter system was required again in that case, but much more measurement time has been spent on this part of the measurements. Figure 3 shows results of the lifetime fits

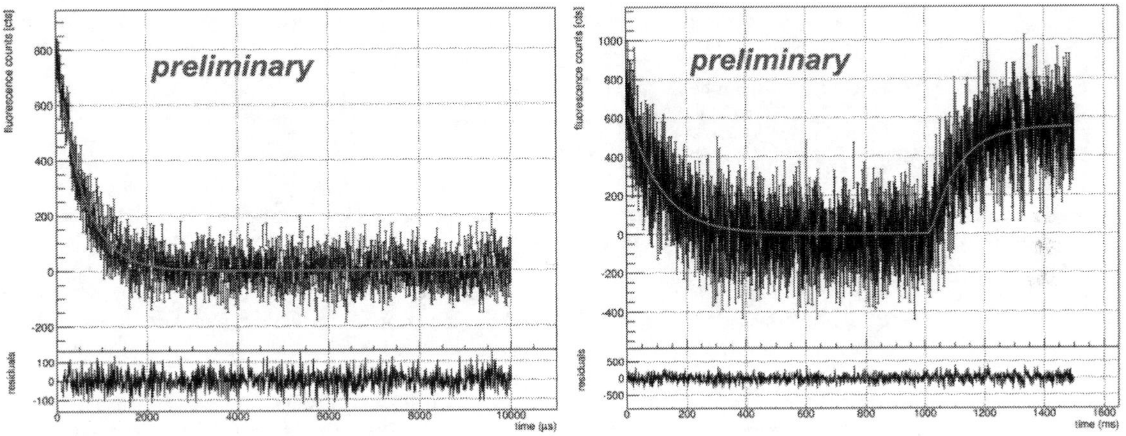

Figure 3. Detected fluorescence photons versus time in ms with preliminary fits of the lifetime data recorded in 2014 for $^{209}Bi^{82+}$and $^{209}Bi^{80+}$.

for H-like and Li-like bismuth yielding preliminary values of

$$\tau_{\rm lab}^{(82+)} = 556(14)\ \mu\text{s} \quad ; \quad \tau_{\rm lab}^{(80+)} = 106.8(4.2)\ \text{ms} \tag{7}$$

in the laboratory frame. The fits currently contain only a subset of the data and systematic effects are not taken into account yet. Analysis of the full dataset and of possible systematic corrections is ongoing.

Acknowledgement

This work is supported by BMBF under contract numbers 05P12PMFAE and 05P12RDFA4 and the Helmholtz International Center for FAIR (HIC for FAIR) within the LOEWE program by the State of Hesse. We thank the operators of the GSI accelerators for technical assistance.

References

[1] Beier T 2000 *Phys. Rep.* **339** 79
[2] Shabaev V M *et al* 2001 *Phys. Rev. Lett.* **86** 3959
[3] Andreev O V *et al* 2012 *Phys. Rev.* A **85** 022510
[4] Volotka A V *et al* 2012 *Phys. Rev. Lett.* **108** 073001
[5] Klaft I *et al* 1994 *Phys. Rev. Lett.* **73** 2425
[6] Shabaev V M *et al* 1998 *Phys. Rev.* A **57** 149
[7] Hannen V M *et al* 2013 *JINST* **8** P09018
[8] Lochmann M *et al* 2014 *Phys. Rev.* A **90** 030501(R)
[9] Andjelkovic Z *et al* 2010 *Hyperfine Interact.* **196** 81
[10] Seelig P *et al* 1998 *Phys. Rev. Lett.* **81** 4824
[11] Lochmann M 2013 PhD thesis, Johannes Gutenberg Universität Mainz
[12] Jöhren R 2013 PhD thesis, Westfälische Wilhelms-Universität Münster

Journal of Physics: Conference Series **583** (2015) 012003 doi:10.1088/1742-6596/583/1/012003

Electron impact excitation of Si II and Fe XIV

K M Aggarwal and F P Keenan

Astrophysics Research Centre, School of Mathematics and Physics, Queen's University Belfast, Belfast BT7 1NN, Northern Ireland, UK

E-mail: K.Aggarwal@qub.ac.uk

Abstract.
Energy levels, radiative rates, lifetimes, collision strengths and effective collision strengths are calculated for two important Al-like ions, namely Si II and Fe XIV. For Si II, the lowest 56 levels of the $3s^23p$, $3s3p^2$, $3p^3$, $3s^23d$, 3s3p3d, $3s^24\ell$ and $3s^25\ell$ configurations are included, whereas for Fe XIV additional 80 levels of $3p^23d$, $3s3d^2$ and $3p3d^2$ are considered, but not of $3s^25\ell$. For the determination of atomic structure GRASP has been adopted and radiative rates are calculated for all E1, E2, M1 and M2 transitions. Electron impact excitation collision strengths are calculated with the DARC code, over a wide energy range, and resonances are resolved in a fine energy mesh to determine effective collision strengths over a wide range of temperatures. Extensive comparisons are made for all atomic parameters with available theoretical and experimental data, and the accuracy of the present results is assessed. Energy levels are estimated to be accurate to ~1% and all other parameters to be better than 20%.

1. Introduction

Emission lines of Si II have been observed at optical and ultraviolet (UV) wavelengths in a variety of plasmas, such as planetary nebulae, quasars and the interstellar medium. Similarly, lines of Fe XIV have been widely observed in solar and other astrophysical plasmas. The strongest observed solar forbidden transition of Fe XIV is from the coronal green line $(3s^23p)\ ^2P^o_{1/2}$ - $^2P^o_{3/2}$ at a wavelength of 5303 Å. Many lines of these ions are density and/or temperature sensitive, and hence provide useful information about physical conditions of the plasmas. Furthermore, these ions are also important for the studies of fusion plasmas, and to reliably analyse observations, atomic data are required for many parameters, such as energy levels, radiative rates (A- values) and excitation rates. Since experimental data are generally not available, except for energy levels, theoretical results are required.

Measured values of energy levels have been compiled by NIST and are available at their website `http://www.nist.gov/pml/data/asd.cfm`. Theoretical energy levels have been determined by several authors, and the most notable results are of [1] and [2] for Si II, and of [3] and [4] for Fe XIV. Similarly, A- values, collision strengths (Ω) and effective collision strengths (Υ) are also available in the literature. However, there are large discrepancies among the available data, particularly for Υ, for many transitions of both ions, and therefore we revisit these ions so that atomic data can be confidently applied for diagnostics and modelling of plasmas.

Journal of Physics: Conference Series **583** (2015) 012003 doi:10.1088/1742-6596/583/1/012003

2. Energy levels

For the calculations of energy levels and A- values we have employed the fully relativistic GRASP (general-purpose relativistic atomic structure package) code available at the website `http://web.am.qub.ac.uk/DARC/`, and the Breit and QED corrections are included in the 'extended average level' option. For Si II 14 configurations are considered, namely ($1s^2 2s^2 2p^6$) $3s^2 3p$, $3s3p^2$, $3p^3$, $3s^2 3d$, 3s3p3d, $3s^2 4\ell$ and $3s^2 5\ell$, which generate 56 levels. However, corresponding calculations for Fe XIV are larger as they include 136 levels of the $3s^2 3p$, $3s3p^2$, $3s^2 3d$, $3p^3$, 3s3p3d, $3p^2 3d$, $3s3d^2$, $3p3d^2$ and $3s^2 4\ell$ configurations.

Our calculated level energies are compared in Table 1 with the NIST compilations for the lowest 40 levels of Si II. Differences with the NIST listings are below $\sim$ 0.1 Ryd, but the orderings are slightly different in a few instances – see for example levels 17–22. The MCHF energies of Tayal [1] (not included in Table 1) are in closer agreement with those of NIST, because he has used non-orthogonal orbitals. However, his level ordering differs with that of NIST in a few instances, such as for 6–7 and 21–22. We have also performed calculations with additional CI, but differences with the NIST compilations remain up to $\sim$ 0.1 Ryd, and there is no change in level orderings.

Our energies for Fe XIV are also listed in Table 1 for the lowest 40 levels, and these agree with the NIST values, in magnitude and orderings. Differences, if any, are within 0.3 Ryd – see levels 37-40, and only for level 18 ($3p^3\ ^2P^o_{1/2}$) the ordering is notably different. To assess the effect of additional CI, we have also performed a larger calculation (GRASP2) with 332 levels, which arise from the $3s^2 3p$, $3s3p^2$, $3s^2 3d$, $3p^3$, 3s3p3d, $3p^2 3d$, $3s3d^2$, $3p3d^2$, $3d^3$, $3s^2 4\ell$, $3s3p4\ell$, $3s^2 5\ell$ and $3s3p5\ell$ configurations. Almost all the additional 196 levels of the $3d^3$, $3s3p4\ell$, $3s^2 5\ell$ and $3s3p5\ell$ configurations yield energies above those of the 136 included earlier. Both sets of energies agree within 0.05 Ryd and the orderings are also nearly the same. Therefore, as for Si II, there is no overall advantage of extensive CI for the determination of energy levels for Fe XIV.

3. Radiative rates

Using the GRASP code we have calculated A- values for four types of transitions, namely electric dipole (E1), electric quadrupole (E2), magnetic dipole (M1) and magnetic quadrupole (M2). In general, E1 transitions dominate in magnitude, but sometimes other types of transitions are also important and hence are (preferably) required for a complete plasma model. Our f- values for Si II agree in general with those of [1] and [2], but differ by up to a factor of two for a few. Discrepancies for a few transitions are because of varying amounts of CI and methods/codes as discussed in detail by [2] – see their Figs. 2 and 3, and particularly Table 3. Similarly, discrepancies for some weak transitions (such as 1– 4, 2 – 4/7 and 4 – 11) are up to an order of magnitude. Weaker transitions are more susceptible to varying amount of CI, due to cancellation or additive effects, and hence such discrepancies are very common. In Table 2 we compare our A- values for a few transitions of Fe XIV with those of Tayal [3] and Liang et al [4]. Also included in this table are the A- values from our larger (GRASP2) calculations as well as the ratio of the velocity and length form of A- values, as it also indicates the accuracy estimate. Based on this comparison and the one discussed for Si II, our results are expected to be accurate to $\sim$ 20%, for a majority of strong transitions.

4. Lifetimes

The measurement of lifetime (τ) can directly lead to the assessment of accuracy of the A- values, particularly if a single transition dominates. Our calculated results for τ are listed in Table 1 for both ions. Schectman et al [5] have measured τ corresponding to the $3s^2 3p\ ^2P^o_{3/2}$ – $3s^2 4s\ ^2S_{1/2}$ transition of Si II to be 0.91±0.04 ns, which compares well with our result of 0.996 ns. Similarly, their measured τ for the $3s^2 5s\ ^2S_{1/2}$ level is 1.99±0.12 ns, in close agreement with our result

Table 1. Energy levels of Si II and Fe XIV, their threshold energies (in Ryd) and lifetimes (s). ($a{\pm}b \equiv a{\times}10^{\pm b}$).

	Si II					Fe XIV				
No.	Config.	Level	NIST	GRASP	τ (s)	Config.	Level	NIST	GRASP	τ (s)
1	$3s^23p$	$^2P^o_{1/2}$	0.0000	0.0000	....	$3s^23p$	$^2P^o_{1/2}$	0.0000	0.0000	....
2	$3s^23p$	$^2P^o_{3/2}$	0.0026	0.0023	6.693+03	$3s^23p$	$^2P^o_{3/2}$	0.1718	0.1707	1.692-02
3	$3s3p^2$	$^4P_{1/2}$	0.3902	0.3531	2.510−04	$3s3p^2$	$^4P_{1/2}$	2.0514	2.0306	3.139-08
4	$3s3p^2$	$^4P_{3/2}$	0.3912	0.3541	8.004−03	$3s3p^2$	$^4P_{3/2}$	2.1213	2.0999	1.675-07
5	$3s3p^2$	$^4P_{5/2}$	0.3928	0.3557	3.391−03	$3s3p^2$	$^4P_{5/2}$	2.2088	2.1876	4.429-08
6	$3s3p^2$	$^2D_{3/2}$	0.5040	0.5257	5.629−06	$3s3p^2$	$^2D_{3/2}$	2.7269	2.7375	4.145-10
7	$3s3p^2$	$^2D_{5/2}$	0.5042	0.5258	4.254−06	$3s3p^2$	$^2D_{5/2}$	2.7472	2.7572	5.269-10
8	$3s^24s$	$^2S_{1/2}$	0.5969	0.6433	6.705−10	$3s3p^2$	$^2S_{1/2}$	3.3233	3.3747	5.101-11
9	$3s3p^2$	$^2S_{1/2}$	0.6986	0.7668	7.363−10	$3s3p^2$	$^2P_{1/2}$	3.5404	3.6010	2.673-11
10	$3s^24p$	$^2P^o_{1/2}$	0.7399	0.7782	8.891−09	$3s3p^2$	$^2P_{3/2}$	3.6133	3.6767	2.295-11
11	$3s^24p$	$^2P^o_{3/2}$	0.7404	0.7787	8.835−09	$3s^23d$	$^2D_{3/2}$	4.3123	4.3939	2.133-11
12	$3s^23d$	$^2D_{3/2}$	0.7230	0.7991	3.078−10	$3s^23d$	$^2D_{5/2}$	4.3304	4.4150	2.341-11
13	$3s^23d$	$^2D_{5/2}$	0.7231	0.7993	3.092−10	$3p^3$	$^2D^o_{3/2}$	5.2524	5.2536	2.147-10
14	$3s3p^2$	$^2P_{1/2}$	0.7637	0.8910	1.850−10	$3p^3$	$^2D^o_{5/2}$	5.2875	5.2859	2.548-10
15	$3s3p^2$	$^2P_{3/2}$	0.7655	0.8927	1.847−10	$3p^3$	$^4S^o_{3/2}$	5.3674	5.3881	2.715-11
16	$3s^25s$	$^2S_{1/2}$	0.8928	0.9219	1.865−09	$3s3p(^3P)3d$	$^4F^o_{3/2}$		5.8524	2.872-09
17	$3s3p(^3P)3d$	$^2D^o_{3/2}$	0.9913	0.9556	4.397−07	$3s3p(^3P)3d$	$^4F^o_{5/2}$	5.8867	5.8891	3.185-09
18	$3s3p(^3P)3d$	$^2D^o_{5/2}$	0.9917	0.9560	4.319−07	$3p^3$	$^2P^o_{1/2}$	5.8532	5.8962	5.830-11
19	$3s^24d$	$^2D_{3/2}$	0.9206	0.9614	6.419−10	$3p^3$	$^2P^o_{3/2}$	5.8814	5.9250	5.910-11
20	$3s^24d$	$^2D_{5/2}$	0.9206	0.9614	6.456−10	$3s3p(^3P)3d$	$^4F^o_{7/2}$	5.9410	5.9428	3.547-09
21	$3s^24f$	$^2F^o_{7/2}$	0.9437	0.9665	3.881−09	$3s3p(^3P)3d$	$^4F^o_{9/2}$	6.0168	6.0179	1.927-02
22	$3s^24f$	$^2F^o_{5/2}$	0.9467	0.9665	3.886−09	$3s3p(^3P)3d$	$^4P^o_{5/2}$	6.2905	6.3145	3.335-11
23	$3s^25p$	$^2P^o_{1/2}$	0.9465	0.9748	1.596−08	$3s3p(^3P)3d$	$^4D^o_{3/2}$	6.3120	6.3392	2.734-11
24	$3s^25p$	$^2P^o_{3/2}$	0.9467	0.9750	1.597−08	$3s3p(^3P)3d$	$^4D^o_{1/2}$	6.3257	6.3564	2.328-11
25	$3s3p(^3P)3d$	$^4F^o_{3/2}$	1.0413	1.0088	7.061−06	$3s3p(^3P)3d$	$^4P^o_{1/2}$	6.4130	6.4344	3.300-11
26	$3s3p(^3P)3d$	$^4F^o_{5/2}$	1.0418	1.0093	3.746−06	$3s3p(^3P)3d$	$^4P^o_{3/2}$	6.4172	6.4414	2.838-11
27	$3s3p(^3P)3d$	$^4F^o_{7/2}$	1.0426	1.0101	3.601−06	$3s3p(^3P)3d$	$^4D^o_{7/2}$	6.4098	6.4403	2.318-11
28	$3s3p(^3P)3d$	$^4F^o_{9/2}$	1.0437	1.0112	2.129+01	$3s3p(^3P)3d$	$^4D^o_{5/2}$	6.4164	6.4446	2.502-11
29	$3s^25d$	$^2D_{3/2}$	1.0242	1.0540	2.146−09	$3s3p(^3P)3d$	$^2D^o_{3/2}$	6.5356	6.5870	2.629-11
30	$3s^25d$	$^2D_{5/2}$	1.0242	1.0540	2.157−09	$3s3p(^3P)3d$	$^2D^o_{5/2}$	6.5416	6.5916	2.700-11
31	$3s^25f$	$^2F^o_{7/2}$	1.0367	1.0584	9.149−09	$3s3p(^3P)3d$	$^2F^o_{5/2}$	6.7886	6.8725	5.516-11
32	$3s^25f$	$^2F^o_{5/2}$	1.0367	1.0584	9.164−09	$3s3p(^3P)3d$	$^2F^o_{7/2}$	6.9239	7.0080	5.175-11
33	$3s^25g$	$^2G_{7/2}$	1.0405	1.0618	1.397−08	$3s3p(^3P)3d$	$^2P^o_{3/2}$	7.3550	7.4769	1.820-11
34	$3s^25g$	$^2G_{9/2}$	1.0405	1.0618	1.397−08	$3s3p(^3P)3d$	$^2P^o_{1/2}$		7.5502	2.024-11
35	$3s3p(^3P)3d$	$^4P^o_{5/2}$		1.0974	4.540−10	$3s3p(^1P)3d$	$^2F^o_{7/2}$	7.4505	7.5703	1.823-11
36	$3s3p(^3P)3d$	$^4P^o_{3/2}$		1.0984	4.520−10	$3s3p(^1P)3d$	$^2F^o_{5/2}$	7.4779	7.5985	1.771-11
37	$3s3p(^3P)3d$	$^4P^o_{1/2}$		1.0990	4.545−10	$3s3p(^1P)3d$	$^2P^o_{1/2}$	7.6500	7.7995	1.422-11
38	$3s3p(^3P)3d$	$^4D^o_{1/2}$		1.1061	2.465−10	$3s3p(^1P)3d$	$^2D^o_{3/2}$	7.6617	7.8086	1.311-11
39	$3s3p(^3P)3d$	$^4D^o_{3/2}$		1.1064	2.477−10	$3s3p(^1P)3d$	$^2P^o_{3/2}$	7.6880	7.8355	1.341-11
40	$3s3p(^3P)3d$	$^4D^o_{5/2}$		1.1067	2.479−10	$3s3p(^1P)3d$	$^2D^o_{5/2}$	7.6954	7.8391	1.185-11

NIST: `http://physics.nist.gov/PhysRefData/ASD_form.html`
GRASP: Present results including the QED effects

Journal of Physics: Conference Series **583** (2015) 012003 doi:10.1088/1742-6596/583/1/012003

Table 2. Comparison of A- values (s^{-1}) for some transitions of Fe XIV. ($a{\pm}b \equiv a{\times}10^{\pm b}$).

I	J	GRASP1		GRASP2	MCHF	AS	R
		A	f	A	A	A	
1	3	2.313+07	6.985−4	2.313+07	2.33+07	2.29+07	8.9−1
1	4	4.779+05	2.698−5	5.133+05	5.68+05	5.06+05	1.2−0
1	6	2.334+09	7.754−2	2.341+09	2.41+09	2.40+09	1.1−0
1	8	1.812+10	1.981−1	1.820+10	1.70+10	1.78+10	9.7−1
1	9	1.515+10	1.454−1	1.511+10	1.49+10	1.49+10	1.0−0
1	10	8.171+09	1.505−1	8.212+09	7.80+09	7.95+09	9.9−1
1	11	3.836+10	4.937−1	3.804+10	3.69+10	3.81+10	1.0−0
2	3	8.724+06	1.570−4	8.788+06	9.16+06	8.88+06	8.1−1
2	4	5.491+06	1.837−4	5.496+06	5.43+06	5.45+06	9.5−1
2	5	2.258+07	1.036−3	2.322+07	2.33+07	2.27+07	1.0−0
2	6	7.886+07	1.490−3	8.179+07	8.79+07	8.82+07	1.2−0
2	7	1.898+09	5.298−2	1.908+09	1.98+09	1.97+09	1.1−0
2	8	1.484+09	8.999−3	1.521+09	1.63+09	1.58+09	8.8−1
2	9	2.226+10	1.177−1	2.231+10	2.14+10	2.19+10	9.7−1
2	10	3.541+10	3.586−1	3.534+10	3.41+10	3.46+10	9.9−1
2	11	8.535+09	5.946−2	8.518+09	8.26+09	8.47+09	1.0−0

GRASP1: Present results for 136 levels with the GRASP code
GRASP2: Present results for 332 levels with the GRASP code
MCHF: Tayal [3]
AS: Liang et al [4]
R: Ratio of velocity/length form of f- values from the GRASP1 calculations

of 1.865 ns. The other measurements of τ available are by [6] for the $3s3p^2\ ^4P_{1/2,3/2,5/2}$ levels, which are 104±16, 811±77 and 406±33 μs, respectively. However, our corresponding theoretical results for these levels (251, 8004 and 3391 μs) are overestimated by up to an order of magnitude. The dominant contributing E1 transitions for these levels are 2 – 3/4/5 for which the f- values are $1.2{\times}10^{-6}$, $1.2{\times}10^{-7}$ and $4.4{\times}10^{-7}$, respectively, i.e. all transitions are very *weak* and for these there are large variations among different calculations. Finally, [7] have measured τ for the ($3s^2$) 4p and 4f levels of Si II to be 9.1±0.5 and 3.4±0.3 ns, which compare well with our calculations of 8.9 and 3.9 ns. For levels of Fe XIV, there are no discrepancies between theory and measurement.

5. Collision strengths and effective collision strengths

For the calculations of Ω, we have employed the *Dirac Atomic R-matrix Code* (DARC) of P H Norrington and I P Grant, available at the website `http://web.am.qub.ac.uk/DARC/`. The *R*-matrix radius (Ra) adopted for Si II is 36.8 au, and 56 continuum orbitals have been included for each channel angular momentum in the expansion of the wave function. The corresponding size of the Hamiltonian matrix is 15,154 and values of Ω are calculated up to an energy of 13 Ryd. The maximum number of channels for a partial wave is 268 and all partial waves with angular momentum $J \leq 40$ are included. For Fe XIV, the Ra is 4.44 au and 30 continuum orbitals have been included for each channel angular momentum. The corresponding size of the Hamiltonian matrix is 21,970 and Ω are calculated up to an energy of 260 Ryd, i.e. about 250 Ryd above the highest threshold considered in the work. The maximum number of channels for a partial

Journal of Physics: Conference Series **583** (2015) 012003 doi:10.1088/1742-6596/583/1/012003

wave is 728 and all partial waves with angular momentum $J \leq 40$ are included. This large J range ensures convergence of Ω for all forbidden and inter-combination transitions, and at all energies. However, for some allowed transitions, particularly at higher energies, Ω are not fully converged within this range, for which a top-up based on the Coulomb-Bethe approximation has been included. Furthermore, a little change in Ra does not affect the values of Ω.

Since Ω does not vary smoothly within the thresholds region, because of the closed-channel (Feshbach) resonances, calculations are performed in a fine energy mesh to account for their contribution. In most astrophysical (and fusion) plasmas, electrons have a *Maxwellian* distribution of velocities, and thereofore an averaged value, known as *effective* collision strength (Υ) is obtained. Depending on the type of transition and temperature, the contribution of resonances may greatly enhance the values of Υ over those of the background collision strengths (Ω_B). For Si II we have calculated Ω in a narrow energy mesh of 0.001 Ryd at over ~1500 energies in the thresholds region. The corresponding calculations for Fe XIV have been performed at over 10,500 energies, because of the wider thresholds energy range.

Bautista et al [2] have reported values of Υ for only 8 resonance transitions of Si II, at three temperatures of 5000, 10,000 and 20,000 K. However, Tayal [1] has reported Υ for a wider range of transitions and temperatures. For a majority of transitions, the Υ of Tayal are the highest, and differences with other calculations are up to 50%. In Fig. 1 we compare our Υ with those of [1] and [2] for three transitions, namely $3s^23p\ ^2P^o_{1/2}$ – $3s^23p\ ^2P^o_{3/2}$ (1–2), $3s^23p\ ^2P^o_{1/2}$ – $3s3p^2\ ^2D_{3/2}$ (1–6) and $3s^23p\ ^2P^o_{1/2}$ – $3s3p^2\ ^2S_{1/2}$ (1–9). For the 1–6/9 transitions the Υ of [2] are comparable to our results but are lower for 1–2, whereas the corresponding results of Tayal are significantly higher for all these three (and many more) transitions, and at (almost) all temperatures. For a few transitions, differences between the two calculations are even larger. For example, for the 2 – 3 transition the Υ of Tayal are lower by a factor of 70 towards the lower end of the temperature range, and are higher by ~ 50% for 2 – 5. Such a behaviour by the Tayal Υ data appears to be anomalous in comparison with our results, particularly when there is a considerably closer agreement towards the higher end of the temperature range.

For the most important transition of Fe XIV, i.e. $(3s^23p)\ ^2P^o_{1/2}$ – $^2P^o_{3/2}$ (1 – 2), our results of Υ are comparable with those of Tayal [3] (within ~20%), but of Liang et al. [4] are significantly higher (by up to a factor of two), particularly towards the lower end of the temperature range. In fact, at the lowest common temperature of $10^{3.593}$ K, the Υ of Liang et al. differ with our calculations by over 20% for ~70% of the transitions, and in a majority of cases their results are higher. For some transitions, such as 1 – 62/66/70/82/88/91/92/93, their data are greater by over an order of magnitude. Similarly, at the highest common temperature of $10^{6.991}$ K, the discrepancies between our results and those of Liang et al. are over 20% for ~46% of the transitions, and as at lower temperatures their Υ values are higher for almost all transitions. Discrepancies of up a factor of 2 are common for many transitions, but are over an order of magnitude for some, such as 1 – 94/112/131, 2 – 94/132 and 3 – 83/106/122. The most likely reasons for such large discrepancies are one or more of the following. (i) They included electron exchange only up to $J = 12$, and for $J \geq 13$ adopted a coarse mesh of energy of ~ 0.34 Ryd. (ii) They performed calculations of Ω in a limited range of energy (~ 90 Ryd), insufficient for the accurate determination of Υ up to $T_e = 4 \times 10^8$ K (~ 2533 Ryd), although they did take into account the high energy expansion of Ω. (iii) An error in the version of the code adopted, because some versions of the R- matrix code are known to contain errors, particularly related to the Breit-Pauli part which accounts for the relativistic effects. (iv) Possibility of the presence of pseudo resonances, because in the generation of wave functions [4] have included levels of additional 77 configurations which are not spectroscopic. If the pseudo resonances are not properly smoothed over then the calculated values of Υ may be abnormally high.

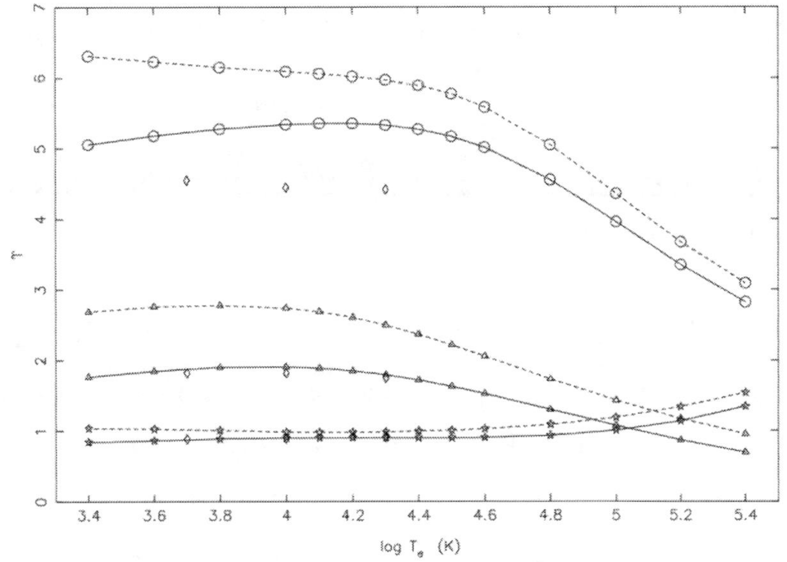

Figure 1. Comparison of effective collision strengths for the $3s^23p\ ^2P^o_{1/2}$ – $3s^23p\ ^2P^o_{3/2}$ (1–2: circles), $3s^23p\ ^2P^o_{1/2}$ – $3s3p^2\ ^2D_{3/2}$ (1–6: triangles) and $3s^23p\ ^2P^o_{1/2}$ – $3s3p^2\ ^2S_{1/2}$ (1–9: stars) transitions of Si II. Continuous lines – present results, broken lines are of Tayal [1] and results of Bautista et al [2] are shown as diamonds.

6. Conclusions

Based on comparisons with measurements and available theoretical results, our energy levels are estimated to be accurate to ~1%. However, scope remains for improvement, although the inclusion of extensive CI is not very helpful in improving the accuracy further, as also demonstrated by [2] for the levels of Si II. Similarly, for a majority of (strong) transitions our listed A- values (and other related parameters including lifetimes) are assessed to be accurate to better than 20%. For Υ the earlier results of Tayal [1] for Si II are assessed to be overestimated for a majority of transitions, and are also anomalous for a few. Similarly, the earlier calculations of [4] for Fe XIV, athough available for a larger range of transitions and over a wider range of temperature, appear to have significantly overestimated the values of Υ for almost all transitions and over an entire range of temperature. We believe the present set of complete results for radiative and excitation rates, for both Al-like ions, are the most reliable to date, and hence should be useful for the modelling of a variety of plasmas. Finally, complete results for all atomic parameters along with detailed comparisons and discussion for Si II are available in [8] and for Fe XIV in [9].

References

[1] Tayal S S 2008 *ApJS* **179** 534
[2] Bautista M A, Quinet P, Palmeri P, Badnell N R, Dunn J, Arav N 2009 *A&A* **508** 1527
[3] Tayal S S 2008 *ApJS* **178** 359
[4] Liang G Y et al 2010 *ApJS* **190** 322
[5] Schectman R M, Povolny H S, Curtis L J 1998 *ApJ* **504** 921
[6] Calamai A G, Smith P L, Bergeson S D 1993 *ApJL* **415** L59
[7] Bashkin S et al 1980 *Phys. Scr.* **21** 820
[8] Aggarwal K M and Keenan F P 2014 *MNRAS* **442** 388
[9] Aggarwal K M and Keenan F P 2014 *MNRAS* **445** 2015

Journal of Physics: Conference Series **583** (2015) 012004 doi:10.1088/1742-6596/583/1/012004

Plasmon excitation in valence shell photoelectron spectroscopy for PAHs

P M Mishra[1], L Avaldi[2], P Bolognesi[2], K C Prince[3], R Richter[3] and U Kadhane[1]

[1] Indian Institute of Space Science and Technology, Thiruvananthapuram 695547, Kerala, India

[2] CNR-Istituto di Metodologie Inorganiche e dei Plasmi, Area della Ricerca di Roma 1, CP10-00016 Monterotondo, Italy

[3] Elettra-Sincrotrone Trieste, Strada Statale 14, km 163.5, Area Science Park, I-34149 Basovizza, Italy

E-mail: `mishra.preetimanjari@gmail.com`

Abstract. The photon energy dependence of photoelectron spectra (PES) for two members of the polycyclic aromatic hydrocarbon (PAH) family namely pyrene and fluorene is studied in FUV regime (15 - 40 eV) using high-resolution synchrotron photoelectron spectrometer. The difference in outer (π dominated) and inner (σ dominated) valence relative photoelectron emission cross sections as a function of photon energy identifies the region of plasmon excitation ($\sim$ 15 - 27 eV). This excitation mode is present in the same region for both the molecule irrespective of the difference in their structure and symmetry. The feature is observed to be independent of the details of the molecular orbital associated with the outgoing electron. The results are in contradiction to the observed in benzene for inner valence bands. With the help of OVGF/cc-pVDZ calculations, the experimental bands are assigned as per their binding energy and symmetry. The first ionization potentials are estimated to be 7.436 ± 0.015 and 7.944 ± 0.055 eV for pyrene and fluorene, respectively.

1. Introduction

The photon collision with PAHs are significant to study because of their role in interstellar physics and chemistry [1] as well as on earth [2]. PAHs are efficient absorber of UV radiation coupling the electronic energy to vibrational excitation and emit in IR. The unidentified IR bands emitted from different galactic sources have resemblance with the spectroscopic signature of PAHs. The IR emission spectrum at 3.3, 6.2, 7.7, and 11.3 μm are attributed to PAHs [3]. A large fraction of interstellar carbon is locked in the form of PAHs and their cations , hydrogenated derivatives are considered as the possible carrier of diffuse interstellar bands [4, 5]. At the same time on earth they are considered as the mutagenic and carcinogenic [2, 6] element formed upon incomplete combustion of carbon (C) rich compounds like fuel, coal, crude oil.

PAHs are a family of hydrocarbons consisting of molecules where C atoms are arranged in a honeycomb lattice type structure of fused six-member aromatic rings with H atoms located at the periphery of the ring. The smallest member of this family is benzene and the family expands by attaching more number of benzene rings linearly or otherwise. The inherent stability for PAHs and aromatic character is due to the existence of delocalized electrons above and below the plane of molecule due to the sp^2 hybridized C atoms.

PAHs are associated with a multiparticle excitation mode upon interaction with external radiation known as plasmon excitation. This mode of excitation is already established for two dimensional graphite (at $\sim$ 25 eV) to three dimensional C_{60} (at $\sim$ 21 eV) [7, 8]. For PAHs it was observed to be at $\sim$ 17 eV in electron energy loss spectroscopy studies of PAH films and photo ion yield studies in the gas phase [7, 9]. For other PAHs like, phenanthrene, anthracene, fluoranthene, pyrene, triphenylene, plasmon resonance is observed in the range of 7 - 21.2 eV for gas phase VUV photoionization mass spectrometry studies [10] and it was found that plasmon excitation leads to autoionization [9]. Hence this common mode of excitation at $\sim$ 17 eV for PAHs [10, 7] is attributed to the collective or plasmon excitation as well as to a mixture of various inner shell transition like $\pi \rightarrow \sigma*, \sigma \rightarrow \pi*, \sigma \rightarrow \sigma*$ transitions. This mode of excitation is shown to play a major role in PAHs in studies like photon impact [10, 7] and ion impact [11].

So far the PES experiments for PAHs are done using He - I and II discharge sources [12, 13, 14] and the theoretical measurements also rely on it. In the present study, synchrotron radiation has been used to investigate the systematic variation of the state selective PE cross sections over a wide range of photon energies. This state selective photoemission intensities have never been measured for PAHs except benzene [15]. In addition to this a detail study of photon energy dependence of PES for a wide range of photon energy in UV regime is performed for the first time for PAHs.

2. Experimental details

The experiment was performed at the GASPHASE photoemission beamline at the Elettra-Sincrotrone Trieste laboratory (Trieste, Italy). The details about the beamline (resolving power 2000 at the energies used here) can be found elsewhere [16, 17]. Pyrene and fluorene were obtained from Sigma-Aldrich at >99% purity. The photon beam crosses the target placed in a copper container having a 3.3 cm length nozzle with a 3 mm opening. The container for pyrene was heated resistively to $\sim$74°C (for obtaining the vapor pressure), which was kept constant with the help of a temperature controller. For fluorene, the vapor pressure was sufficiently high to avoid any heating. PES for a wide range of photon energies from 15 to 40 eV were measured using a hemispherical electron energy analyzer with six channel electron multiplier (CEM) detectors with a resolution down to 50 meV. The spectrometer acceptance angle is a cone of $\pm 4°$ with the axis of the cone oriented at the magic angle with respect to the direction of polarization.

3. Experimental data analysis

The relative photoelectron cross sections were obtained by including photon flux normalization and spectrometer efficiency correction (assuming a constant target density). These relative cross sections are directly proportional to the absolute cross sections, by a constant factor unique to each target molecule. Xenon ionization cross section from the same set up (with identical experimental set up parameters) were used to obtain the normalization to the photon flux after comparing with the previously published absolute cross sections for Xe 5p photoionization [18, 19]. This procedure is followed because of the change in photodiode efficiency with photon energy which makes the photon beam intensity measurement with this photodiode unreliable for normalization. For the spectrometer efficiency correction, Xe Auger electron spectra at 95.1 ev photon energy (figure 1) with present set up is compared with the Auger intensity distribution reported in previous investigations [20]. The efficiency curve was obtained by fitting a polynomial to the measured factors for each individual CEM (figure 2).

4. Experimental results

The HOMO of PES is associated with vibrational structure resolved up to $3 \leftarrow 0$ progression in the case of pyrene due to symmetry as compare to fluorene. The first ionization potential

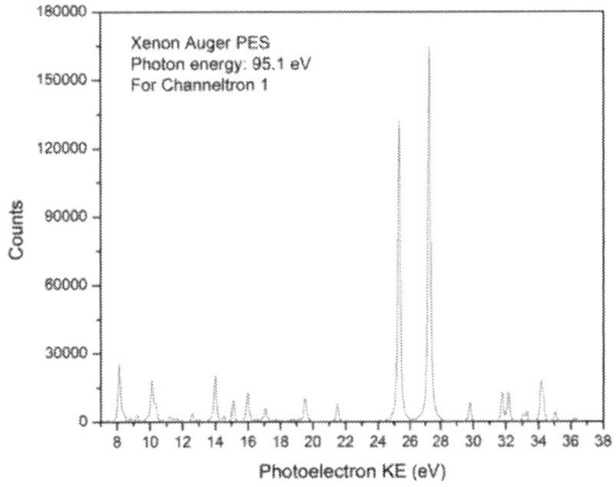

Figure 1. Xenon Auger electron spectrum at 95.1 eV photon energy.

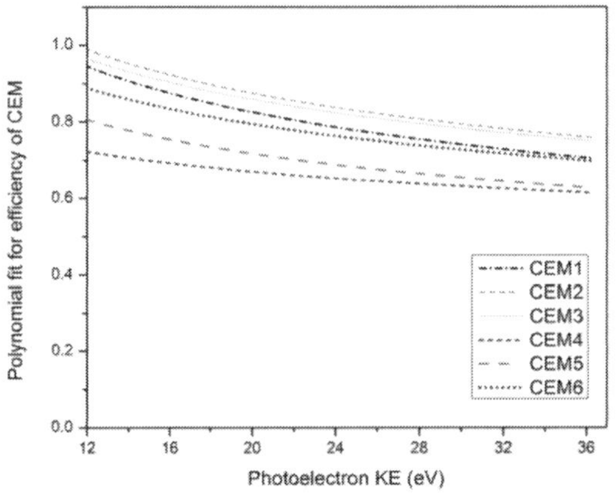

Figure 2. The efficiency variation for different CEMs as a function of photoelectron kinetic energy.

agrees very well with the highly accurate results of studies like REMPI-ZEKE and TPEPICO. The inner valence band onwards in PES are broadened due to the breakdown of MO picture as described by Deleuze [21] which involves the strong electron-electron correlation leading to redistribution of intensity around the main peak.

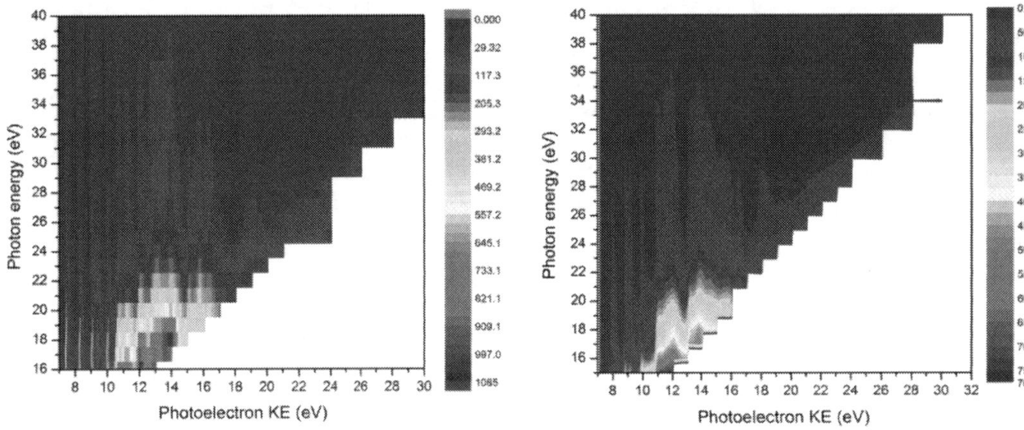

Figure 3. Contour plot for PES as a function of photon energy for pyrene (left) and fluorene (right).

The relative photoelectron cross sections are plotted as a function of photoelectron kinetic energy and photon energy in figure 3. Despite the fundamental differences in the structure of the two molecules, such as symmetry, the presence of sp^3-hybridized carbon and cyclopentane ring, etc., the photon energy dependences are markedly similar. This figure shows the enhanced intensity for the inner valence band in the photon energy region < 27 eV which is attributed to the plasmon excitation. The cross section again follows an almost uniform variation for the photon energy > 27 eV for the studied range. The trend in outer valence band (pure π MOs) and inner bands (mixture of π and σ as well as pure σ bands) are almost same for all photon energy for the case of pyrene. Whereas for fluorene the PE cross section for entire band differs except the outer valence band. Hence it is concluded that the inner valence band couples to plasmon excitation more efficiently. The most distinct characteristic we observe is the peak of cross sections at $\sim$ 17 eV particularly for the outer valence bands of pyrene (a plateau in the case of fluorene).

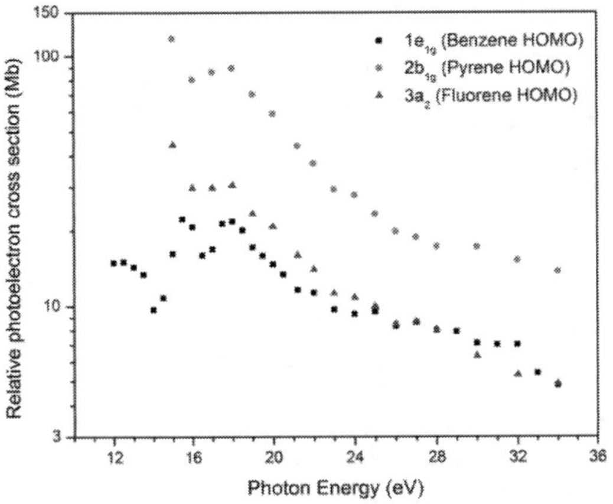

Figure 4. Comparison of HOMO photoelectron cross section with that of benzene study.

This study for the first time reports the quantitative intensity variation showing the parallel trend for π orbital photoelectron cross section for different photon energy which is theoretically studied by Carravetta *et al* [22] for large polyenes. However for the benzene case [23], particularly at a lower photon energy the individual π MO cross sections show a different trend which is possibly due to the role of autoionization in large PAHs. Another contradictory measurement with benzene is the increasing evolution of inner bands for benzene as a function of photon energy [23, 15]. Figure 4 shows no difference in behavior of PE cross section from HOMO for benzene, pyrene, fluorene for the studied range of photon energy. The detail discussion can be found elsewhere [24].

5. Computational results and analysis

The outer valence Green's function calculations were performed using the nonlocal hybrid Becke three parameter Lee-Yang-Parr functional (B3LYP) and Dunning's correlation consistent polarized valence basis set of double-ζ quality (cc-pVDZ) incorporated in GAUSSIAN09 [25] to estimate molecular orbital ionization energies with respective pole strengths. The ground-state geometry optimization was done within the constraint of D_{2h} and C_{2v} symmetry point groups for pyrene and fluorene respectively, using density functional theory (DFT).

The convoluted spectrum obtained from OVGF/cc-pVDZ calculations using Lorentzian and Gaussian function reproduced the experimental PES binding energy positions as well as their intensity to a reasonably good accuracy for the valence MOs. Orbital symmetry assignments are made by comparing the binding energies of experimental and theoretical peaks obtained by OVGF calculations.

6. Conclusions

PAHs due to their significance in nature form a very intriguing system to investigate. Taking Pyrene and fluorene as two examples of this family, a two dimensional photoionization measurement was performed with binding energy and photon energy as variable. The binding energy structure was well reproduced with the help of OVGF calculations particularly in the outer valence region where the shakeup and shakeoff processes are less dominant. The evolution of PES as a function of photon energy demonstrated clearly that at the lower photon energies the inner valence bands contribute substantially compared to the outer valence bands. But beyond 27 eV photon energy the outer valence bands dominate. Over all, the cross sections for

Journal of Physics: Conference Series **583** (2015) 012004 doi:10.1088/1742-6596/583/1/012004

both inner and outer valence orbital are found to peak at around 17 eV photon energy. This observation along with the previously studied photoion yield studies and EELS studies indicate the plasmon like excitation which appears to be independent of individual valence bands. The observations are explained in terms of this plasmon excitation process which couples with various ionization channels after photoexcitation.

References

[1] Tielens A G 2008 *Annu. Rev. Astron. Astrophys.* **46** 289–337
[2] Yu H 2002 *J. Environ. Sci. Heal. C* **20** 149–83
[3] Tielens A G 2005 *The physics and chemistry of the interstellar medium* (Cambridge University Press, New York)
[4] Salama F, Bakes E, Allamandola L and Tielens A 1996 *Astrophys. J.* **458** 621–36
[5] Dabestani R and Ivanov I N 1999 *Photochem. Photobiol.* **70** 10–34
[6] Dong S, Hwang H M, Harrison C, Holloway L, Shi X and Yu H 2000 *B. Environ. Contam. Tox.* **64** 467–74
[7] Jochims H, Baumgärtel H and Leach S 1999 *Astrophys. J.* **512** 500–10
[8] Bertsch G F, Bulgac A, Tománek D and Wang Y 1991 *Phys. Rev. Lett.* **67** 2690
[9] Ling Y and Lifshitz C 1996 *Chem. Phys. Lett.* **257** 587–91
[10] Jochims H, Rühl E, Baumgärtel H, Tobita S and Leach S 1997 *Int. J Mass Spectrum.* **167/168** 35–53
[11] Kadhane U, Kelkar A, Misra D, Kumar A and Tribedi L C 2007 *Phys. Rev. A* **75** 041201
[12] Boschi R, Clar E and Schmidt W 1974 *J. Chem. Phys.* **60** 4406–18
[13] Schmidt W 2008 *J. Chem. Phys.* **66** 828–45
[14] Klasinc L *et al.* 1980 *Pure Appl. Chem.* **52** 1509–24
[15] Baltzer P, Karlsson L, Wannberg B, Öhrwall G, Holland D, MacDonald M, Hayes M and Von Niessen W 1997 *Chem. Phys.* **224** 95–119
[16] Blyth R, Delaunay R, Zitnik M, Krempasky J, Krempaska R, Slezak J, Prince K, Richter R, Vondracek M, Camilloni R *et al.* 1999 *J. Electron Spectrosc. Relat. Phenom.* **101** 959–64
[17] Plekan O, Coreno M, Feyer V, Moise A, Richter R, de Simone M, Sankari R and Prince K 2008 *Phys. Scripta* **78** 058105
[18] Fahlman A, Krause M O, Carlson T A and Svensson A 1984 *Phys. Rev. A* **30** 812–19
[19] Kutzner M, Radojević V and Kelly H 1989 *Phys. Rev. A* **40** 5052–57
[20] Kivimäki A, Pfeiffer L, Aksela H, Nommiste E and Aksela S 1999 *J. Electron Spectrosc. Relat. Phenom.* **101** 43–47
[21] Deleuze M S 2002 *J. Chem. Phys.* **116** 7012–26
[22] Carravetta V, Yang L and Ågren H 1997 *Phys. Rev. B* **55** 10044–50
[23] Carlson T A, Gerard P, Krause M O, Grimm F A and Pullen B 1987 *J. Chem. Phys.* **86** 6918–26
[24] Mishra P M, Avaldi L, Bolognesi P, Prince K C, Richter R and Kadhane U R 2014 *J. Phys. Chem. A* **118** 3128–35
[25] Frisch M J, Trucks G W, Schlegel H B, Scuseria G E, Robb M A, Cheeseman J R, Scalmani G, Barone V, Mennucci B and et al G A P 2009 Gaussian 09 Revision D.01 gaussian Inc. Wallingford CT 2009

Calculation of differential cross section for dielectronic recombination with one-electron uranium

K N Lyashchenko and O Yu Andreev

Department of Physics, St. Petersburg State University, Ulyanovskaya 3, Petergof, St. Petersburg, 198504, Russia

E-mail: laywer92@mail.ru

Abstract. Calculation of the differential cross section for the dielectronic recombination with one-electron uranium within the framework of QED is presented. The contribution of the QED corrections and the interfernce of the photon multipoles is investigated.

1. Introduction

The electron recombination, in particular, the dielectronic recombination with one-electron uranium is under intensive investigation during the last decade [1, 2, 3, 4, 5]. The present paper is devoted to calculation of the differential cross section of the dielectronic recombination with one-electron uranium within the framework of QED.

We study a process, where an ion captures an electron and goes into a single excited state with emission of one photon

$$e^- + U^{91+}(1s) \quad \to \quad U^{90+}(r) + \gamma \to U^{90+}(1s1s) + \gamma + \gamma' , \tag{1}$$

where r denotes one of the single excited states $(1s2s)$, $(1s2p)$. The first emitted photon (γ) can be registered in the experiment [5]. The interference between the first (γ) and the second emitted (γ') photons can be neglected [2].

If the energy of the initial state is close to the energy of a double excited state, the cross section shows resonances. The resonances are explained by the dielectronic recombination

$$e^- + U^{91+}(1s) \quad \to \quad U^{90+}(d) \to U^{90+}(r) + \gamma \to \dots , \tag{2}$$

where d is one of the double excited states $(2s2s)$, $(2s2p)$, $(2p2p)$.

2. Electron recombination with a bare nucleus

We consider first the electron recombination with a bare nucleus. In the zeroth order of the QED perturbation theory the amplitude of electron recombination is written as [6]

$$U \quad = \quad e \int d^3\mathbf{r}\, \overline{\psi}_b(\mathbf{r}) \gamma^\nu A_\nu^{(k,\lambda)*}(\mathbf{r}) \psi_{\mathbf{p}\mu}(\mathbf{r}) , \tag{3}$$

where e is the electron charge; $\psi_{\mathbf{p},\mu}$ is a wave function of the incident electron with certain momentum $\mathbf{p}$ and polarization μ; $\psi_b = \psi_{njlm}$ is the wave function of the bound electron; 4-vector $A_\nu^{(k,\lambda)*}$ describes the emitted photon with wave vector $k = (\omega, \mathbf{k})$ and polarization λ ($A_0^{(k,\lambda)} = 0$, the transverse gauge is used); γ^ν – Dirac gamma matrices.

In the transverse gauge the photon wave function reads (the relativistic units are used)

$$\mathbf{A}^{(k,\lambda)}(\mathbf{r}) = \sqrt{\frac{2\pi}{\omega}} e^{i\mathbf{kr}} \mathbf{e}^{(\lambda)} . \tag{4}$$

This function can be presented as [7]

$$\mathbf{A}^{(k,\lambda)} = \sqrt{\frac{2\pi}{\omega}} \sum_{jlm} i^l g_l(\omega r)(\mathbf{e}^{(\lambda)}, \mathbf{Y}^*_{jlm}(\mathbf{k}))\mathbf{Y}_{jlm}(\mathbf{r}) , \tag{5}$$

where $g_l(x) = 4\pi j_l(x)$ and $j_l(x)$ is the spherical Bessel function.

The wave function $\psi_{\mathbf{p},\mu}(\mathbf{r})$ describes the incident electron with certain momentum ($\mathbf{p}$), energy ($\epsilon = \sqrt{1+\mathbf{p}^2}$) and polarization ($\mu$). It is convenient to expand the wave function $\psi_{\mathbf{p},\mu}(\mathbf{r})$ in series over the wave functions of electron with certain energy, total angular momentum (j), its projection (m), and parity (l) [6]

$$\psi_{\mathbf{p}\mu}(\mathbf{r}) = \int d\epsilon \sum_{jlm} a_{\mathbf{p}\mu,\varepsilon jlm} \psi_{\varepsilon jlm}(\mathbf{r}) , \tag{6}$$

$$a_{\mathbf{p}\mu,\varepsilon jlm} = \frac{(2\pi)^{3/2}}{\sqrt{p\epsilon}} i^l e^{i\phi_{jl}} [\Omega^+_{jlm}(\mathbf{p}) v_\mu(\mathbf{p})] \delta(\epsilon - \varepsilon) , \tag{7}$$

where $\Omega_{jlm}(\mathbf{p})$ is the spherical spinor and $v_\mu(\mathbf{p})$ is the spinor with certain projection on the electron momentum ($\mathbf{p}$), the phase ϕ_{jl} is the Coulomb phase shift. The wave function $\psi_{\varepsilon jlm}$ are normalized as

$$\int d^3\mathbf{r}\, \psi^+_{\varepsilon jlm}(\mathbf{r})\, \psi_{\varepsilon' j' l' m'}(\mathbf{r}) = \delta(\varepsilon - \varepsilon')\delta_{jj'}\delta_{ll'}\delta_{mm'} . \tag{8}$$

Accordingly, the electron recombination amplitude can be written as

$$\begin{aligned} U_{if} = & -e \int d^3\mathbf{r} \sum_{j_p l_p m_p} \sum_{jlm} \psi^+_{n_b j_b l_b m_b}(\mathbf{r}) \vec{\alpha} \sqrt{\frac{2\pi}{\omega}} (-i)^{l_p} g_{l_p}(\omega r)(\mathbf{e}^{(\lambda)*} \mathbf{Y}_{j_p l_p m_p}(\mathbf{k}))\mathbf{Y}^*_{j_p l_p m_p}(\mathbf{r}) \\ & \times \frac{(2\pi)^{3/2}}{\sqrt{p\epsilon}} i^l e^{i\phi_{jl}} [\Omega^+_{jlm}(\mathbf{p}) v_\mu(\mathbf{p})] \psi_{\varepsilon jlm}(\mathbf{r}) , \end{aligned} \tag{9}$$

where $\vec{\alpha}$ are Dirac alpha matrices.

3. Electron recombination with a one-electron ion

The initial state (the bound ($1s$) electron and the incident electron) is described by

$$\Psi_i = \frac{1}{\sqrt{2}} \det\{\psi_{\mathbf{p},\mu}, \psi_{n_b j_b l_b m_b}\} . \tag{10}$$

This function can be expanded over two-electron functions with certain total momentum (in the j-j coupling scheme)

$$\Psi_i = \sum_{JMj_1j_2l_1l_2} \langle \Psi^{(0)}_{JMj_1j_2l_1l_2} | \Psi_i \rangle \, \Psi^{(0)}_{JMj_1j_2l_1l_2} , \tag{11}$$

where

$$\Psi^{(0)}_{JMj_1j_2l_1l_2} = N \sum_{m_1m_2} C^{j_1j_2}_{JM}(m_1m_2)\det\{\psi_{n_1j_1l_1m_1}, \psi_{n_2j_2l_2m_2}\}. \tag{12}$$

Here, n – the principle quantum number, j – total momentum of electron, l – parity, m – total momentum projection, N – normalization constant, $C^{j_1j_2}_{JM}(m_1m_2)$ – Clebsch-Gordan coefficients.

The interaction with the quantized electromagnetic and electron-positron fields leads to various correction to the amplitude: interelectron interaction correction, electron self-energy and vacuum polarization corrections. For consideration of these corrections the line-profile approach (LPA) was employed [3, 4]. Within the LPA we introduce the set of two-electron configurations (g) which includes all the two-electron configurations composed by $1s, 2s, 2p, 3s, 3p, 3d$ electrons and the incident electron e.

Within the LPA there is also introduced the matrix V which is defined by the one and two-photon exchange, electron self-energy and vacuum polarization matrix elements. Matrix $V = V^{(0)} + \Delta V$ is considered as a block matrix

$$V = \begin{bmatrix} V_{11} & V_{12} \\ V_{21} & V_{22} \end{bmatrix} = \begin{bmatrix} V^{(0)}_{11} + \Delta V_{11} & \Delta V_{12} \\ \Delta V_{21} & V^{(0)}_{22} + \Delta V_{22} \end{bmatrix}. \tag{13}$$

Matrix V_{11} is defined on set g, which contains configurations mixing with the reference state $n_g \in g$. Matrix $V^{(0)}$ is a diagonal matrix: sum of Dirac energies. Matrix ΔV_{11} is not a diagonal matrix, but it contains a small parameter of the QED perturbation theory. Matrix V_{11} is a finite matrix and can be diagonalized numerically:

$$V^{diag}_{11} = B^{+}V_{11}B, \qquad B^{+}B = I. \tag{14}$$

Taking into account the interaction with the quantized fields leads to the substitution of

$$\Phi_{n_g} = \sum_{k_g \in g} B_{k_g n_g}\Psi^{(0)}_{k_g} + \sum_{k \notin g, l_g \in g} [\Delta V_{21}]_{kl_g} \frac{B_{l_g n_g}}{E^{(0)}_{n_g} - E^{(0)}_{k}} \Psi^{(0)}_{k} + \cdots. \tag{15}$$

for $\Psi^{(0)}_{n_g}$ functions, where $n_g = (JMj_1j_2l_1l_2)$ is a reference state. Here, $E^{(0)}$ is the energy of the corresponding two-electron configuration.

The amplitude of the electron recombination is given as a matrix element of the photon emission operator ($\Xi^{(0)}$) with the bra and ket-vectors given by the corresponding Φ_{n_g} functions:

$$U_{if} = \langle \Phi_f | \Xi^{(0)} | \Phi_i \rangle, \qquad \Xi^{(0)}_{ab} = -\int d^3\mathbf{r}\, \psi^{+}_{a}(\mathbf{r})(\vec{\alpha}\mathbf{A}^{(\mathbf{k},\lambda)*}(\mathbf{r}))\psi_i(\mathbf{r}). \tag{16}$$

4. Results

We preset calculation of the diferential cross section for the electron recombination with one-electron ion of uranium being in the ground state ($1s$) as a function of the incident electron energy (ϵ_e) and the angle of the emitted photon in the rest frame of the ion. The calculation is performed for the kinetic energy of the incident electron ($\epsilon_e - m_e c^2$) within the interval $[62.5, 64.2]$ keV. It supposed that the process of electron recombination is registered by registration of emitted photons with energy near the value $\omega \approx \epsilon_e + 1s - r$, where r denotes a single excited configuration: $(1s2s)$, $(1s2p)$. The higher orders of the multipole expansion of the photon ($0 < l_0 < 10$) were taken into account, what led to changes of $2 - 6\%$ in the peaks area and up to 30% in the rest area of the differential cross section. The total cross section of the electron recombination was also calculated and found good agreement with the previous works [4, 5].

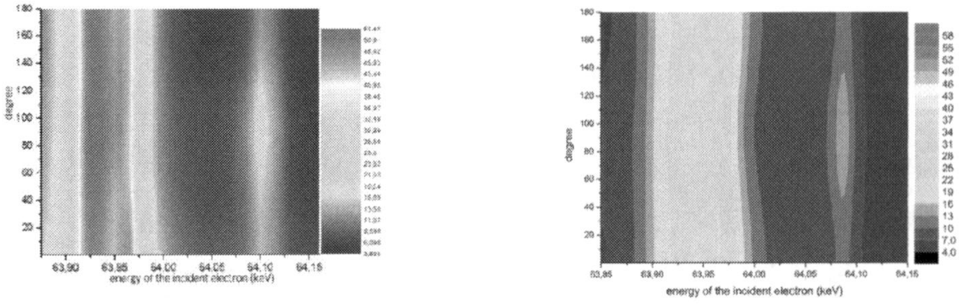

Figure 1. The differential cross section for recombination of non-polarized electrons with non-polarized ions. The left graph corresponds to the full QED calculation and the right graph represents the calculation with disregard of the Breit interaction.

Results for the differential cross section (in barn/str) for recombination of non-polarized electron with non-polarized ions are presented in Fig.(1).

Results for the differential cross section (in barn/str) for recombination of polarized ($\mu = 1/2$) electron with polarized ($m = 1/2$) ions are presented in Fig.(2).

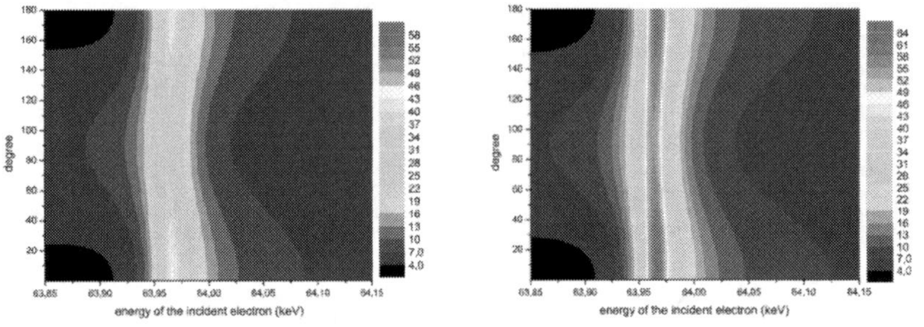

Figure 2. The differential cross section for recombination of polarized electron with polarized ions. The left graph corresponds to the full QED calculation and the right graph represents the calculation with disregard of the Breit interaction.

The presented calculation shows that the differential cross section of the electron and dielectronic recombination is very sensitive to the QED corrections and to the interference of the photon multipoles.

Acknowledgments

The authors acknowledge the support by RFBR grants 14-02-20188 and St. Petersburg State University for a research Grant No. 11.38.227.2014.

References

[1] Karasiov V V, Labzowsky L N, Nefiodov A V, and Shabaev V M 1992 *Phys. Lett. A* **161** 453
[2] Zakowicz S, Scheid W and Grün N 2004 *J. Phys. B: At. Mol. Opt. Phys.* **37** 131
[3] Andreev O Yu, Labzowsky L N, Plunien G and Solovyev D A 2008 *Phys. Rep.* **455** 135
[4] Andreev O Yu, Labzowsky L N and Prigorovsky A V 2009 *Phys. Rev. A* **80** 042514
[5] Bernhardt D *et al.* 2011 *Phys. Rev. A* **83** 020701
[6] Akhiezer A I 1965 *Quantum Electrodynamics* (New York: Wiley Interscience)
[7] Labzowsky L N 1996 *Theory of atoms. Quantum electrodynamics of the electron shells and the processes of radiation (in Russian)* (Moscow: Nauka)

Journal of Physics: Conference Series **583** (2015) 012006 doi:10.1088/1742-6596/583/1/012006

New spectroscopic data for atomic tungsten XIV

M Raineri [13], J Reyna Almandos [1], M Gallardo [1], C J B Pagan [2]
[1]Centro de Investigaciones Opticas (CIOp), CC 3, 1897 Gonnet, La Plata, Argentina
[2]School of Electrical and Computer Engineering, University of Campinas (UNICAMP), 13083-970 Campinas, SP, Brazil

[3]E-mail: monicar@ciop.unlp.edu.ar

Abstract. The thirteen times ionized tungsten is isoelectronic with PmI. Wavelengths and transition probabilities for the 5s-5p and 5p-5d transitions of WXIV, identifying the ground state as $4f^{13}\ 5s^2\ {}^2F_{7/2}$ were calculated. Both, a relativistic Hartree Fock approach, including core-polarization effects, and a purely relativistic multiconfiguration Dirac-Fock method were used for the calculations. Particularly, 5s-5p transitions were compared with experimental results obtained with VUV electron beam ion trap (EBIT) spectroscopy.

1. Introduction

The thirteen times ionized tungsten (WXIV) is isoelectronic with PmI. Reliable spectroscopic parameters in the tungsten atom are of fundamental importance for the study of fusion plasmas, but some discrepancies remain unsolved. On one hand, Curtis and Ellis [1], using a nonrelativistic single-configuration model, identified the collapse of the 4f orbital and predicted the wavelengths for 5s-5p transitions setting the ground state as [Kr] $4d^{10}\ 4f^{14}\ 5s\ {}^2S_{1/2}$. On the other, Theodosiou and V. Raftopoulos [2] used a more sophisticated multiconfigurational relativistic approach to identify the ground state as $4f^{13}\ 5s^2\ {}^2F_{5/2}$, but predicting the wavelengths of 5s-5p transitions far away from the experimental values [3].

In this work we made HFR[4] and GRASP [5] calculations for wavelength and transition probabilities for the 5s-5p (table 1) and 5p-5d (table 2) transitions, identifying the ground state as $4f^{13}\ 5s^2\ {}^2F_{7/2}$ in accordance with the paper of Kramida [6], showing good agreement with the observed values of the reference [3]. We also took into account the core polarization effects that refer to the deformation of the internal atomic orbitals due to the orbit of the active electron, which repels the remaining electrons (section 2.2)

2. Theory and discussions

Three calculation methods were used in this paper to obtain the wavelengths gA and gf. For the first two methods, Cowan's package [4] was used, and the third method used the GRASP relativistic program [5].

2.1. The Relativistic Hartree–Fock method

In Cowan's package, wavefunctions are calculated in a Hartree–Fock approximation with relativistic corrections. The wavefunctions are used to calculate a multiconfigurational energy matrix. Both the eigenvalues and the eigenvectors of the matrix are functions of the Slater parameters, i.e., functions of the average configuration energy E_{ave}, electrostatic direct F^k and exchange G^k integrals,configuration interaction integrals R^k, and spin–orbit parameters ζ_{nl}, as well as the coefficients of the multiconfigurational expansion (Ref. [4, Eq. (16.7)]). This method was used in several previous paper (see for example M.Raineri et.al Ref [7])

In this calculation the following configurations were included: $4f^{14}5s$, $4f^{14}5d$, $4f^{13}5s5p$, $4f^{13}5p5d$, $4f^{14}5p$, $4f^{13}5s^2$, $4f^{12}5s^25p$, $4f^{13}5p^2$ and $4f^{13}5s5d$ for both parities.

Figure 1 shows the average energy values (E_{ave}) along the promethium isoelectronic sequence of all configurations considered in our Relativistic Hartree-Fock calculation.

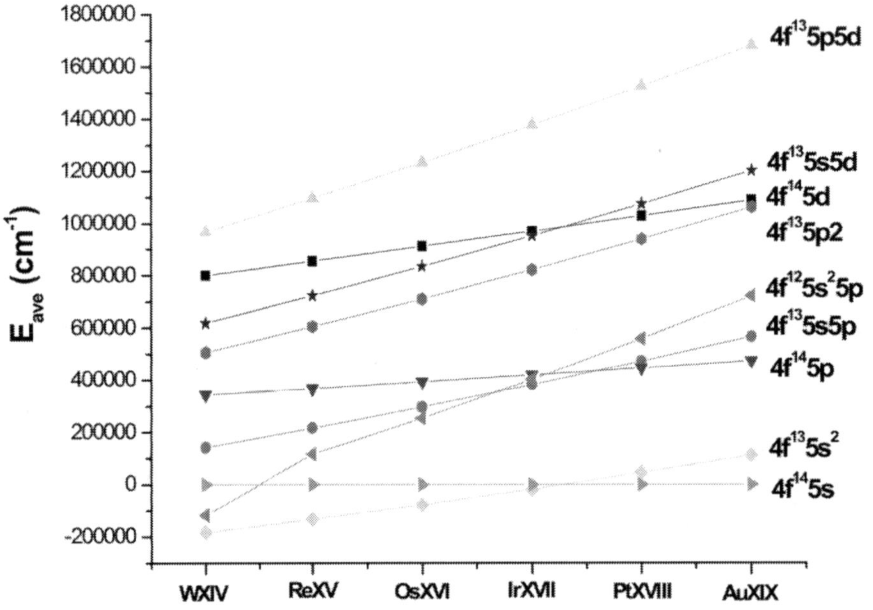

Figure 1. The average energy values (E_{ave}) along the promethium isoelectronic sequence of all configurations considered in our Relativistic Hartree-Fock calculation

In figure 1 we can observe that the ground state of Pm-like ions starting with W XIV, is the odd-parity $4f^{13}5s^2$ and in our relativistic approach it predicts the change to happen at Ir XVII. In accordance with the paper of Kramida [6] our HFR[4] and GRASP[5] calculations identify the ground state as $4f^{13}\,5s^2\ {}^2F_{7/2}$ separated from the first excited state $4f^{13}\,5s^2\ {}^2F_{5/2}$ by approximately 18000 cm $^{-1}$.

2.2 Hartree–Fock plus core polarization

We included the Core Polarization (CP) effects (see, for example, Curtis L.J. [8] and Biémont [9]) just by replacing the dipole integral

$$\int_0^\infty P_{nl}(r) r P_{n'l'}(r) dr \quad \rightarrow \int_0^\infty P_{nl}\, r \cdot \left[1 - \frac{\alpha_d}{(r^2+r_c^2)^{3/2}}\right] P_{n'l'}(r) dr - \frac{\alpha_d}{r_c^3} \int_0^\infty P_{nl}(r) P_{n'l'}(r) dr$$

Here α_d is the electric dipole polarizability of the core, and r_c is the cut-off radius, which defines the boundaries of the atomic core. This is the same modification used by Quinet [10] to correct transition matrix elements when including CP effects. In our case, the radial functions were obtained from the

Journal of Physics: Conference Series **583** (2015) 012006 doi:10.1088/1742-6596/583/1/012006

single configuration Hartree-Fock Method with relativistic corrections, and no modification was done to include CP effects in the Hamiltonian. As suggested by Migdalek & Baylis [11], core polarization approximately describes the core-valence correlation effects. As it is shown in table 1 and table 2 we present calculations with two values of the polarizability using different ion cores: one when the core has a [Kr]$4d^{10}5s^2$ (WXXVII) (α= 0.233a_0^3 [12] structure, and another for transitions that preserve the [Kr]$4d^{10}$ (WXXIX) core (α= 0.0716 a_0^3). For this last value of polarizability we considered the expression [4]:

$$\alpha_d = 4Ry^2 \sum_{yJ}^{M} f_{y'J',yJ}\ \lambda^2_{y'J',yJ}$$

where f is the oscillator strength of the transition, λ its corresponding wavelength, and Ry is the Rydberg constant.

2.3 Relativistic Dirac–Fock calculations

The GRASP package solves the Dirac equations within the framework of relativistic quantum theory [5]. This program offers energy levels wavelength and dipole transition rates from a multiconfigurational, relativistic approach. The configurations included in the calculation were: $4f^{14}$ 5s, $4f^{14}$ 5d, $4f^{13}$ 5s5p, $4f^{14}$ 5p, $4f^{13}$ $5s^2$, $4f^{13}$ $5p^2$ and $4f^{13}$ 5s5d.

3. Results and conclusion

In table1 and table 2 we present the calculations described above for transitions 5s-5p, 5p-5d respectively. The Slater parameters (section 2.1) were scaled at 85% of their RHF values. In table 1 we compared our results with the Reference [3] showing good accordance between λ_{HFR} for 5s $^2S_{1/2}$ - $5p^2P_{1/2}$ to experimental wavelength.

Table 1 Wavelengths, transition rates and oscillator strengths for 5s – 5p transition of WXIV

Transition	λ_{HFR} [Å]	λ_{MCDF} [Å]	$\lambda_{Ref[3]}$ [Å]	Experimental Wavelength[3]	Log gf	gA	gA CP α = 0.0716 a_0^3	gA CP α = 0.233 a_0^3
5s $^2S_{1/2}$-5p$^2P_{1/2}$	365.9	372.1	366.2	365.3	-0.288	2.564 10^{10}	2.203 10^{10}	1.843 10^{10}
5s $^2S_{1/2}$-5p$^2P_{3/2}$	264.2	263.2	263.7	258.2	0.145	1.332 10^{11}	1.141 10^{11}	9.571 10^{10}

Table 2 Wavelengths, transition rates and oscillator strengths for 5p–5d transition of WXIV

Transition	λ_{HFR} [Å]	λ_{MCDF} [Å]	Log gf	gA	gA CP α = 0.0716 a_0^3	gA CP α = 0.233 a_0^3
5p $^2P_{1/2}$-5d $^2D_{3/2}$	194.1	192.8	0.322	3.711 10^{11}	3.231 10^{11}	2.955 10^{11}
5p $^2P_{3/2}$-5d $^2D_{3/2}$	243.9	243.5	-0.433	4.131 10^{10}	3.655 10^{10}	3.284 10^{10}
5p $^2P_{3/2}$-5d $^2D_{5/2}$	233.9	233.2	0.532	4.151 10^{11}	3.680 10^{11}	3.3020 10^{11}

The difference between the various calculations can be summarized as follows: GRASP took into account relativistic effects by means of a more complete approach than Cowan's package by the HFR method; HFR took into account correlation effects in a more complete way. Finally, in the HFR + CP, not only were correlation effects considered more deeply, but core polarization effects were taken into account in the gA calculations for the different values of polarizabilities.

Acknowledgment

This research was supported by the Consejo Nacional de Investigaciones Científicas y Técnicas (CONICET), Argentina, by the Coordenação de Aperfeiçoamento de Pessoal de Nível Superior (CAPES) and by the Conselho Nacional de Pesquisa (CNPq), Brazil. The Comisión de Investigaciones Científcas de la Provincia de Buenos Aires (CIC), where J. G. Reyna Almandos and M. Raineri are researchers, is also gratefully acknowledged

References

[1] Curtis L J, and Ellis D G 1980 *Phys. Rev. Lett.* **45** 2099

[2] Theodosiou C E, and Raftopoulos V 1982 *Phys. Rev.* A **28** 1186

[3] Hutton R, Zou Y,. Reyna Almandos J, Biedermann C, Radtke R R, Greier A, and Neu R 2003 *Nuc. Instrum. Methods Phys. Res.* B **205** 114

[4] Cowan R D 1981 *The Theory of Atomic Structure and Spectra* (Berkeley, CA: University of California Press)

[5] Dyal KG, Grant IP, Johnson C T, Parpia F A and Plummer E P 1989 *Comp. Phys. Commun.* **55** 425

[6] Kramida A E and Shirai T 2009 *At. Data Nucl. Data Tables* **95** ,305

[7] Raineri M, Gallardo M, Pagan C J B,. Trigueiros AG and Reyna Almandos 2012 *J. Quantum. Spectrosc. Radiat. Transfer* **113** 1612

[8] Curtis L J 2003 *Atomic structure lifetimes: a conceptual approach* (Berkeley, CA: University of California Press)

[9] Biémont, E, Buchard V, Garnir H -P, Lefébvre, P –H and Q uinet P 2005 *Eur. Phys. J.* D **33** 181

[10] Quinet, P, Palmeri, P, Biémont, E, McCurdy, M M, Rieger, G, Pinnington, E H, Wickliffe M E, Lawler J E 1999 *Mon. Not. R. Astron. Soc.* **307** 934

[11] Migdalek, J., Baylis, W. E. 1978 *J. Phys.* B *: At. Mol. Opt .Phys.* **11**, L497

[12] Fraga S, Muszyñska J 1981 *Atoms in external fields* (Elsevier Science and Technology)

Journal of Physics: Conference Series **583** (2015) 012007 doi:10.1088/1742-6596/583/1/012007

EUV spectra from highly charged terbium ions in optically thin and thick plasmas

C Suzuki[1], F Koike[2], I Murakami[1], N Tamura[1], S Sudo[1], E Long[3], J Sheil[3], E White[3], F O'Reilly[3], E Sokell[3], P Dunne[3] and G O'Sullivan[3]

[1] National Institute for Fusion Science, 322-6 Oroshi-cho, Toki 509-5292, Japan
[2] Sophia University, 7-1 Kioi-cho, Chiyoda-ku, Tokyo 102-8554, Japan
[3] University College Dublin, Belfield, Dublin 4, Ireland

E-mail: csuzuki@nifs.ac.jp

Abstract. We have observed extreme ultraviolet (EUV) spectra from terbium (Tb) ions in optically thin and thick plasmas for a comparative study. The experimental spectra are recorded in optically thin, magnetically confined torus plasmas and dense laser-produced plasmas (LPPs). The main feature of the spectra is quasicontinuum emission with a peak around 6.5–6.6 nm, the bandwidth of which is narrower in the torus plasmas than in the LPPs. A comparison between the two types of spectra also suggests strong opacity effects in the LPPs. A comparison with the calculated line strength distributions gives a qualitative interpretation of the observed spectra.

1. Introduction

Extreme ultraviolet (EUV) emission from highly charged heavy ions in laser-produced plasmas (LPPs) has widely been investigated in terms of their potential applications as EUV light sources. For example, quasicontinuum emission, referred to as unresolved transition array (UTA), from tin plasmas, the peak of which is around 13.5 nm, is now being industrialized for the EUV lithography of semiconductor devices [1]. The expected future trend in the semiconductor feature size has recently prompted the survey of deep EUV emission from highly charged lanthanide ions such as gadolinium (Gd) and terbium (Tb) at around 6.5–6.7 nm [2,3]. However, theoretical models of EUV spectra in dense LPPs are complicated because of opacity effects such as line broadening and self-absorption. Therefore, it would be worthwhile to make a comparative study of EUV spectra between LPPs and optically thin plasmas for benchmarking.

In this article, we present EUV spectra from highly charged Tb ions observed in optically thin and thick plasmas for a comparative study. Optically thin, high-temperature plasmas are produced in the Large Helical Device (LHD) at the National Institute for Fusion Science, while dense LPPs are generated using Nd:YAG lasers at University College Dublin. Though we have already performed a similar study for Gd [4], Tb should also be investigated because the bandwidth of a candidate multilayer mirror lies between the emission peaks of Gd and Tb [5]. In addition, signal-to-noise ratios of the data in the LPPs have been improved since the previous study by utilizing a new grazing incidence spectrometer with better throughput. Therefore, we have observed not only Tb but also Gd spectra with the new experimental setup. Also,

Journal of Physics: Conference Series **583** (2015) 012007 doi:10.1088/1742-6596/583/1/012007

Table 1. Some of the major parameters of the experimental setup for the two lasers with ns and ps pulses.

	ns-laser	ps-laser
Model	Continuum Surelite III	EKSPLA SL312P
Maximum energy	$\simeq$ 670 mJ	$\simeq$ 250 mJ
Pulse width	$\simeq$ 7 ns	$\simeq$ 180 ps
Maximum power density on targets	$\simeq 2\times10^{13}$ W/cm^2	$\simeq 5\times10^{14}$ W/cm^2

some qualitative interpretations are given based on the calculations of line strengths using a Hartree-Fock with configuration interaction (HFCI) Cowan suite of codes [6].

2. Experimental

The LPP experiment was performed using two Q-switched Nd:YAG lasers with 7 ns and 180 ps pulses at the wavelength of 1064 nm. The laser beam is guided into a chamber and focused onto pure Gd and Tb metal targets, placed on a movable stage, by a plano-convex lens (76 mm focal length at 1064 nm). Some of the major parameters of the experimental setup for these ns- and ps-lasers are summarized in table 1.

EUV spectra in LPPs were recorded with a 0.25 m grazing incidence spectrometer connected to the target chamber via a 75 μm entrance slit. The spectrometer is equipped with a flat field 2400 lines mm^{-1} grating (Shimadzu) and a CCD camera (Andor, DX436-BN) for direct detection of EUV photons. The wavelength range of this setup is 2–9 nm. The absolute wavelength was calibrated using known emission lines of silicon and nitrogen ions from LPPs using silicon nitride targets. Space- and time-integrated single spectrum can only be recorded for each laser shot. In order to obtain enough signal intensity, we usually accumulate 5–20 identical laser shots reproduced by moving the target stage.

Since the experimental setup used in the LHD has already been described elsewhere [7], only a brief explanation is given here. Small amounts of Gd and Tb powders are introduced into high-temperature hydrogen plasmas, the density of which are of the order of 10^{19} m^{-3}, using tracer encapsulated solid pellets (TESPEL) [8]. Spatial profiles of electron density and temperature are measured precisely by a laser Thomson scattering diagnostic system [9]. EUV spectra are recorded by a 2 m Schwob-Fraenkel grazing incidence spectrometer [10] equipped with a 600 lines mm^{-1} grating. The temporal evolutions of EUV spectra in the 5.6–9 nm range were recorded with a frame rate of 0.1 or 0.2 s.

3. Results and discussion

EUV spectra from Tb ions observed in LPPs for the ns- and ps-lasers (referred hereafter as ns-LPP and ps-LPP) at their maximum energies are shown in figure 1 (a) in red and green, respectively. A sharp drop of the emissivity at 4.3 nm seems to arise from the K-edge absorption of carbon in the deposition on the surface of the grating or the CCD chip. Apart from this absorption feature, LPP spectra are characterized by broad quasicontinuum emission with a main peak around 6.6 nm and rugged structure with several noticeable peaks found on the shorter wavelength side of the main peak. As is well known, the main peak is caused by the UTA of the n=4–4 transitions of open N shell ions. The rugged structure is more pronounced in the ns-LPP than in the ps-LPP. The small dip found at 6.5 nm in the ns-LPP is not observed in the ps-LPP. Indeed, this dip is more pronounced in a Gd plasma (not shown in the figure).

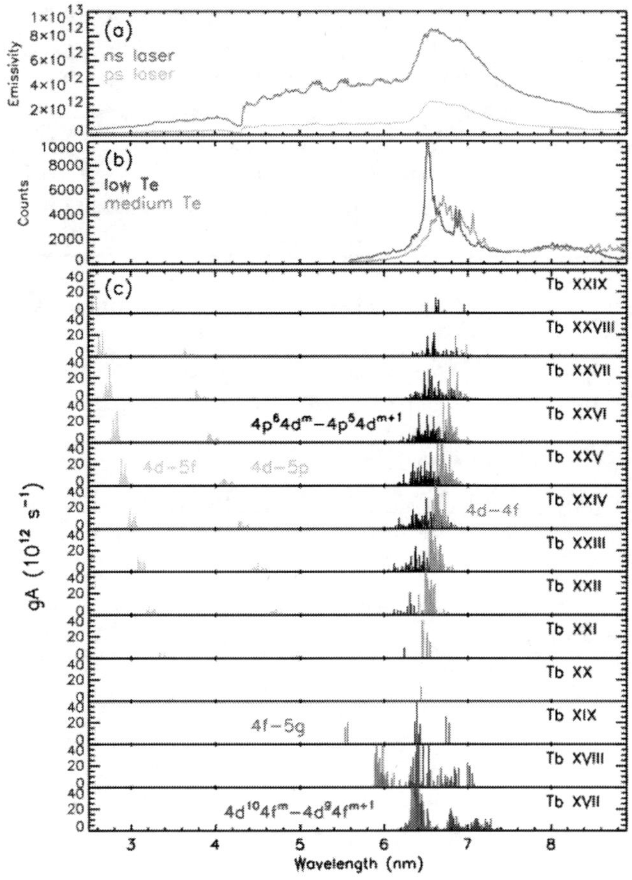

Figure 1. Tb spectra observed in (a) LPPs and (b) LHD plasmas compared with (c) the weighted probabilities of spontaneous transitions for Tb XVII–XXIX calculated by the HFCI Cowan suite of codes.

For a comparison, two different types of EUV spectra from Tb ions observed in LHD are shown in figure 1 (b) in the range of 5.6–9 nm. The spectra in blue and orange are recorded when the peak electron temperature is around 400 eV (low) and 1100 eV (medium), respectively. It is apparent that the UTA bandwidth of the n=4–4 transitions in the LHD plasmas is narrower than that in the LPPs. In particular, the spectrum is drastically narrowed with a sharp peak near 6.5 nm, and a clear double peak is observed near 6.9 nm under the low-temperature condition in the LHD. The positions of these peaks agree well with the previously identified lines of Ag-like Tb XIX and Pd-like Tb XX [11,12], which will be discussed elsewhere in detail. Interestingly, the position of the small dip in the ns-LPP spectrum completely agrees with that of the sharp peak in the low temperature LHD plasma, which is also the case in Gd plasmas. This indicates a strong effect of self-absorption by Ag-like and Pd-like ions in the ns-LPP.

The n=4–4 and n=4–5 transitions of Tb ions have recently been calculated [13,14] using the Cowan code and Flexible Atomic Code (FAC) [15]. In this study, we have compared the measured spectra with the line strengths calculated by the Cowan code to interpret them qualitatively. The weighted probabilities of spontaneous transitions for all of the open 4d and some of the open 4f subshell Tb ions with simple electronic configurations are shown in figure 1 (c). The minimum number of excited states and configuration interactions among them are included

Journal of Physics: Conference Series **583** (2015) 012007 | doi:10.1088/1742-6596/583/1/012007

in the calculations. Following the general procedure in the preceding studies [2,13,16], Slater-Condon and configuration interaction parameters were empirically scaled to 90% of their *ab initio* values, while the spin orbit integrals were unchanged. Ion stages lower than In-like Tb XVII have more complex ground-state configurations because of an interplay of 4f and 5s or 5p electrons [17]. It is obvious that the spectra in the medium temperature LHD plasma can be well explained by the spectral bands of n=4–4 transitions for open 4d subshell ions. On the other hand, the LPP spectra have broad tail structure on the longer wavelength side, which seems to include the contributions from open 4f ions.

The present calculation also indicates the possibility of the contributions from n=4–5 transitions to the rugged feature in the shorter wavelength side of the LPP spectra because the positions of n=4–5 (Δn=1) transitions move to shorter wavelength as the ion stage increases. Indeed, similar structures for tungsten and gold have been previously identified as n=4–5 transitions [16,18]. The detailed analysis of these features is underway.

4. Summary

In terms of the application to semiconductor lithography, we have observed EUV spectra from highly charged Tb ions in dense LPPs and optically thin LHD plasmas by grazing incidence spectrometers. A comparative analysis of EUV spectra between optically thick and thin plasmas clearly shows the differences in the spectral feature and opacity effects. Though the rugged feature observed on the shorter wavelength side in the LPP spectra seems to arise from n=4–5 transitions, further studies are necessary to make detailed identifications of these structures.

Acknowledgments

The main part of this work was carried out during one of the authors' (CS) stay at UCD funded by SOKENDAI Young Researcher Overseas Visit Program. This work was partly supported by the NIFS-collaboration research program (NIFS12KLPF025). The UCD group was supported by Science Foundation Ireland International Co-operation Strategic Award No. 13/ISCA/2846.

References

[1] Bakshi V 2006 *EUV Sources for Lithography* (Bellingham: SPIE Press)
[2] Churilov S S, Kildiyarova R R, Ryabtsev A N and Sadovsky S V 2009 *Phys. Scr.* **80** 045303
[3] Otsuka T, Kilbane D, White J, Higashiguchi T, Yugami N, Yatagai T, Jiang W, Endo A, Dunne P and O'Sullivan G 2010 *Appl. Phys. Lett.* **97** 111503
[4] Suzuki C, Koike F, Murakami I, Tamura N, Sudo S, O'Gorman C, Li B, Harte C S, Donnelly T and O'Sullivan G 2013 *Phys. Scr.* **T156** 014078
[5] Makhotkin I, Louis E, Zoethout E, Yakunin A, Banine V, Müllender S and Bijkerk F 2012 *Proc. Int. Workshop on EUV and soft X-ray Sources* (Dublin)
[6] Cowan R D 1968 *J. Opt. Soc. Am.* **58** 808
[7] Suzuki C, Koike F, Murakami I, Tamura N and Sudo S 2012 *J. Phys. B: At. Mol. Opt. Phys.* **45** 135002
[8] Sudo S and Tamura N 2012 *Rev. Sci. Instrum.* **83** 023503
[9] Narihara K, Yamada I, Hayashi H and Yamauchi K 2001 *Rev. Sci. Instrum.* **72** 1122
[10] Schwob J L, Wouters A W, Suckewer S and Finkenthal M 1987 *Rev. Sci. Instrum.* **58** 1601
[11] Sugar J and Kaufman V 1981 *Phys. Scr.* **24** 742
[12] Sugar J and Kaufman V 1982 *Phys. Scr.* **26** 419
[13] Kilbane D and O'Sullivan G 2010 *J. Appl. Phys.* **108** 104905
[14] Li B, Otsuka T, Higashiguchi T, Yugami N, Jiang W, Endo A, Dunne P and O'Sullivan G 2012 *Appl. Phys. Lett.* **101** 013112
[15] Gu M F 2008 *Can. J. Phys.* **86** 675
[16] Li B, Higashiguchi T, Otsuka T, Yugami N, Dunne P, Kilbane D, Sokell E and O'Sullivan G 2014 *J. Phys. B: At. Mol. Opt. Phys.* **47** 075001
[17] Kilbane D and O'Sullivan G 2010 *Phys. Rev. A* **82** 062504
[18] Harte C S, Higashiguchi T, Otsuka T, D'Arcy R, Kilbane D and O'Sullivan G 2012 *J. Phys. B: At. Mol. Opt. Phys.* **45** 205002

Journal of Physics: Conference Series **583** (2015) 012008 doi:10.1088/1742-6596/583/1/012008

Laboratory study supporting the interpretation of Solar Dynamics Observatory data

E Träbert[1,2] and P Beiersdorfer[1]

1 Physics Division, Lawrence Livermore National Laboratory, Livermore, CA 94550-9234, U.S.A.

2 Astronomisches Institut, Ruhr-Universität Bochum, D-44780 Bochum, Germany

E-mail: traebert@astro.rub.de

Abstract. High-resolution extreme ultraviolet spectra of ions in an electron beam ion trap are investigated as a laboratory complement of the moderate-resolution observation bands of the AIA experiment on board the Solar Dynamics Observatory (SDO) spacecraft. The latter observations depend on dominant iron lines of various charge states which in combination yield temperature information on the solar plasma. Our measurements suggest additions to the spectral models that are used in the SDO data interpretation. In the process, we also note a fair number of inconsistencies among the wavelength reference data bases.

1. Introduction

The *Solar Dynamics Observatory (SDO)* spacecraft [1] aims to study the highly variable Extreme Ultraviolet (EUV) emission of the solar corona (which fluctuates by several orders of magnitude), combining high spatial resolution observations at a high exposure frequency with determinations of the actual plasma temperature in the range from 1 MK to 20 MK. The Atmospheric Imaging Assembly (AIA) experiment [2] on board the *SDO* spacecraft observes in seven, mostly iron-dominated, EUV channels (centered at wavelengths 94, 131, 171, 193, 211, 304, 355 Å) with moderate wavelength selection by multilayer reflective optics. Supported by extensive modeling [3, 4], the observations also yield the temperature in the solar coronal field of view. The seven AIA data channels have spectral band passes from 1 Å to about 20 Å wide. In parallel, a grating spectrograph of the Extreme ultraviolet Variability Experiment (EVE) [5] monitors the EUV spectrum with a band width of 1 Å.

The solar spectrum reflects the chemical composition of the Sun with elements at their natural abundances and charge state distributions depending on the local temperature. In the laboratory, the spectra of elements can be studied individually, and the charge state balance in a plasma can be adjusted. Hence it is possible to recognize the origin of spectral lines, possibly resolve spectral blends that originate from different elements, and check spectral models for consistency.

Among terrestrial experiments, an electron beam ion trap (EBIT) offers the laboratory environment that is closest to the solar corona. Various elements of coronal interest (He, C, N, O, F, Ne, Mg, Si, S, Ar, Ca, Fe, Ni) are introduced into EBIT and ionized and excited by an energetic electron beam. The EUV spectra in the vicinity of the *SDO*/AIA observation channels are studied with spectrographs of resolving powers 1100 to 3000. We thus check the

Journal of Physics: Conference Series **583** (2015) 012008 doi:10.1088/1742-6596/583/1/012008

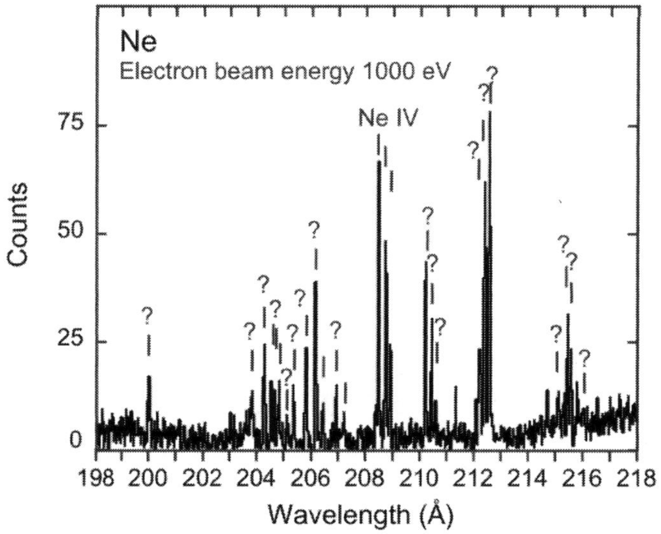

Figure 1. EBIT spectrum of Ne in the vicinity of the SDO/AIA 211 Å band. Only three of the Ne lines are known from the literature (NIST on-line data base and CHIANTI). The newly found and as yet unidentified Ne lines carry question marks.

completeness of the spectral data that are used in the collisional-radiative modeling necessary for the interpretation of the *SDO*/AIA raw data, but with deliberate control over the elements introduced as well as the charge state distribution, and at a much higher spectral resolution than is available on board of the *SDO* spacecraft.

We mainly compare our observations with the NIST on-line data base [6] and with the CHIANTI data base (v. 7.1) [7]; the latter compiles wavelength data mostly from solar observations and models the relative line intensities. We note quite a number of inconsistencies among the wavelength reference data tables. Overall, the CHIANTI wavelengths are better compatible with our observations of the three (out of six) AIA iron channels covered so far. However, in our work in progress [8, 9, 10] we also find a number of spectral lines that may have been under-appreciated in the spectral models.

2. Experiment

The experiment was performed at the EBIT-I electron beam ion trap [11] at the Lawrence Livermore National Laboratory. The device has been optimized for spectroscopic studies of highly charged ions [12], including the study of spectra needed for diagnosing high energy density plasmas [13]. We employ flat-field EUV spectrographs with cryogenically cooled CCD detectors [14, 15] that exceed the resolving power of the grating spectrograph on board *SDO* by one to two orders of magnitude, in spectral slices 15 to 20 Å wide and thus exceeding the widths of the AIA detection bands. For each injected gas (CO_2, N_2, SF_6, Ne, Ar, ironpentacarbonyl, nickelocene, and so on) spectra were recorded at several electron beam energies from 200 to 1000 eV. This procedure enabled a rough estimate of the charge state range of emitting ions. All spectra contained emission from low charge states, because neutral gas was bled into the trap continually.

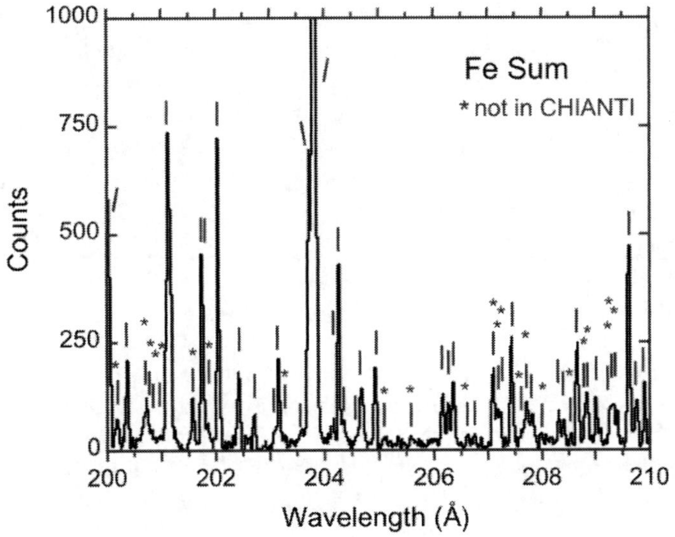

Figure 2. Sum of EBIT spectra of Fe recorded at various electron beam energies in the vicinity of the SDO/AIA 211 Å band. While most of the stronger Fe lines are identified, the lines marked with an asterisk are not listed in the CHIANTI data base.

3. Wavelength calibration

Calibration lines are needed within the field-of-view of each spectrograph setting. Our wavelength calibrations are based on sequential observations of several elements (C, N, O, F, Ne, Mg, Si, S, Ar, Ca, Fe, Ni), the combination of individual data sets yielding an average calibration curve. Fortunately, lines of oxygen, which is an indigeneous background material in EBIT-I, appear in many of the spectra we measure.

Reliable reference lines are often scarce; from historic measurements of light elements, one would expect the best reference data from ions such as those of O, F, or Ne (see [16]). However, even the spectra of the lighter ions are incompletely known, as we see for the case of Ne. Near 94 Å (one of the AIA channels), neon has lines of various charge states, and some of these lines also appear in second diffraction order and then appear near 196 Å, useful for calibrating the spectra in the vicinity of the 193 Å AIA channel [10]. Ne lines are not only wanted for the present laboratory astrophysics work, but they are important also for the calibration in our measurements on Li- and Be-like ions of specific isotopes of rare earth elements (such as $^{141}_{59}$Pr [17]).

Most of the neon lines listed in the neighbourhood of the seven SDO/AIA bands belong to charge states 2+ to 8+, and none of them are known with high accuracy. Actually, some 90 Ne lines are expected from the NIST and CHIANTI tables in the neighbourhood of the seven AIA bands, but only about 60 of these are seen in our EBIT spectra. Of the predictions, about 40 lines appear in only one of the two tables and not in the other. Some 60 Ne lines are seen in our EBIT spectra that have no equivalent in the tables. In the special example of the 211 Å SDO/AIA band, three known Ne IV lines are seen in the EBIT spectrum as well as 26 unidentified Ne lines [9] (Fig. 1). The Ne wavelength listings of the two tables perfectly agree in some cases and differ by up to 80 mÅ in others. Consequently most of our line positions are limited in precision by statistics and in accuracy by the calibration problem, combining to 10 to 20 mÅ in most cases.

4. Discussion

The complexity of the spectra causes identification problems. The charge states of elements Fe and Ni that are of particular interest feature an open n=3 valence shell, with a multitude of lines. Around 200 Å, Fe has many more than just one line per 1 Å interval (see the sample spectrum in Fig. 2). In order to identify known lines, the wavelength calibration has to be much better established than 0.1 Å. Correspondingly, calculations of wavelengths that are not better than 100 mÅ may be easy to obtain, but are largely useless in practice. In the present study, parts of the Fe spectrum are so crowded with spectral lines that other elements would be difficult to recognize, unless their relative abundance exceeds the natural one. Fe is the element with the ninth highest natural abundance. Most elements beyond H, He, C, N, O, Ne, Mg, and Si are clearly less abundant and thus their spectra are correspondingly less bright in the Sun, and their lines are progressively difficult to discern in the presence of line-rich Fe.

The present example of Ne reveals many previously unknown lines of Ne in the vicinity of the SDO/AIA 211 Å band. Although Ne is more abundant than Fe in the Sun, the Ne lines in this band are likely hidden in the solar spectra by the dense line forest of Fe. However, now knowing that Ne has these lines that are not yet listed in models such as CHIANTI points out the necessity for refining the model, so that the pedestal under the Fe lines can be treated more correctly in the interpretation of the SDO observations.

Our investigation reveals a number of cases in which the modeling of the solar signal of the seven AIA data channels should benefit from the inclusion of additional spectral lines that we identify. Hence the reliability of the interpretation of the SDO data for establishing a temperature scale can be expected to improve.

Acknowledgements

This work was performed under the auspices of the U.S. Department of Energy by Lawrence Livermore National Laboratory under Contract DE-AC52-07NA27344 and was supported by the Solar and Heliospherical Physics Program of the National Aeronautics and Space Administration under award NNH10AN31I. E.T. acknowledges support by the German Research Association (DFG) (grants Tr171/18 and Tr171/19).

References

[1] Pesnell W D, Thompson B J and Chamberlin P C 2012 *Sol. Phys.* **275** 3 (special issue)
[2] Lemen J R, Title A M, Akin D J, *et al* 2012, SoPh. 275, 17
[3] O'Dwyer B, Del Zanna G, Mason H E, Weber M A and Tripathi D 2010 *Astron. Astrophys.* **521** A21
[4] Parenti S, Schmieder B, Heinzel P and Golub L 2012, arXiv:1205.5460v1 [astro-ph.SR] 24 May 2012
[5] Woods T N, Eparvier F G, Hock R, *et al* 2012 *Sol. Phys.* **275** 115
[6] Kramida A, Ralchenko Yu, Reader J andNIST ASD team (2012). NIST Atomic Spectra Database (ver. 5.0), (Online). Available: http://physics.nist.gov/asd [2012, October 3]. National Institute of Standards and Technology, Gaithersburg, MD, U.S.A.
[7] Landi E, Del Zanna G, Young P R, Dere K P and Mason H E 2012 *Astrophys. J. Suppl. Ser.* **744**
[8] Träbert E, Beiersdorfer P, Brickhouse N S and Golub L 2014 *Astrophys. J. Suppl. Ser.* **211** 14
[9] Beiersdorfer P, Träbert E, Lepson J K, Brickhouse N S and Golub L 2014 *Astrophys. J.* **788** 25
[10] Träbert E, Beiersdorfer P, Brickhouse N S and Golub L 2014 *Astrophys. J. Suppl. Ser.* (to be published)
[11] Levine M A, Marrs R E, Henderson J R, Knapp D A and Schneider M B 1988 *Phys. Scr.* T **22** 157
[12] Beiersdorfer P 2008 *Can. J. Phys.* **86** 1
[13] Schneider M B, Mancini R, Widmann K, *et al* 2008 *Can. J. Phys.* **86** 259
[14] Beiersdorfer P, Magee E W, Träbert E, *et al* 2004 *Rev. Sci. Instrum.* **75** 3723
[15] Beiersdorfer P, Magee E W, Brown G V, Träbert E and Widmann K 2014 *Rev. Sci. Instrum.* **85** 11E422
[16] Kelly R L 1987 *J Phys. Chem. Ref. Data* **16** Suppl. No. 1
[17] Beiersdorfer P, Träbert E, Brown GV, *et al* 2014 *Phys. Rev. Lett.* **112** 233003

17th International Conference on the Physics of Highly Charged Ions IOP Publishing
Journal of Physics: Conference Series **583** (2015) 012009 doi:10.1088/1742-6596/583/1/012009

X-ray Measurements of Highly Charged Europium

K Widmann, P Beiersdorfer, G V Brown, N Hell, E W Magee, E Träbert

Lawrence Livermore National Laboratory, P.O. Box 808, Livermore, CA 94550, USA

E-mail: widmann1@llnl.gov

Abstract. We present spectroscopic measurements of the M-shell emission of highly charged europium performed at the Livermore SuperEBIT electron beam ion trap facility using the EBIT Calorimeter Spectrometer (ECS). There is significant blending among the emission lines from the different charge states but despite the complexity of the observed spectra we have successfully identified the ten brightest $n = 4 \rightarrow 3$ transitions from sodium-like Eu^{52+} utilizing the Flexible Atomic Code (FAC). We find that the difference between the calculated and measured transition energies for these ten Eu^{52+} lines does not exceed 3 eV. In fact, for four of the identified lines we find agreement within the measured uncertainties. Additional comparison with semi-empirical transition-energy predictions for sodium-like ions from laser-generated plasmas is included and shows that overall the semi-empirical predicted values for the transition energies are slightly higher than the measured values, while the FAC values that didnt agree with the measured transition energies are almost 1 eV lower than the measured values.

1. Introduction

M-shell emission from highly charged ions has been observed in a variety of different laboratory and astrophysical plasma sources, spanning a wide range of densities and temperatures. The complexity of the spectra originating from the $4 \rightarrow 3$ transitions of ions that have a partially filled M-shell makes it difficult to utilize M-shell spectroscopy for measurements of plasma parameters, such as electron and ion temperatures, electron densities, and the charge state distribution, respectively [1]. The richness of the M-shell spectra of highly charged ions opens the window on radiative cooling of high-temperature low-Z plasmas by introducing high-Z impurities. For example, using tungsten coated tiles in the divertor region of tokamaks significantly reduces the temperature of the plasma in the divertor region and, thus, extends the lifespan of the walls in the tokamak vessel [2, 3]. In contrast, for indirectly driven ICF plasmas the presence of M-shell emission from laser-heated high-Z hohlraum walls is an unwanted side effect. As the X-ray conversion efficiency significantly increases with the Z of the hohlraum material, so does the energy of M-shell radiation. The few-keV X rays are hard enough to penetrate the capsule and preheat the fuel which leads to a decrease in the achievable velocity of the imploding capsule. In both cases, detailed spectroscopic measurements are needed in order to quantitatively assess the impact of the M-shell emission on radiative cooling and fuel preheat, respectively. Such detailed studies of the M-shell spectra also pave the way for a reliable utilization of this emission regime as a plasma diagnostic tool.

The Na-like iso-electronic sequence is the simplest of the M-shell ions and has already received some experimental and theoretical attention. In most cases, however, the object of interest is the $3p_{3/2} \rightarrow 3s_{1/2}$ resonance transition of the sodium like ions and not the $4 \rightarrow 3$ transitions.

Journal of Physics: Conference Series **583** (2015) 012009 doi:10.1088/1742-6596/583/1/012009

Interestingly, a comparison of transition energy measurements of this intra-M-shell resonance line in Na-like ions obtained from laser-produced plasmas versus accelerators and ion traps indicates a systematic difference between these two experimental approaches to measure M-shell emission. In particular, spectroscopic measurements on laser-produced plasmas involving elements as high as Gd (Z=64) have featured a deviation from theory toward shorter wavelengths [4], while high-Z measurements ($Z \geq 78$) on accelerators and ion traps have yielded wavelengths that are longer than semi-empirical predictions based on measurements of sodium-like ions with $Z \geq 50$ [5, 6, 7, 8]. In order to overlap in Z with the laser-produced plasma data, we have performed measurements at the Livermore high-energy electron beam ion trap SuperEBIT of the $3p_{3/2} \rightarrow 3s_{1/2}$ resonance transition in sodium-like Eu^{52+} [9]. Here, we present measurements of the $n = 4 \rightarrow 3$ transitions that we made concurrently in the X-ray region utilizing a cryogenic microcalorimeter as the energy-dispersive detector.

2. Experimental Setup

The spectroscopic measurements were performed at the SuperEBIT facility [10, 11, 12] utilizing the EBIT Calorimeter Spectrometer (ECS) [13, 14]. For obtaining the M-shell spectra of highly charged europium, the high-energy SuperEBIT was operated at rather low beam energies, i.e., near 10 keV.

The core feature of the ECS is a pixel array, made of HgTe absorbers, which is cooled to 60 mK. The heat generated by absorption of X rays in the HgTe absorbers is being measured by an intrinsic thermometer. The array contains two types of absorbers that are sensitive to low- and high-energy X rays, respectively. Thus, the ECS can be used to detect X rays in the 0.1 to 100 keV range [15, 16], with a line width (in the low-energy range) of a few eV.

Europium injection into SuperEBIT required the development of a new injection scheme, because europium cannot readily be machined, as it tends to combust in air. For the measurements, we utilized the fact that its vapor pressure is rather high when the material is heated to about 300 ^{o}C [17].

3. Measurements and Results

A spectrum of the $4 \rightarrow 3$ emission array of europium obtained with the ECS is shown in Fig. 1. The spectrum exhibits a multitude of (often blended) lines, as is readily seen from the figure. A benefit of the energy-dispersive detection is the observation of whole transition arrays, which gives a wider range and offers opportunities for cross-checks to the ensuing spectrum analysis.

Despite the complexity of the measured spectrum and the significant blending among the emission lines from the different charge states, we were able to identify the ten brightest lines emitted by Na-like Eu^{52+}. A series of measurements with varying electron beam energy was conducted to observe the change in the emission spectra as a function of charge state distribution in the ion trap which subsequently allowed us to perform a coarse identification of the Na-like Eu^{52+} emission. The final identification of the observed transitions was done by comparing the measured lines with calculations using the Flexible Atomic Code [18, 19]. Figure 1 shows an overlap of the measured spectrum with the FAC calculations and the obtained values are listed in Table 1.

The data set presented in Table 1 show that for most transitions the energy values obtained with FAC are about 1 eV lower than the semi-empirical predictions by Seely *et al* [4] and that the measured values are in between the two predictions. In particular, four of the ten identified lines show agreement with the FAC calculations, and three of the ten lines agree with the semi-empirical predictions by Seely *et al* .

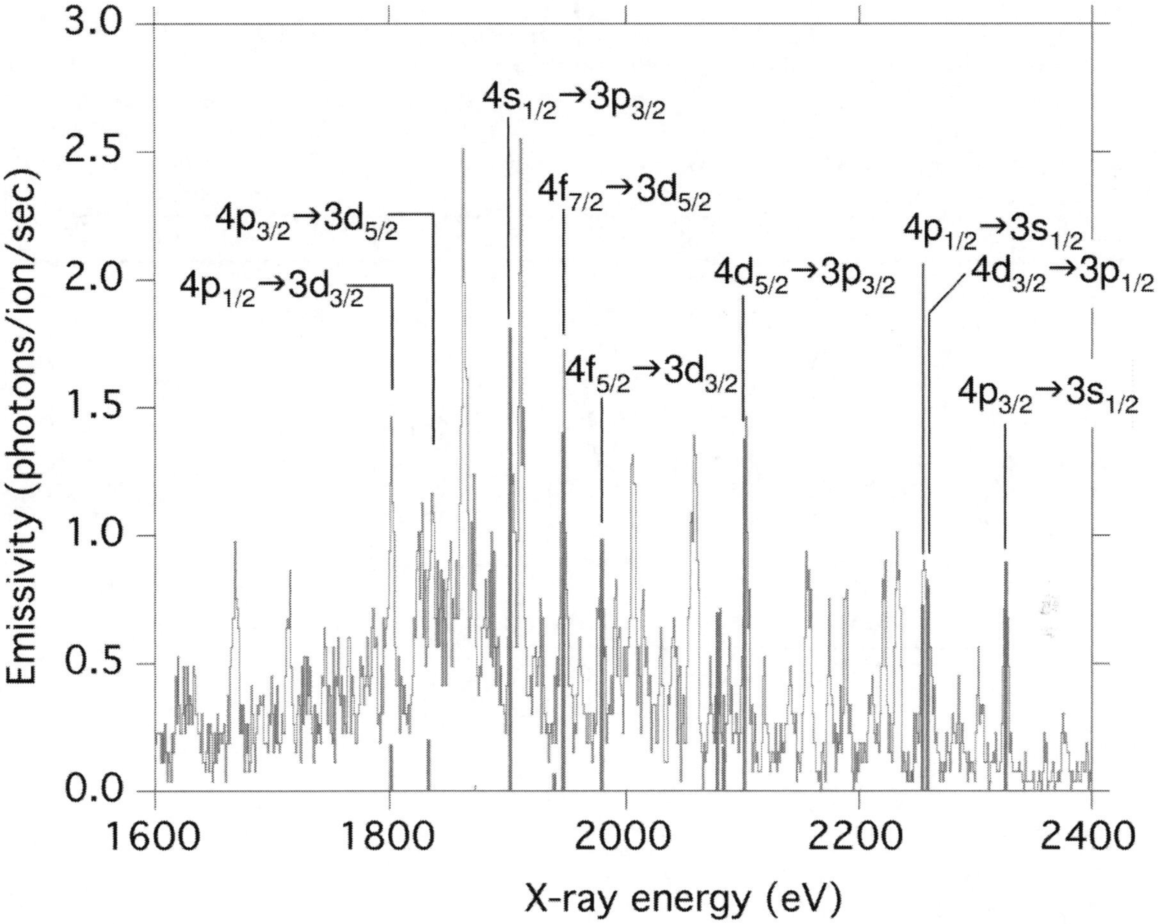

Figure 1. M-shell emission spectrum, $4 \rightarrow 3$ transitions, from highly charged europium. The measured spectrum (solid, red trace) was obtained using the EBIT Calorimeter Spectrometer (ECS). The green bars indicate the result obtained with the Flexible Atomic Code (FAC) [18] and the transition labels indicate the lines that we have identified as belonging to Na-like Eu^{52+}. The vertical axis is the output of the FAC calculations; the data have been scaled for display.

Acknowledgments
This work was performed under the auspices of the U.S. Department of Energy by Lawrence Livermore National Laboratory under Contract DE-AC52-07NA27344.

Table 1. Measured and calculated values for the energies of the $n = 4 \rightarrow 3$ transitions in Na-like Eu^{52+}. The "ECS" column shows the values measured at the Livermore SuperEBIT with the EBIT Calorimeter Spectrometer (ECS), the column "FAC" lists the calculated values using the Flexible Atomic Code (FAC)[18], and the "Seely" values are from semi-empirical calculations [4]. The three right columns show the difference between the measured and calculated values. All values are in eV.

Transition	Measured ECS	Calculated FAC	Calculated Seely	Difference Seely - FAC	Difference ECS - Seely	Difference ECS - FAC
$4p_{1/2} \rightarrow 3d_{3/2}$	1801.4(5)	1800.75	1750.07	−50.68	51.4	0.7
$4p_{3/2} \rightarrow 3d_{5/2}$	1836.0(5)	1832.78	1832.89	0.11	3.1	3.2
$4s_{1/2} \rightarrow 3p_{3/2}$	1902.7(5)	1902.00	1903.59	1.59	−0.9	0.7
$4f_{7/2} \rightarrow 3d_{5/2}$	1946.8(5)	1946.34	1947.79	1.45	−1.0	0.5
$4f_{5/2} \rightarrow 3d_{3/2}$	1978.8(5)	1978.87	1980.56	1.69	−1.8	−0.1
$4s_{1/2} \rightarrow 3p_{1/2}$	2080.1(5)	2077.76	2079.40	1.64	0.7	2.3
$4d_{5/2} \rightarrow 3p_{3/2}$	2102.0(5)	2100.97	2102.14	1.17	−0.1	1.0
$4p_{1/2} \rightarrow 3s_{1/2}$	2255.2(10)	2254.95	2255.48	0.53	−0.3	0.2
$4d_{3/2} \rightarrow 3p_{1/2}$	2260.9(10)	2259.38	2260.01	0.63	0.9	1.5
$4p_{3/2} \rightarrow 3s_{1/2}$	2326.5(5)	2326.71	2327.47	0.76	−1.0	−0.2

References

[1] Griem H R 1989 *Plasma Spectroscopy* McGraw-Hill, New York
[2] Clementson J *et al* 2010 *Rev. Sci. Instrum.* **81**, 10E326
[3] Clementson J *et al* 2010 *J. Phys. B: At. Mol. Opt. Phys.* **43**, 144009
[4] Seely J F *et al* 1991 *At. Data Nucl. Data Tables* **47**, 1
[5] Cowan T E *et al* 1991 *Phys. Rev. Lett.* **66** 1150
[6] Simionovici A *et al* 1993 *Phys. Rev.* A **48** 3056
[7] Beiersdorfer P and Wargelin B J 1994 *Rev. Sci. Instrum.* **65**, 13
[8] Beiersdorfer P *et al* 2003 *Phys. Rev.* A **67**, 052103
[9] Träbert E *et al* 2014 *submitted to Phys. Rev.* A
[10] Widmann K *et al* 1997 *Hyperfine Inter.* **108** 73
[11] Beiersdorfer P *et al* 2003 *AIP CP652* 131
[12] Beiersdorfer P 2008 *Can. J. Phys.* **86**, 1
[13] Porter F S *et al* 2004 *Rev. Sci. Instrum.* **75** 3772
[14] Porter F S *et al* 2008 *Can. J. Phys.* **86** 231
[15] Porter F S 2008 *Rev. Sci. Instrum.* **19** 10E307
[16] Porter F S 2008 *J. Low Temp. Phys.* **151** 1061
[17] Magee E W *et al* 2014 *Rev. Sci. Instrum.* **85** 11E820
[18] Gu M-F *et al* 2004 *AIP APP Conference Proceedings* **730** 127
[19] Gu M-F 2008 *Can. J. Phys.* **86**, 675

Journal of Physics: Conference Series **583** (2015) 012010 doi:10.1088/1742-6596/583/1/012010

The role of multiple electron capture in the x-ray emission process following charge exchange collisions with neutral targets

S. Otranto[1], N. D. Cariatore[1] and R. E. Olson[2]

[1]IFISUR and Departamento de Física, Universidad Nacional del Sur, 8000 Bahía Blanca, Argentina.

[2]Physics Department, Missouri University of Science and Technology, Rolla MO 65409, USA.

sotranto@uns.edu.ar

Abstract. In this work we theoretically study photonic spectra that follow charge exchange processes between highly charged ions and neutral argon and CO targets. The range of collision energies studied is 5 eV/amu-10 keV/amu, covering typical EBIT-traps and Solar Wind energies. Our studies are based on multiple electrons schemes within the classical trajectory Monte Carlo method. Electrons are sorted with the sequential binding energies for the target under consideration. The role played by the multiple electron capture process for the different collision systems under consideration is explicitly analyzed and its contribution separated as arising from double radiative decay and autoionizing multiple capture. Present studies are stimulated by the upcoming launch of the Astro-H mission in 2015, which will provide high resolution spectra in the 0.3 keV-12keV band.

1. Introduction

The study of photonic spectra following charge exchange collisions has been a field with growing interest during the past 15 years. The unexpected detection of the X-ray emission from comet P/Hyakutake by ROSAT in 1996 [1], later on ascribed to the charge exchange mechanism between the Solar Wind ions and the neutrals targets in gas phase in the cometary coma [2], triggered a deep search of cometary x-ray sources in other astrophysical environments like planetary atmospheres, the heliosphere, the interestellar medium and even other galaxies. As recently pointed out by Dennerl [3], it seems now strange that such emission had remained unnoticed to us and had been detected even with surprise by the astrophysical community, provided that all the elements to predict such an emission were already at hand by 1968.

By the year 2000, and with the Chandra X-ray Observatory (CXO) already in orbit providing cometary spectra with a much improved resolution of about 100 eV, several laboratories worldwide worked upon the study of the collision processes and the possible recreation of cometary spectra based on laboratory data [4-7]. EBIT traps married to calorimeter spectrometers provided *in-situ* spectra with

Journal of Physics: Conference Series **583** (2015) 012010 doi:10.1088/1742-6596/583/1/012010

an unprecedented resolution of about 8 eV at collision energies which are two orders of magnitude below those corresponding to the Solar Wind. Linear accelerators, on the other hand, reached Solar Wind velocities but missed the dominant forbidden lines $2^3S \rightarrow 1^1S$ which are long lived and cannot take place within the typical dimensions of the devices under use. Since then, a limited but nonetheless important set of data has been collected by several experimental groups [8].

Theoretical studies in this context, on the other hand, have been mainly conducted by means of the classical overbarrier model [5], the Landau-Zener multichannel model [9] and the classical trajectory Monte Carlo (CTMC) method [10-15]. In addition, electron capture cross sections for hydrogen targets obtained by means of atomic orbitals close coupling codes have been used by Bodewits and Hoekstra to simulate cometary spectra [7] and line emission studies have been also recently made within the two center-basis generator model [16].

A significant problem in this context is that of multiple-electron capture by highly charged ions colliding with multielectronic targets. As multiple capture probabilities are expected to increase with the projectile charge, a key point during the analysis of laboratory and astrophysical spectra is the accurate determination of the fraction of X-rays which originate in multiple electron transfer [13].

In this work, we explicitly evaluate the role of the multiple electron capture process in the photonic spectra which follow charge exchange collisions involving highly charged ions and argon and CO targets. The former target has been largely explored at different laboratories [8,17,18], while the latter is a significant component in cometary environments and can be also found in the atmospheres of Venus and Mars. In section 2, we describe the CTMC models used for the different targets under consideration and the scheme used to account for the Auger cascading processes which ultimately lead to single or double radiative decay. In section 3, we show our main results. Conclusions and outlook are drawn in Section 4.

2. Theoretical Method

Theoretical electron capture and line emission cross sections have been calculated using the CTMC method [19]. In the following we detail how this method has been adapted for the two targets under consideration and the decay scheme used to account for the Auger cascading processes.

2.1 Argon

Hamilton's equations are solved for a system composed by the projectile, the ionic core, and eight non-interacting electrons (the whole M shell) which are sorted according to the sequential binding energies for the target. Although electron-electron correlations are not accounted in detail, the energy deposition needed for many electron removal is properly considered. Electrons are sorted according to the quantum mechanical momentum distribution and their interaction with the ionic core is considered to be Coulomb-like with an effective charge that is set in order to provide the best possible agreement with the quantum mechanical radial distribution. To further exemplify, in figure 1a) we show the radial distribution obtained for the electron initialized with the first ionization potential of 15.7 eV. Interesting to notice, the above detailed procedure correctly leads to the nodal structure in the radial distribution.

2.2 CO

For CO we consider a many-electron-multicenter approach. Hamilton's equations are solved for a system composed by the projectile, two ionic centers (fixed in space at the equilibrium distance R = 2.13 a.u.), and eight non-interacting electrons again sorted according to their sequential binding energies.

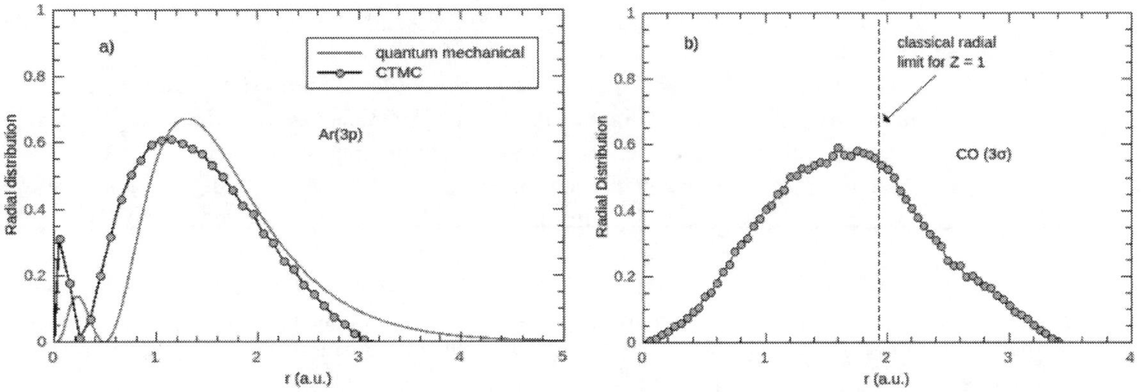

Figure 1. CTMC radial distributions for a) Ar(3p) and b) the 3σ orbital of CO.

The molecular axis is randomly oriented and the eight electrons interact with the ionic centers via the pseudopotential developed by Gabas et al. [20]:

$$V_{CO^+}(r_{eC}, r_{eO}) = V_C(r_{eC}) + V_O(r_{eO}),$$

$$V_O(r_{eO}) = -\frac{8-N_O}{r_{eO}} - \frac{N_O}{r_{eO}}(1+\alpha_O r_{eO})\, e^{-2\alpha_O r_{eO}},$$

$$V_C(r_{eC}) = -\frac{6-N_C}{r_{eC}} - \frac{N_C}{r_{eC}}(1+\alpha_C r_{eC})\, e^{-2\alpha_C r_{eC}}.$$

The corresponding coefficients are N_o=*7.609*, N_c=*5.391*, α_o=*1.501* and α_c=*0.85*.

In figure 1b) we show the radial distribution obtained for the 3σ orbital (ionization potential of 14.01 eV). Compared to simpler single-center approaches which model the target as an hydrogenic core of Z = 1 with the ionization potential of the target under consideration, the present multicenter treatment leads to a more extended radial distribution and is expected to provide a more accurate description of the target.

2.3 Decay scheme for the multiple-electron capture

The Auger decay scheme used in this work is based in the decay scheme developed by Ali et al [21] for collisions involving argon ions with argon atoms and can be summarized as follows:

(i) Multiply excited states dominantly stabilize via multiple Auger processes.
(ii) Two-electron Auger processes are considered only.
(iii) Transitions involving electrons in the same shell proceed first. If several electrons are in different shells, the Auger process involves the two electrons which are energetically closer.
(iv) Each Auger transition proceeds with the unit probability to the nearest continuum limit. The decaying electron falls to a well-established *n* level according to the energy equation.

Journal of Physics: Conference Series **583** (2015) 012010 doi:10.1088/1742-6596/583/1/012010

(v) If the new configuration still provides a multiple excited state involving more than two electrons, those rules are applied again until only two electrons remain bound to the projectile.

(vi) If a cascading process leads to an asymmetric doubly excited state ($|n_2-n_1| \geq 2$), the event is characterized as radiative decay (R2CX). Otherwise, a final Auger process takes place and the event is characterized as single charge exchange. As a result, the single charge exchange channel includes the contributions arising from the single electron capture (SEC) and the autoionizing multiple capture (AutoMC) channels.

Finally, we note that the electronic angular momenta of the decaying electrons are determined throughout the Auger process by requiring the preservation of their respective orbital eccentricities. For the sake of simplicity, hydrogenic energy levels are used to determine the projectile population after multiple charge exchange.

3. Results

In figure 2 we show the line emission cross sections following Ar^{18+} + Ar charge exchange collisions at an impact energy of 4 keV/amu. Such impact energy is very close to that corresponding to the fast Solar Wind (~700 km/sec). Present theoretical results are compared to the experimental data of Tawara *et al* [8] which are normalized to our theoretical absolute values. Although the reported energy resolution of the detector is roughly 130 eV for Ti Ly-α x-rays at 4.5keV, we convolute our line emission cross sections by means of 160 eV FWHM Gaussian functions in order to match the experimental widths. In figure 2a) we show the separate contribution of the single-electron-capture (SEC) channel. In this case, the fraction of x-rays originating in multiple capture amounts up to 55%.

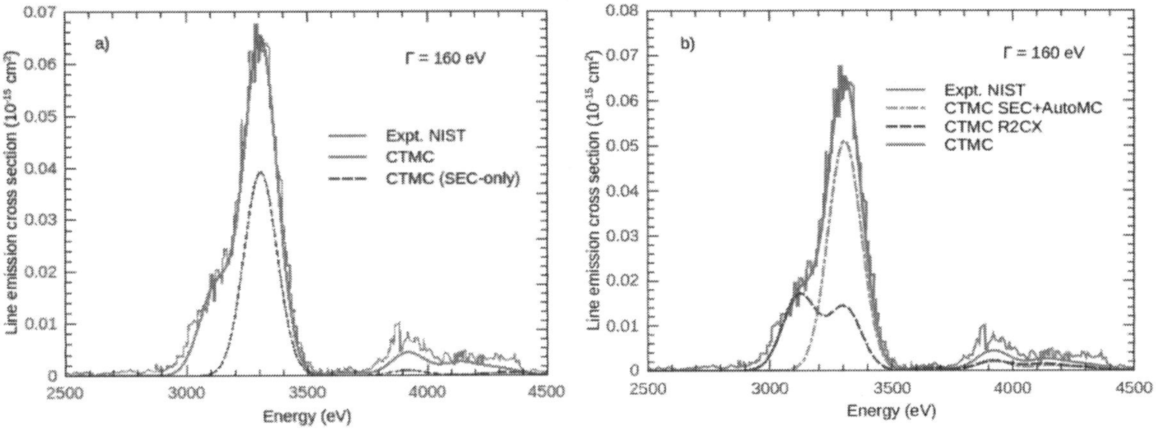

Figure 2. CTMC line emission cross sections for 4 keV/amu Ar^{18+}+ Ar collisions. Present results are compared to the NIST experimental data of Tawara *et al* [8]. The partial contributions of the SEC channel and the radiative decay (R2CX) are explicitly shown.

In Figure 2b) the autoionizing multiple capture and double radiative decay contributions are individually shown. It can be easily seen that the satellite peak originating in double radiative decay is shifted compared to the hyper-satellite position and provides a low-energy shoulder to the Ly-α peak. Also, the CTMC estimates for the fraction of x-rays due to radiative double decay amount to about

71.4% of the multiple capture channel. Our preliminary results indicate that the fraction of x-rays associated to double radiative decay increases with the projectile charge.

Now moving to the CO target, in figure 3 we show the line emission cross sections following charge exchange reactions in 3.5 keV/amu Ne^{10+} + CO collisions. The impact energy under consideration belongs to the Solar Wind range and has been widely considered by the JPL group during the last decade [4,5]. Present results are again convoluted over an energy resolution of 160 eV FWHM. In this case, the fraction of x-rays originating in multiple capture amounts up to 33%. In figure 3b) the autoionizing multiple capture and double radiative decay contributions are individually shown. The CTMC estimates for the fraction of x-rays due to radiative double decay amount to about 63.5% of the multiple capture channel.

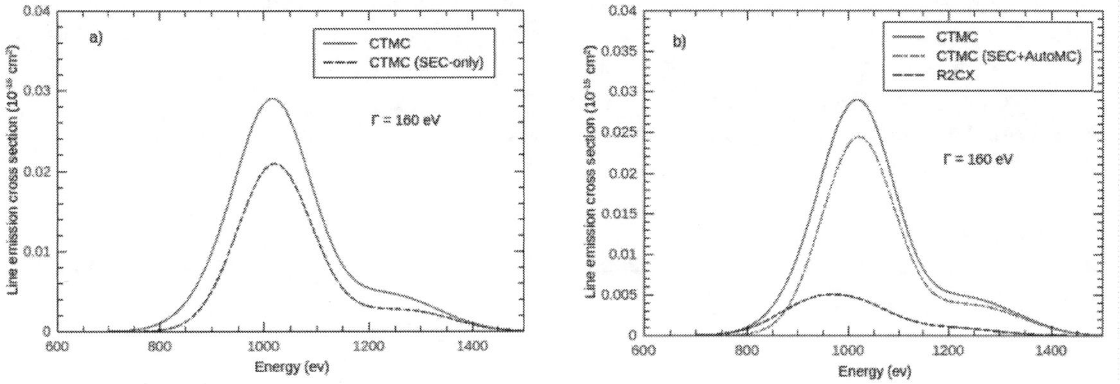

Figure 3. CTMC line emission cross sections for 3.5 keV/amu Ne^{10+}+ CO. The partial contributions of the SEC channel and the radiative decay (R2CX) are explicitly shown.

In Table 1 we consider the state selective relative populations to $n = 3$ and 4 in 2.25 keV/amu O^{6+} + CO collisions. The JPL experimental data from Miller *et al* [22] obtained from line emission spectra are shown for comparison. It can be seen that present results are in good agreement with the experiments. Nevertheless, we point out that we have detected a non negligible amount of capture into the 5g state that would entirely feed the 3d + 4f line. For that reason, we have incorporated our 5g state selective cross sections in the results shown.

Table 1. State selective relative populations to n= 3 and n = 4 in 2.25 keV/amu O^{6+} +CO collisions. The experimental data corresponds to Miller *et al* [22].

	Expt	*mmCTMC*
4p	0.13 ± 0.03	0.169
4d	0.13 ± 0.03	0.121
4s	0.23 ± 0.05	0.174
3p	0.07 ± 0.02	0.089
3d+4f	0.41 ± 0.09	0.360
3s	0.03 ± 0.01	0.085

4. Conclusions

In this work we have theoretically studied the role of the multiple electron capture channel in the x-ray spectra following charge exchange collisions between highly charged ions and neutral targets (argon and CO). A classical trajectory Monte Carlo model based on eight non-interacting electrons sorted with sequential binding energies has been married with a decay scheme which allows the explicit separation of the multiple electron capture events in autoionizing multiple capture (leading to single electron capture) and double radiative emission. Present results for Ar^{18+}+Ar are found in very good agreement with the x-ray spectra of Tawara et al obtained at NIST. Moreover, our calculations nicely reproduce the low energy shoulder in the Ly-α structure which corresponds to the satellite transition and originates in double radiative decay. For Ne^{10+}+CO, on the other hand, our results suggest a not so prominent role for the multiple capture channel. These results indicate that the fraction of x-rays originating in multiple electron capture increases for larger projectile charges. Finally, we contrasted our theoretical results with the relative populations of the n = 3 and 4 levels measured by Miller et al at JPL for 2.25 keV/amu O^{6+}+CO collisions.

New complementary laboratory studies would be welcome at this point to further test our present theoretical capabilities.

Acknowledgments

Work at Universidad Nacional del Sur is supported by Grants No. PGI 24/F059 and No. PIP 112-201101-00749 of CONICET (Argentina).

References

[1] Lisse C M et al 1996 *Science* **274** 205
[2] Cravens T E 1997 *Geophys. Res. Lett.* **24** 105
[3] Dennerl K 2010 *Space Sci. Rev.* **143** 427
[4] Greenwood J B et al 2000 *Astrophys. J.* **533** L175
[5] Greenwood J B et al 2001 *Phys. Rev. A* **63** 062707
[6] Beiersdorfer P et al 2003 *Science* **300** 1558
[7] Bodewitz D et al 2006 *Astrophys. J.* **642** 593
[8] Tawara H et al 2006 *Phys. Rev. A* **73** 012704
[9] Rigazio M, Kharchenko V and Dalgarno A 2002 *Phys. Rev. A* **66** 064701
[10] Otranto S, Olson R E and Beiersdorfer P 2006 *Phys. Rev. A* **73** 022723
[11] Otranto S Olson R E and Beiersdorfer P 2007 *J. Phys. B* **40** 1755
[12] Otranto S and Olson R E 2008 *Phys. Rev. A* **77** 022709
[13] Otranto S and Olson R E 2011 *Phys. Rev. A* **83** 032710
[14] Simcic J et al 2010 *Phys. Rev. A* **81**, 062715
[15] Cariatore N D and Otranto S 2013 *Phys. Rev. A* **88**, 012714
[16] Salehzadeh A and Kirchner T 2013 *J. Phys. B* **46** 025201
[17] Beiersdorfer P et al 2000 *Phys. Rev. Lett.* **85**, 5090
[18] Allen F I et al 2008 *Phys. Rev. A* **78** 032705
[19] Olson R E and Salop A 1977 *Phys. Rev. A* **16** 531
[20] Gabas P M M et al 2012 *Phys. Rev. A* **85** 012702
[21] Ali R et al 1994 *Phys. Rev. A* **49** 3586
[22] Miller K A et al 2011 *Astrophys. J.* **742** 130

Journal of Physics: Conference Series **583** (2015) 012011 doi:10.1088/1742-6596/583/1/012011

Ion-Induced Reactivity in Pyrene Clusters

M Gatchell[1], R Delaunay[2,3], S Maclot[2,3], T Chen[1], M H Stockett[1], A Domaracka[2], A Méry[2,3], L Adoui[2,3], H Zettergren[1], P Rousseau[2,3], H Cederquist[1] and B A Huber[2]

[1]Department of Physics, Stockholm Univeristy, SE-106 91 Stockholm, Sweden
[2]CIMAP, CEA-CNRS-ENSICAEN-UCBN, Bd Henri Becquerel, 14070 Caen Cedex 05, France
[3]Université de Caen Basse-Normandie, Esplanade de la Paix, F-14032 Caen, France

E-mail: `gatchell@fysik.su.se`

Abstract. We report experimental indications of chemical reactions inside clusters of pyrene ($C_{16}H_{10}$) molecules following collisions with 11.25 keV He^+- and 12.0 keV Ar^{2+} ions. It appears that bond-forming reactions are more likely with the heavier projectile. We have also performed classical molecular dynamics simulations of these processes where we treat the interaction between the projectiles and all atoms in the cluster as well as non-dispersive and dispersive forces between all atoms in the cluster before, during and after the collision. The time step is typically 10^{-17} s and the total simulation time 1 picosecond. The simulations were performed for a fixed cluster size with 36 pyrene molecules, although there is a broad range of cluster sizes in the experiment. Still, there is good qualitative agreement between the experimental and the simulated mass spectra exhibiting reaction products with masses between those of the $C_{16}H_{10}$-monomer and dimer in both cases. Additional studies of the influence of the projectile charge and mass is planned as well as simulations on longer time scales and as functions of cluster size.

1. Introduction

Polycyclic Aromatic Hydrocarbon molecules, or PAHs for short, attract considerable interest in several fields of research. PAHs, which typically consists of three or more aromatic rings of carbon atoms, are common combustion products and are believed to play key roles in soot formation processes although the exact mechanisms for this is not yet clear [1, 2]. As PAHs are formed in combustion processes they are also of great interest for environmental and health research. Some PAHs are toxic [3] and some may even be cancerogenic [4, 5]. It has been suggested that PAHs are involved in the formation of fullerenes in sooting flames [1] and that large dehydrogenated PAH molecules lead to the formation of C_{60} and C_{70}-fullerenes in space [6]. It is generally believed that PAHs or closely related molecules are responsible for the characteristic infrared emission features of a very large number of astrophysical objects including the interstellar medium [7, 8] and it is generally believed that they play important roles in astrochemical reactions. The IR-emission from astrophysical PAHs are most likely caused by comparatively large PAHs containing at least 50-100 carbon atoms which are heated mainly in UV-absorption processes, but it is also possible that they are heated in collisions with electrons, ions or other particles. Interestingly, ion collisions may lead to other fragmentation processes than photo-absorption and as discussed below these fragments are more reactive and effective in molecular growth processes. The latter of course requires that a fragment collides with other molecules or atoms before it is destroyed.

This may happen when the PAH is part of a cluster. Here, we will study cases where molecular growth processes seem to occur inside clusters of only pyrene molecules.

Recently, growth pathways for fullerenes have been discovered in collisions between keV ions and clusters of fullerene molecules [9, 10, 11]. In short range collisions between 400 keV Xe^{20+} ions and mixed clusters of C_{60} and C_{70} (containing 4 % C_{70}) the ions induced significant damage to molecules along the projectiles path, creating a hot gas/plasma of C atoms and small carbon molecules [9]. Fullerene molecules containing up to 180 carbon atoms were then formed through either the coalescence of the small fragments with neighbouring fullerene molecules in the cluster, or through self-assembly of the fragments. These growth processes were only detected when the initial clusters were very large, containing several tens of molecules, showing that the cluster environment is crucial for these mechanisms to take place. In later experiments with collisions between 22.5 keV He^{2+} ions and clusters of C_{60} molecules, the dumbbell shaped C_{119}^+ and C_{118}^+ molecules were detected [10]. These molecules are formed when the projectile ion transfers sufficient energy to one (or two) C atom(s) in a molecule to knock out the atom(s) on the femtosecond timescale. The so formed C_{59} (C_{58}) radical forms covalent bonds with a neighbouring intact C_{60} molecule on the picosecond timescale with a reaction barrier close to 0 eV [10, 12]. This is much lower than the $\sim$ 60 eV kinetic energy required to induce covalent bond formation between two intact C_{60} molecules [13, 14, 15]. Dumbbell fullerene formation was also observed in collisions between 22.5 keV He^{2+} and mixed clusters of C_{60} and the polycyclic aromatic hydrocarbon (PAH) coronene ($C_{24}H_{12}$) [16]. However, equivalent covalent bond formation reactions involving (intact or fragmented) coronene molecules were not detected in that experiment [16].

Here we present early results showing the possible growth of PAH molecules induced by collisions with keV ions. We have studied collisions between 11.25 keV He^+ or 12 keV Ar^{2+} projectile ions and loosely bound clusters of pyrene ($C_{16}H_{10}$) and found that the latter collisions lead to the formation of molecules larger than pyrene. The experimental results are compared to those from simulated collisions and there is a good agreement in the distribution of molecular masses formed by the collisions and subsequent bond-forming reactions.

2. Experimental Details

The experiments are performed at the ARIBE beamline at GANIL in Caen, France. Clusters of pyrene are formed in a liquid nitrogen cooled gas aggregation source [17]. Ions are produced in a Electron Cyclotron Resonance (ECR) ion source, accelerated to 11.25 keV for He^+ and 12 keV for Ar^{2+} and pulsed into 500 ns bunches. The ions collide with the neutral clusters in the interaction region of a linear time-of-flight (TOF) mass spectrometer. Following $\sim$ 100 ns delay, a voltage is applied to the extraction electrodes and the cationic products from individual collision events are detected in coincidence by a microchannel plate (MCP) detector. A schematic of the experimental setup is shown in Fig. 1. The clusters in our target follow a log-normal intensity distribution as a function of their size [18].

3. Simulation Details

We model the collisions between projectile ions/atoms and pyrene clusters containing 36 molecules (Fig. 2) using classical Molecular Dynamics (MD) simulations. The present MD-simulations are performed for neutral systems as it has been shown that the location of the charge does not influence the bond formation or fragmentation processes to any large extent for systems with many delocalized electrons [12, 19]. The molecules and their mutual interactions are modelled using the Adaptive Intermolecular Reactive Empirical Bond Order (AIREBO) potential [20] as implemented in the LAMMPS molecular dynamics suite [21] using a microcanonical (constant energy and volume) ensemble. Interactions with the projectile are modelled using the Ziegler-Biersack-Littmark (ZBL) potential [22] in order to simulate nuclear

Journal of Physics: Conference Series **583** (2015) 012011 doi:10.1088/1742-6596/583/1/012011

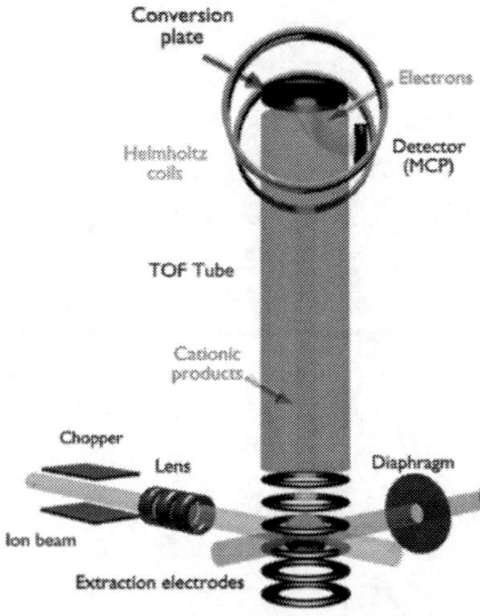

Figure 1. Schematic of the experimental setup. The cluster target is produced by sublimation of a pyrene powder in a liquid nitrogen cooled gas aggregation source. This gives a log-normal size distribution. The cluster target is crossed with a pulsed beam of keV ions (see text).

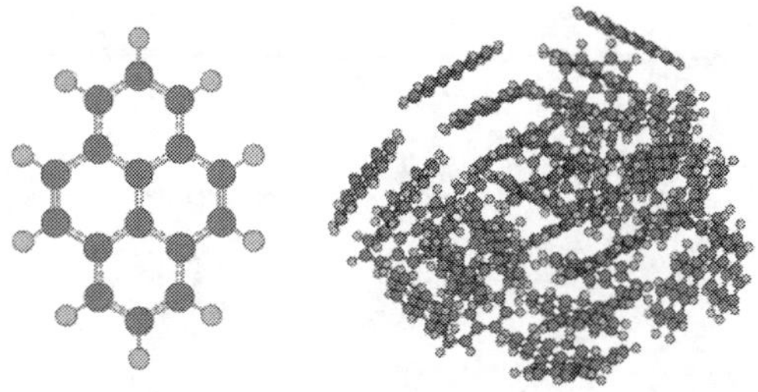

Figure 2. A pyrene molecule (left) and cluster of 36 pyrene molecules (right) from the present MD simulations.

scattering of colliding atoms and does not include electronic stopping. The exclusion of electronic stopping is compensated for by assuming a temperature of 2000 Kelvin for the individual molecules at the beginning of the simulations. This corresponds to an internal energy of 480 eV distributed evenly over the whole $[C_{16}H_{10}]_{36}$ cluster. This is a reasonable approximation in view of estimates of energy transfers to the electronic degrees of freedom in the individual molecules (see e.g. ref. [19]). Further, it has been shown that PAH clusters rapidly redistribute internal energy and charge evenly across the cluster [23]. Time steps of 5×10^{-17} s and 1×10^{-17} s are used for simulation with 12 keV Ar and 11.25 keV He projectiles, respectively. In a simulation, the cluster and projectile are randomly positioned and oriented. Then the projectile is given a velocity corresponding to the desired kinetic energy. The collision and subsequent bond-forming reactions are then followed for a few picoseconds. For the results presented here, 10^4 trajectories have been simulated for each projectile species and energy.

4. Results and Discussion

We show mass spectra from experiments and simulations in Fig. 3. All four panels of Fig. 3 show a mass range which includes the pyrene monomer (mass 202 amu) and the dimer (mass 404 amu). With 12 keV Ar^{2+} projectiles we see that a series of peaks corresponding to masses between that of the monomer and dimer are visible in the experimental mass spectrum. Each subsequent peak in this range is due to an additional C atom, and a distribution of the number of H atoms, above the mass of the intact pyrene molecule. These additional features are largely absent from the mass spectrum with the 11.25 keV He^+ projectile. Here, only a few weak peaks are visible in the mass range between the pyrene monomer and dimer. Also visible in this case is a tail to the right of the monomer caused by the delayed fragmentation of clusters during the extraction of the charged products in time-of-flight mass spectrometer. This feature is present in the case with Ar^{2+} projectiles as well, but is masked by the additional peaks in this mass spectrum.

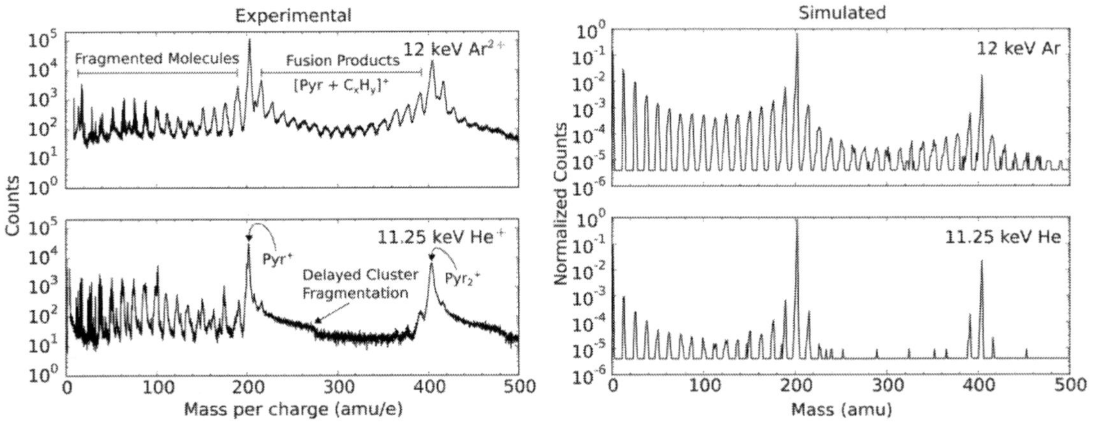

Figure 3. Mass spectra from experiments (left panels) and MD simulations (right panels) for 11.25 keV He^+ (bottom row) and 12 keV Ar^{2+} (top row) projectiles colliding with clusters of pyrene molecules.

The right panels of Fig. 3 show calculated mass spectra from MD simulations. The simulations show results very similar to the experimental ones with respect to the distribution of fusion products in the mass range between the monomer and the dimer. Molecules with these masses are created when the projectile atom promptly knocks one or several atoms out of molecules along the trajectory - this is the so called knockout mechanism which has been shown to be dominant at lower energy collisions with H and He atoms and to lead to highly reactive fragments [19, 24, 25]. These fragments may then easily form covalent bonds with neighboring molecules, either with intact pyrene molecules or other pyrene fragments. We see from the simulations that the primary source of masses between the intact monomer and dimer is from bond coalescence between two rather than several molecules/fragments, typically between an intact pyrene and a single fragment. The region of the mass spectrum around the dimer mass contains contributions from both H knockout, giving reactive radicals which form covalent bonds with neighboring molecules, and intact dimers bound by the dispersion forces. The Ar projectiles are capable of causing more severe damage to the pyrene molecules than the lighter He projectiles, with projectiles colliding with edge-on molecules creating more small fragments than trajectories where the molecules are closer to face-on where the projectile may pass through the molecular rings without transferring significant amounts of energy. The result of this is the lower intensity of fusion products and fragmented molecules in the case of He projectiles. This is expected based on previous studies with fullerenes [26].

The mechanisms driving the growth of pyrene molecules observed here appear to be similar to those seen in the growth of fullerenes. As is the case with fullerenes, atom-knockout in the collisions is key to forming radical fragments which in a second step form bonds with neighboring molecules within picoseconds. For these processes to occur, the cluster environment is crucial as it both supplies the radical with a reactant and also allows the energy transfered in the collision to be redistributed among many degrees of freedom, increasing the probability of the new molecular species surviving the experimental microsecond timescales and longer.

5. Conclusions

We have seen indications of covalent bond forming molecular growth processes inside medium sized clusters of the PAH molecule pyrene due to keV collisions with Ar^{2+} and He^+. The igniting process seems to be knockout of individual atoms from one or several molecules in the clusters which is followed by bond forming reactions with another molecule or fragment from the cluster. The effect is much stronger with the Ar projectile which may be explained by the larger probability for multiple atom knockout in this case. The qualitative agreement between the experimental and simulation results are remarkable and reproduces the difference between reactions following He^{+} and Ar^{2+} collisions. These results suggests efficient pathways to molecular growth in space due to interactions between ions from supernova shock waves and stellar winds with PAH molecules in clusters and grains [27]. It is of considerable interest to extend the study to other projectile masses and charge states and to study the influence of the cluster size both in the experiment and in the simulations. It is expected that there is a smallest possible, critical, cluster size for which the molecular growth processes become important.

Acknowledgements

This work was supported by the Swedish Research Council (Contract Nos. 621-2012-3662, 621-2012-3660, and 621-2011-4047). We acknowledge the COST action CM1204 "XUV/X-ray light and fast ions for ultrafast chemistry (XLIC)" and the European Associated Laboratory DYNAMO.

References

[1] Richter H and Howard J B 2000 *Progress in Energy and Combustion Science* **26** 565–608
[2] Shukla B and Koshi M 2011 *Combustion and Flame* **158** 369 – 375
[3] de Kok T M C M, Driece H A L, Hogervorst J G F and Briedé J J 2006 *Mutation Research/Reviews in Mutation Research* **613** 103–122
[4] Bostrom C E, Gerde P, Hanberg A, Jernstrom B, Johansson C, Kyrklund T, Rannug A, Tornqvist M, Victorin K and Westerholm R 2002 *Environ Health Perspect* **110 Suppl 3** 451–488
[5] Armstrong B, Hutchinson E, Unwin J and Fletcher T 2004 *Environ Health Perspect* **112** 970–978
[6] Berné O and Tielens A G G M 2012 *Proceedings of the National Academy of Sciences* **109** 401–406
[7] Tielens A 2008 *Annual Review of Astronomy and Astrophysics* **46** 289–337
[8] Tielens A G G M 2013 *Rev. Mod. Phys.* **85**(3) 1021–1081
[9] Zettergren H, Johansson H A B, Schmidt H T, Jensen J, Hvelplund P, Tomita S, Wang Y, Martín F, Alcamí M, Manil B, Maunoury L, Huber B A and Cederquist H 2010 *The Journal of Chemical Physics* **133** 104301
[10] Zettergren H, Rousseau P, Wang Y, Seitz F, Chen T, Gatchell M, Alexander J D, Stockett M H, Rangama J, Chesnel J Y, Capron M, Poully J C, Domaracka A, Méry A, Maclot S, Schmidt H T, Adoui L, AlcamíM, Tielens A G G M, Martín F, Huber B A and Cederquist H 2013 *Phys. Rev. Lett.* **110**(18) 185501
[11] Seitz F, Zettergren H, Rousseau P, Wang Y, Chen T, Gatchell M, Alexander J D, Stockett M H, Rangama J, Chesnel J Y, Capron M, Poully J C, Domaracka A, Méry A, Maclot S, Vizcaino V, Schmidt H T, Adoui L, Alcamí M, Tielens A G G M, Martín F, Huber B A and Cederquist H 2013 *The Journal of Chemical Physics* **139** 034309
[12] Wang Y, Zettergren H, Rousseau P, Chen T, Gatchell M, Stockett M H, Domaracka A, Adoui L, Huber B A, Cederquist H, Alcamí M and Martín F 2014 *Phys. Rev. A* **89**(6) 062708
[13] Rohmund F, Glotov A V, Hansen K and Campbell E E B 1996 *Journal of Physics B: Atomic, Molecular and Optical Physics* **29** 5143
[14] Knospe O, Glotov A V, Seifert G and Schmidt R 1996 *Journal of Physics B: Atomic, Molecular and Optical Physics* **29** 5163
[15] Jakowski J, Irle S and Morokuma K 2010 *Phys. Rev. B* **82**(12) 125443
[16] Gatchell M, Rousseau P, Domaracka A, Stockett M H, Chen T, Schmidt H T, Chesnel J Y, Méry A, Maclot S, Adoui L, Huber B A, Zettergren H and Cederquist H 2014 *Phys. Rev. A* **90**(2) 022713
[17] Bergen T, Biquard X, Brenac A, Chandezon F, Huber B A, Jalabert D, Lebius H, Maurel M, Monnand E, Opitz J, Pesnelle A, Pras B, Ristori C and Rocco J C 1999 *Review of Scientific Instruments* **70** 3244–3253
[18] Maul J, Berg T, Marosits E, Schönhense G and Huber G 2006 *Phys. Rev. B* **74**(16) 161406
[19] Chen T, Gatchell M, Stockett M H, Alexander J D, Zhang Y, Rousseau P, Domaracka A, Maclot S, Delaunay R, Adoui L, Huber B A, Schlathölter T, Schmidt H T, Cederquist H and Zettergren H 2014 *The Journal of Chemical Physics* **140** 224306
[20] Stuart S J, Tutein A B and Harrison J A 2000 *The Journal of Chemical Physics* **112** 6472–6486
[21] Plimpton S 1995 *Journal of Computational Physics* **117** 1–19
[22] Ziegler J F, Biersack J P and Littmark U 1985 *In The Stopping and Range of Ions in Matter* vol 1 (New York: Pergamon) ISBN 0-08-022053-3

Journal of Physics: Conference Series **583** (2015) 012011 doi:10.1088/1742-6596/583/1/012011

[23] Holm A I S, Zettergren H, Johansson H A B, Seitz F, Rosén S, Schmidt H T, Ławicki A, Rangama J, Rousseau P, Capron M, Maisonny R, Adoui L, Méry A, Manil B, Huber B A and Cederquist H 2010 *Phys. Rev. Lett.* **105**(21) 213401

[24] Gatchell M, Stockett M, Rousseau P, Chen T, Kulyk K, Schmidt H, Chesnel J, Domaracka A, Méry A, Maclot S, Adoui L, Støchkel K, Hvelplund P, Wang Y, Alcamí M, Huber B, Martín F, Zettergren H and Cederquist H 2014 *International Journal of Mass Spectrometry* **365–366** 260 – 265 special issue: Tilmann Mäerk

[25] Stockett M H, Zettergren H, Adoui L, Alexander J D, Bērziņš U, Chen T, Gatchell M, Haag N, Huber B A, Hvelplund P, Johansson A, Johansson H A B, Kulyk K, Rosén S, Rousseau P, Støchkel K, Schmidt H T and Cederquist H 2014 *Phys. Rev. A* **89**(3) 032701

[26] Tomita S, Hvelplund P, Nielsen S B and Muramoto T 2002 *Phys. Rev. A* **65**(4) 043201

[27] Micelotta, E R, Jones, A P and Tielens, A G G M 2010 *A&A* **510** A36

Journal of Physics: Conference Series **583** (2015) 012012 doi:10.1088/1742-6596/583/1/012012

Charge exchange produced emission of carbon in the extreme ultraviolet spectral region

J K Lepson[1], P Beiersdorfer[2], M Bitter[3], A L Roquemore[3], K Hill[3] and R Kaita[3]

[1] Space Sciences Laboratory, University of California, Berkeley, CA 94720 USA
[2] Physics Division, Lawrence Livermore National Laboratory, Livermore, CA 94550 USA
[3] Princeton Plasma Physics Laboratory, Princeton, NJ 08543 USA USA

E-mail: `lepson@ssl.berkeley.edu`

Abstract.
We used a time-reolving high-resolution grating spectrometer to study extreme ultraviolet emission from plasmas in the National Spherical Tokamak Experiment (NSTX). The NSTX spectral range from 150-250 Å is typically dominated by emission from M-shell iron lines, L-shell transitions of oxygen, or K-shell lines of lithium. However, we also observed several intense emission lines, which we now attribute to transitions in C V and C VI. Collisional-radiative modeling shows that electron-impact excitation is far too weak to account for the features we observed. Instead, these lines appear to be produced by charge exchange with neutral hydrogen.

1. Introduction

Charge exchange occurs when an ion takes an electron from a neutral species or from an ion with lesser charge, thereby reducing its own charge while increasing the charge of the donor species. In recent years, it has been identified as an important process in astrophysics, where it contributes to the soft x-ray background. Charge exchange occurs extensively in the solar system, e.g., in the interaction of neutral atoms of cometary comae and planetary atmospheres with the highly charged solar wind [1, 2, 3], and may also be a significant source of soft x rays in such objects as supernova remnants and starburst galaxies [4, 5]. Charge exchange is also of interest as a diagnostic tool in tokamaks, especially for determining ion temperature and impurity profiles using an energetic neutral beam to enable charge exchange deep within the plasma [6, 7, 8].

Spectral observations of astrophysical charge exchange phenomena have mainly focused on the x-ray emission of K-shell ions. Identifying lines produced by charge exchange in the extreme ultraviolet (EUV) region may aid in ascertaining charge exchange phenomena in cooler plasmas. Both the *Chandra* and *XMM-Newton* x-ray observatories are sensitive to EUV emission from cosmic sources, while the Hinode satellite and the *Solar Dynamics Observatory* observe this spectral band in the sun. Thus, it may be possible to use these currently operating observatories to search for lines indicative of charge exchange.

In recent years we have made measurements of the EUV emission spectra generated by the NSTX spherical torus [9, 10, 11]. The focus of these measurements has been to study the iron L-shell and iron M-shell emission and to calibrate density-sensitive line ratios [11, 12].

Journal of Physics: Conference Series **583** (2015) 012012 doi:10.1088/1742-6596/583/1/012012

While studying the M-shell iron emission in the 170–250 Å region, we observed unexpected enhancements of some of the lines in this region, which initially appeared to be iron lines. Upon additional studies, we attributed the lines to transitions in carbon C V and C VI [15]. In the following, we present spectral data from NSTX discharges that have no noticeable iron emission. This confirms that the lines in question are not from iron. A comparison of the observed carbon emission with spectral calculations rules out that the lines are produced by electron-impact excitation. This leaves charge exchange as the production mechanism.

2. Experiment

We obtained spectra on NSTX using the Long-Wavelength EUV Spectrometer (LoWEUS; [11]) instrument. LoWEUS is a flat-field grating spectrometer with variable line spacing and a mean 1200 ℓ/mm. The spectral resolution is ~0.3 Å, resulting in a resolving power $\lambda/\Delta\lambda$ ~500-800 in the 150-250 Å region we examined ([16]). LoWEUS is capable of a time resolution of ~5 ms, but was operated in time-integrated mode during the shot we present here. We operate a similar instrument on the Alcator tokamak [17, 18].

Figure 1 shows a spectrum from NSTX shot 140324. The strongest line in the spectrum is the Li III $2p \rightarrow 1s$ Lyman-α transition at 133 Å. The $2p \rightarrow 1s$ emission from heliumlike Li II can be seen just below 200 Å. The other strong lines are all from carbon, i.e., the they are K-shell transitions from C V and C VI. Although these lines occur near 41 and 33 Å, respectively, they appear here in higher order diffraction. The fact that these carbon lines are seen proves the abundance of carbon in the plasma.

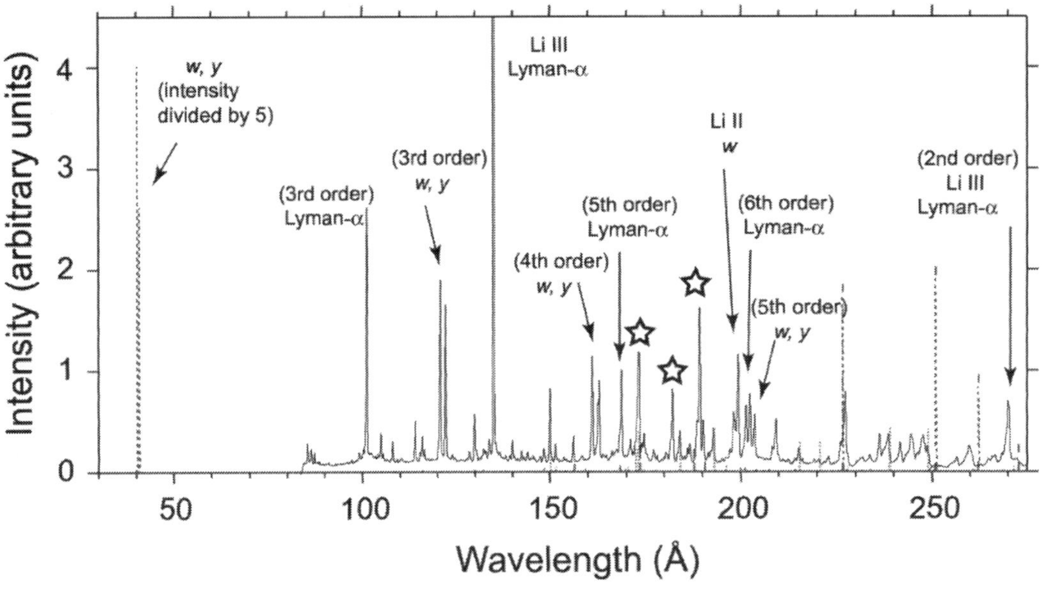

Figure 1. NSTX Shot 140324 (solid line) overlaid with FAC calculations of carbon emission due to collisonal excitation (blue dashed lines) and oxygen emission from the CHIANTI database (red dotted lines). Intensities of carbon IV lines w and y in first order have been decreased by a factor of 5 to fit on the figure. Lines not identified to element are carbon; strong new carbon lines are marked with stars.

It is interesting to note that the the K-shell spectrum of heliumlike C V is comprised of two lines, commonly known as w and y, which emanate from the upper levels $1s2p\ ^1P_1$ and $1s2p\ ^3P_1$,

Journal of Physics: Conference Series **583** (2015) 012012 doi:10.1088/1742-6596/583/1/012012

Table 1. Candidate carbon lines in the 170– 250 Å region.

Ion	Wavelength (Å)	Lower Level	Upper Level
C V	173.281	$1s2s\ ^3P_1$	$1s4p\ ^3P_1$
C VI	182.088	$2p\ ^2P_{1/2}$	$3d\ ^2D_{3/2}$
C VI	182.097	$2p\ ^2P_{1/2}$	$3d\ ^2D_{3/2}$
C VI	182.132	$2p\ ^2P_{1/2}$	$3d\ ^2D_{1/2}$
C VI	182.144	$2p\ ^2P_{1/2}$	$3d\ ^2D_{1/2}$
C VI	182.230	$2p\ ^2P_{3/2}$	$3d\ ^2D_{5/2}$
C VI	182.246	$2p\ ^2P_{3/2}$	$3d\ ^2D_{3/2}$
C VI	182.290	$2p\ ^2P_{3/2}$	$3d\ ^2D_{1/2}$
C V	189.255	$1s2p\ ^3P_1$	$1s4s\ ^3S_1$
C V	189.260	$1s2p\ ^3P_0$	$1s4s\ ^3S_1$
C V	189.304	$1s2p\ ^3P_2$	$1s4s\ ^3S_1$
C V	227.190	$1s2s\ ^3S_1$	$1s3p\ ^3P_2$
C V	227.202	$1s2s\ ^3S_1$	$1s3p\ ^3P_0$
C V	227.203	$1s2s\ ^3S_1$	$1s3p\ ^3P_1$
C V	248.660	$1s2p\ ^3P_1$	$1s3d\ ^3D_2$
C V	248.660	$1s2p\ ^3P_1$	$1s3d\ ^3D_1$
C V	248.672	$1s2p\ ^3P_0$	$1s3d\ ^3D_1$
C V	248.740	$1s2p\ ^3P_2$	$1s3d\ ^3D_3$
C V	248.748	$1s2p\ ^3P_2$	$1s3d\ ^3D_1$
C V	248.748	$1s2p\ ^3P_2$	$1s3d\ ^3D_2$

respectively.

Several weaker lines can also be seen. We have not yet identified all lines, but many of the weaker lines can be attributed to oxygen. The positions and relative intensities of the oxygen lines in this wavelength band are given by the CHIANTI spectral model [19, 20], as indicated by the red dotted lines in Fig. 1.

The spectrum in Fig. 1 also shows three relatively strong features marked with an asterisk. These are the features we have recently attributed to the carbon lines listed in Table 1.

3. Comparison with Theoretical Spectra

Calculations of the carbon line emission from collisional excitation were performed using the Flexible Atomic Code [21]. The line emission was calculated for the temperature of maximum abundance for C V (31 eV) and C VI (87 eV), using the electron density $n_e = 5 \times 10^{13}\mathrm{cm}^{-3}$ measured by Thomson scattering. Calculations with higher and lower densities were also performed, but an effect of the density on the relative line emission could not be seen unless the density was varied by several orders of magnitude.

The results from our calculations for C V are shown in Fig. 1 as blue dashed traces. Here, we also show the w and y lines in 1st order. These lines appear in the spectrum only in 3rd and higher orders. We have normalized the intensity of the calculated emission in 1st order to the observed emission in 3rd order with the help of a measurement of the grating efficiency for measuring the w line in different orders made at the EBIT-I machine [22]. Because the first-order w and y lines are much stronger than the higher orders we observed, the NSTX spectrum and the calculated lines in the region we observed have been expanded vertically by a factor of 5 in order to better see the lines.

Our calculations predict several C V lines that fall into the observed spectral range. The largest predicted lines are near 228 and 251 Å, and they are stronger than any of the observed features. This may mean that our normalization to the 3rd order K-shell lines of C V is off by

Journal of Physics: Conference Series **583** (2015) 012012 doi:10.1088/1742-6596/583/1/012012

several factors, thus resulting in an overprediction of these intensity of these lines. Despite this overprediction, the predicted intensity of the lines in question near 173, 182, and 189 Å is far too weak to account for the intensity of these carbon lines.

Figure 2 shows an expanded view of the spectral region near the three carbon features. The figure clearly demonstrates that the strengths predicted in our FAC model are vanishingly small. This rules out electron-impact excitation as the process forming these lines.

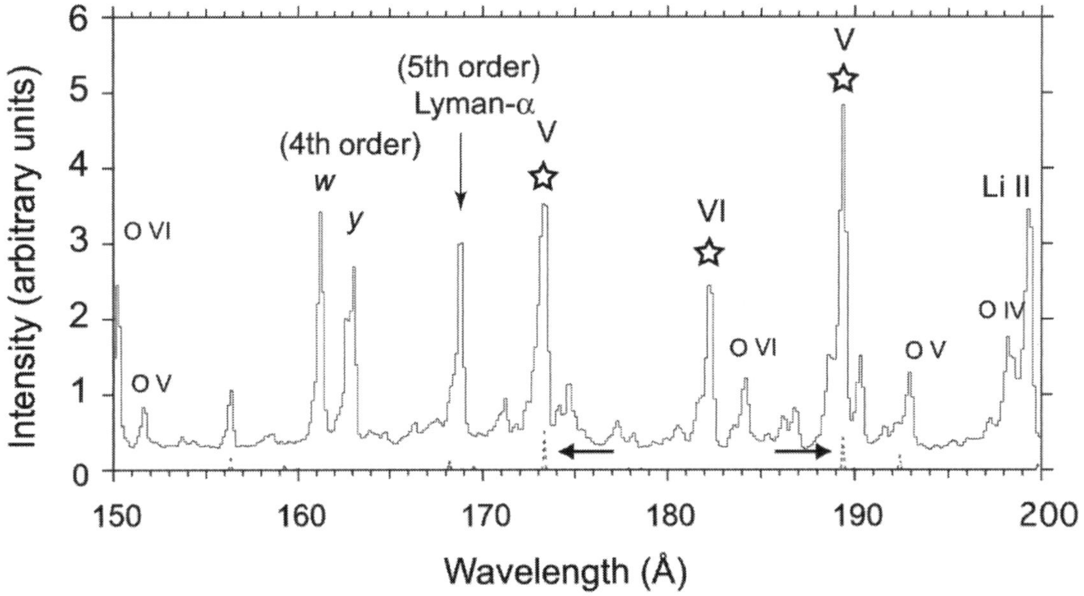

Figure 2. Inset of NSTX Shot 140324 showing comparison between calculated emission strengths via collisional excitation (blue dashed lines) vs. measured (solid line). Lines not identified to element are carbon; strong new carbon lines are marked with stars.

The strength of the new C V and C VI lines in our spectrum is indicative of non-equilibrium plasma processes, in particular, of charge exchange between highly charged carbon with neutral hydrogen. Charge exchange populates the n = 3,4 levels in carbon ions, and our two strongest lines, at 173.19 and 189.40 Å are both 4→2 transitions, while the feature at 182.23Å is a blend of 3→2 transitions. We are now undertaking calculations of the spectral emission due to charge exchange between hydrogen and bare and hydrogenlike carbon utilizing the Charge Exchange Spectral Synthesizer, CHESS [8]. CHESS predicts spectral emission patterns that match the observations well. These calculations are ongoing and will be presented in future work.

Acknowledgments
This work was performed by Lawrence Livermore National Laboratory and Princeton Plasma Physics Laboratory under the auspices of the U. S. Department of Energy under Contracts DEAC52-07NA27344 and DE-AC02-09CH11466 and supported by the U.S. Department of Energy, Office of Science, Office of Fusion Energy Sciences.

References

[1] Cravens T E 2002, *Science*, **296**, 1042
[2] Beiersdorfer P, Boyce K R, Brown G V, Chen H, Kahn S M, Kelley R L, May M, Olson R E, Porter F S, Stahle C K and Tillotson W A 2003, *Science*, **300**, 1558.

[3] Wargelin B J, Beiersdorfer P, Neill P A, Olson R E and Scofield J H 2005, *Astrophys.J.Suppl*, **634**, 637.
[4] Katsuda S, Tsunemi H, Mori K, Uchida H, Kosugi H, Kimura M, Nakajima H, Takakura S, Petre R, Hewitt J W and Yamaguchi H 2011, *Astrophys.J.*, **730**, 24.
[5] Wang Q D and Liu J 2012, *Astron.Nachtrichten*, **333**, 373.
[6] Källne E, Källne J, Dalgarno A, Marmar E S, Rice J E and Pradhan, A K 1984, *Phys.Rev.Lett.*, **52**, 2245
[7] Rice J E, Marmar E S, Terry J L, Källne E and Källne J 1986, *Phys.Rev.Lett.*, **56**, 50.
[8] Beiersdorfer P, Bitter M, Marion M and Olson R E 2005, *PhysRevA*, **72**, 032725.
[9] Beiersdorfer P, Bitter M, Roquemore L, Lepson J K and Gu M F 2006, *Rev.Sci.Instrum.*, **77**, 10F306.
[10] Beiersdorfer P, Lepson J K, Bitter M, Hill K W and Roquemore L 2008, *Rev.Sci.Instrum.*, **79**, 10E313.
[11] Lepson J K, Beiersdorfer P, Clementson J, Gu M F, Bitter M, Roquemore L, Kaita R, Cox P G and Safronova A S 2010, *J.Phys.B*, **43**, 142010.
[12] Lepson J K, Beiersdorfer P, Gu M F, Desai P, Bitter M, Roquemore L and Reinke M L 2012, *AIPConf.Proc.* **1438**, 136.
[13] Beiersdorfer P, Crespo López-Urrutia J R, Springer P, Utter S B and Wong K L 1999, *Rev.Sci.Instrum.*, **70**, 276.
[14] Graf A, Brockington S, Horton R, Howard S, Hwang D, Beiersdorfer P, Clementson J, Hill D, May M, Mclean H, Wood R, Bitter M, Terry J, Rowan W L, Lepson J K and Delgado-Aparicio L 2008, *Can.J.Phys.*, **86**, 307.
[15] Lepson J K, Beiersdorfer P, Bitter M, Roquemore A L, Kaita R 2013, *PhysicaScripta*, **156**, 014075.
[16] Lepson J K, Beiersdorfer P, Bitter M, Roquemore L, Hill K W, Kaita R, Skinner C H and Zimmer G 2012, *Rev.Sci.Instrum.*, **83**, 10D520.
[17] Reinke M L, Beiersdorfer P, Howard N T, Magee E W, Podpaly Y, Rice J E and Terry J L 2010, *Rev.Sci.Instrum.*, **81**, 10D736.
[18] Beiersdorfer P, Brown G V, Kamp J B, Magee E W, Lepson J K, Podpaly Y and Reinke M L 2011, *Can.J.Phys.*, **89**, 653.
[19] Dere K P, Landi E, Mason H E, Monsignori Fossi B C and Young P R 1997, *Astron.Astrophys.Suppl.*, **125**, 149.
[20] Landi E, Del Zanna G, Young P R, Dere K P and Mason H E 2012, *Astrophys.J.*, **S744**, 99.
[21] Gu M F 2008, *Can.J.Phys.*, **86**, 675.
[22] Brown G V, Beiersdorfer P, Clementson J, Dunn J, Kelley, R L, Kilbourne C A, Leutenegger M, Magee E W, Park J, Porter F S, Schneider M and Träbert E 2010, Space Telescopes and Instrumentation 2010: Ultraviolet to Gamma Ray. Ed. by Arnaud, Monique; Murray, Stephen S.; Takahashi, Tadayuki. *Proc.SPIE*, **7732**, 77324Q.

Journal of Physics: Conference Series **583** (2015) 012013 doi:10.1088/1742-6596/583/1/012013

Electron emission in collisions of dressed ions with He atoms

D Fregenal[1], J M Monti[2], J Fiol[1], G Bernardi[1], S G Suárez[1], P D Fainstein[1], R D Rivarola[2], and R Schuch[3]

[1] Centro Atómico Bariloche, CNEA, Av. Bustillo 9500, 8400 Bariloche, Argentina
[2] Laboratorio de Colisiones Atómicas, Facultad de Ciencias Exactas, Ingeniería y Agrimensura and Instituto de Física Rosario, CONICET-UNR, Av. Pellegrini 250, 2000 Rosario, Argentina.
[3] Department of Physics, Stockholm University, AlbaNova University Center, SE-106 91 Stockholm, Sweden

E-mail: fregenal@cab.cnea.gov.ar

Abstract. We present theoretical and experimental results for angular and energy distributions of electrons emitted in collisions of B^{2+} with He atoms. In particular, we analyze the cusp formed by projectile Electron Loss to Continuum (ELC) and Electron Capture to Continuum (ECC) . The cusp shape dependence on the electronic initial state is discussed and compared with our previous measurements on $Li^{+} + He$. The projectile incident energy was about 414 keV/u and the electron detection angle was varied between 0° and 150°. The experimental data is compared with three-body quantum Continuum-Distorted-Wave calculations.

1. Introduction

Electron emission produced by collisions has been studied for decades in order to understand fundamental atomic processes. Of particular interest here are one- and many-electron processes induced by interactions between two partially screened nuclei and their bound electrons. We studied collision systems where the binding energy of the projectile electrons is similar to the binding energy of the target electrons, thus projectile as well as target electrons can no longer be considered as passive (frozen in their orbitals during the reaction). Therefore, target and projectile, and even simultaneous ionisation can give a noticeable contribution to electron emission.

The studies of projectile ionisation (stripping) have also important applications in design parameters for ion accelerators, for the energy loss (range) of ions in matter, and for limiting beam transport losses as well as losses in storage rings. At present, scaling laws, applicable to a wide range of systems, are the best way of extrapolate data [1–5], specially for many-electron collision systems where multiple-electron processes can dominate the observed spectra and one-electron theories are inadequate.

In this work we present experimental angular and energy distributions of electrons emitted in collisions of B^{2+} with He atoms and the corresponding theoretical results of quantum three-body CDW calculations. In particular, data taken at zero degrees for energies in the region of the cusp, formed by projectile Electron Loss to Continuum (ELC) and Electron Capture to Continuum (ECC), are compared with measured data for $Li^{+} + He$ at very similar collision energies.

Journal of Physics: Conference Series **583** (2015) 012013 doi:10.1088/1742-6596/583/1/012013

2. The experimental arrangement

Experimental data were measured at the 1.7 MV Tandem accelerator at Centro Atómico Bariloche. The experimental set-up is described in detail elsewhere [6], thus only details related to the present experiment are reported here. Boron ion beams emerging from the accelerator with approximately 414 keV/u are collimated in the transport section by two sets of four collimators to 0.60×0.60 mm^2, determining a beam divergence of 0.7 mrad (half-angle). Inside the collision chamber the projectiles collide with the effusive target at the focus of a cylindrical mirror spectrometer. The analyzer rotates in a plane perpendicular to the gas flow direction, so it measures energy distributions of the electrons emitted in the collision for any selected angle between 0 and 180 degrees. After the collision region, the beam is collected in a Faraday cup. The collected charge is used to normalize the electronic distributions to a constant number of projectiles.

3. Theory

The experimental data have been compared to quantum three-body CDW calculations. The theoretical model has previously been used by the authors to describe electron emission in Li^{q+} ($q = 1, 2$) [7] and Al^{q+} ($q = 1, 2, 3$) collisions with He targets [8]. The initial-channel distorted wavefunction is written as the product of an initial target bound state, and a Coulomb continuum factor that describes the interaction between the target active electron and the projectile in the entry channel. After the ionisation, the electron evolves in the combined fields of the projectile and the residual target. The final-channel distorted wavefunction is chosen as a target continuum state multiplied by a Coulomb continuum factor. In order to extend this model to the case of dressed projectiles [9], the projectile potential was represented by means of parametric Green-Sellin-Zachor (GSZ) potentials [10].

To estimate the contributions of simultaneous ionisation from the target and the projectile a probabilistic approach was used. It is obtained as the sum of the differential cross sections for projectile ionisation due to the interaction with a He target, and for He target ionisation by B^{2+} impact, each of them weighted by their corresponding total cross sections (see [7, 8]). This approach gives a rough estimate of the contribution by simultaneous ionisation, and does not take into account any interference effects. It is expected to be more suitable when electrons in one center are much more deeply bound than in the other center.

4. Results and Discussions

Doubly differential cross sections (DDCS) for electron emission in collisions between B^{2+} projectiles and He targets were measured. Figure 1 shows the experimental electron distributions at different emission angles for 4.5 MeV B^{2+} ions impinging on He. At 0 degrees the cusp at around 226 eV is most pronounced and its intensity decreases with increasing angle. These electrons are ejected into a small cone of half angle about 5° in the forward direction with a velocity that matches the velocity of the incident ions. At larger angles the so-called Binary-Encounter Peak (BEP) appears at higher electron energies, whose maximum shifts to lower energies with increasing angles.

At large emission angles (backward directions) a small *bump* is observed at energies of about 35 eV. This is the signature of autoionisation processes for the doubly excited state of Helium. At lower angles this mechanism is hidden in the background produced by other direct ionisation mechanisms. A similar structure, but associated with B^{2+} autoionisaton, is shown for electron emission in the forward direction at electron energies of 740 eV [11].

Electrons with laboratory velocities close to the beam velocity are produced in three major processes. (i) Projectile electrons are ionized to continuum states of low momentum with respect to the projectile nucleus (ELC). (ii) Electrons are transferred from a bound state of the target nucleus to a low-lying continuum state of the projectile nucleus (ECC). (iii) Electrons from projectile Auger processes appear near the projectile velocity in the laboratory frame. However, in the experimental spectra we do not observe any of these autoionisation lines from the Li-like B ions in the main peak.

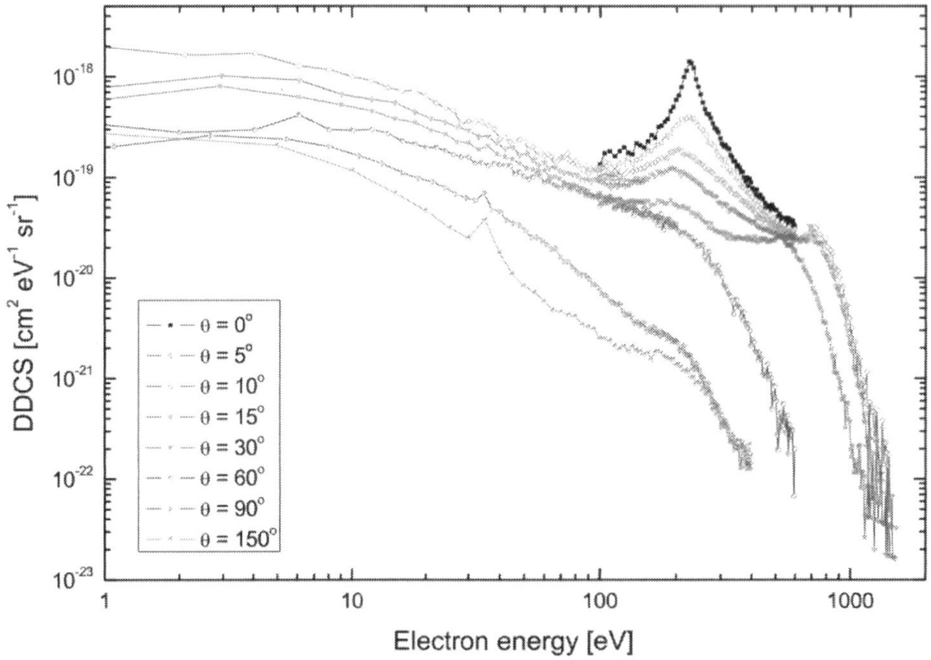

Figure 1. Measured DDCS for electron emission in collisions between 4.5 MeV B^{2+} ions impinging on He as a function of emission energy for several fixed ejection angles. Data were normalized to absolute values to match the theoretical calculations at 400 eV and 0° emission.

The energy distribution of electrons ejected at 0° cusp are compared in Figure 2 with the predictions of the CDW model. Just for comparison reasons the experimental data were normalized to agree with the theory at an energy of 400 eV. A good overall agreement is observed, with marked differences only on the ECC cusp, where the experimental angular and energy finite resolutions smooth the $1/v$ behaviour.

The calculations allow us to distinguish between the different contributions to the differential electron emission cross section. In the neighborhood of the cusp the major contribution comes from electrons emitted from the projectile (ELC). It is observed that both projectile ionisation and simultaneous ionisation contribute similarly to the ELC cusp, and the corresponding cross-sections have similar shapes.

We separate the contributions to projectile ionisation due to the 2s electron from the 1s electrons of the Li-like B ion. In the cusp, the 2s ionisation dominates but in the high energy end, above 700 eV, 1s ionisation takes over. However, this effect is completely hidden in the total emission cross-section because target ionisation becomes strongly dominating above 400 eV. The difference in the cusp by ionisation of the 2s electron from the 1s electron ionisation can be seen when comparing the Li-like B ion with the data taken for He-like Li ions (see Figure 3) [7]. In the cusp from B^{2+}, the 2s ionisation is dominant, whereas for Li^{+} it is only 1s ionisation. Contributions from a possible metastable He-like Li was found to be negligible. The cusp for B^{2+} is significantly wider than for Li^{+} projectiles, even when the binding energies of the 1s electron in Li^{+} (first ionisation potential, 76 eV) almost doubles the one of 2s electrons in B^{2+} (38 eV).

In Figure 4 we compare the experimental spectra with predictions from the CDW model for four different emission angles, 10, 30, 90, and 150 degrees. The data were normalized using the same factor as in figure 1. The overall agreement for emission close to the direction of incidence is quite good. However, for out-of-axis emission the theory overestimates the measured data, and the agreement only improves again for very large angles (150°).

When we analyse the contributions from the different processes: projectile, target and simultaneous

Journal of Physics: Conference Series **583** (2015) 012013 doi:10.1088/1742-6596/583/1/012013

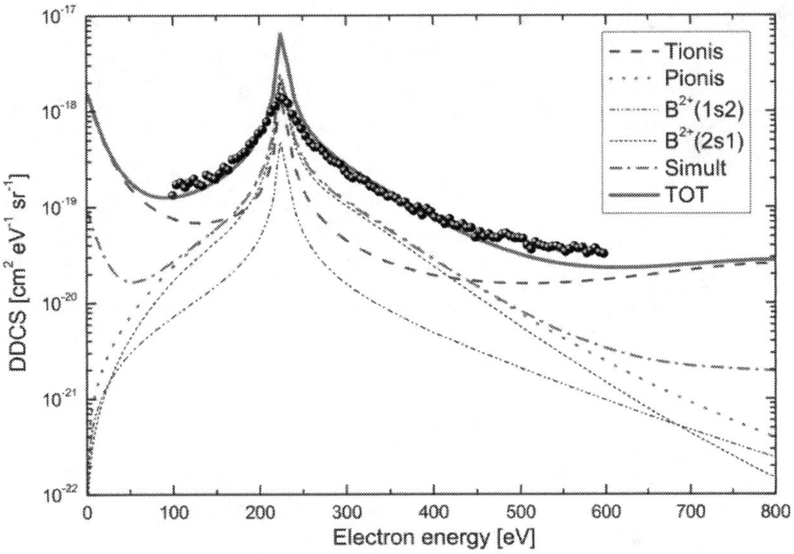

Figure 2. Theoretical and experimental DDCS in the same condition as in Figure 1, for 0° emission angle. The absolute value of the experimental results has been scaled to those of the CDW theory at 400 eV. Full line, total electron emission; dashed line, target ionisation; dotted line, projectile ionisation; dash-dotted line, simultaneous ionisation. The dot-dot-dash line is the contribution to projectile ionisation from the B^{2+}(1s2) orbital, while short-dashed line is the contribution from the B^{2+}(2s1) orbital.

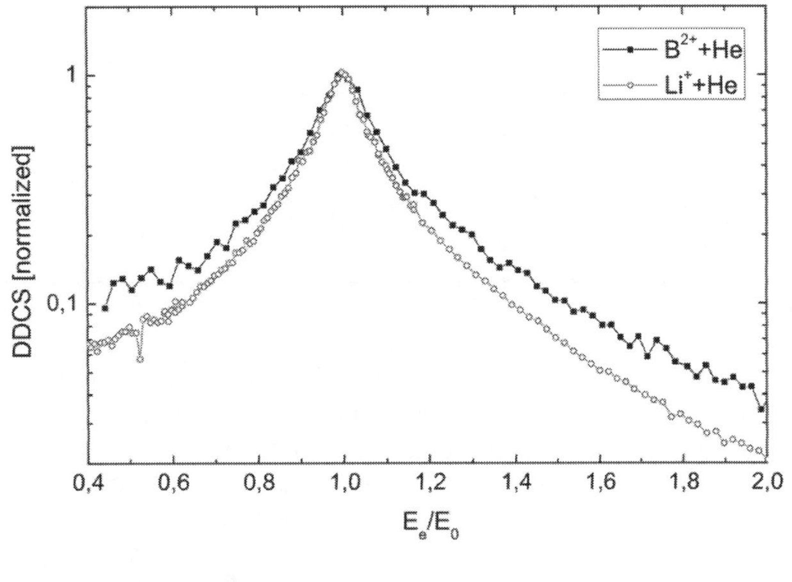

Figure 3. DDCS for electron emission in collision between 414 keV/u ($v_p \approx 4.08$ au) B^{2+} and 440 keV/u ($v_p \approx 4.20$ au) Li^+ with He targets for a fixed 0° emission angle. The cusp top has been normalized to a value of one, and the energy axis has been scaled to the incident energy E_0 for better comparison.

ionisations, we see that the main reason for the discrepancies with the experiment at intermediate angles arises from overestimation of target emission. On the other hand, in the vicinity of the incident velocity, at forward angles, the theoretical values for processes associated with electron projectile emission (projectile and simultaneous ionisation) are slightly overestimated, giving a more pronounced maximum than the experimentally observed.

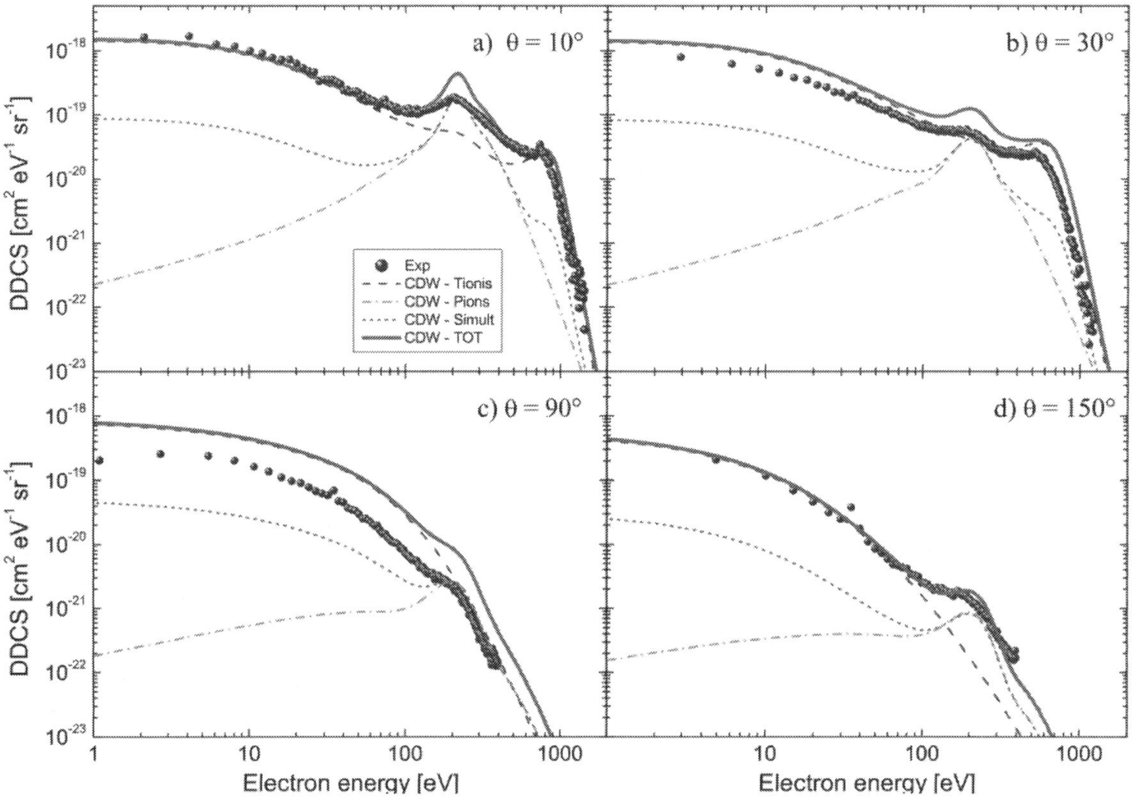

Figure 4. Experimental and theoretical DDCS for electron emission in collisions of 414 keV/u B^{2+} on He. Corresponding emission angles are, a) $\theta = 10°$, b) $\theta = 30°$, c) $\theta = 90°$, and d) $\theta = 150°$.

5. Conclusions

Double differential cross-sections for electron emission in B^{2+} + He collisions were experimentally determined. The measured data has been compared to previous measurements on $Li^{+} + He$ collisions and to quantum-mechanical CDW calculations. By comparing the observed data with the theoretical results, the relative importance of the different processes contributing to the total electron emission has been analysed. Present results show that target ionisation dominates the spectra for all energies and angles emissions, except in the vicinity of the incident velocity (the ELC region). The main contributions of electrons emitted in the ELC region arise from electrons emitted from the projectile, either in pure projectile-ionisation or simultaneous-ionisation processes.

Acknowledgments
One of the authors (R.S.) acknowledge the support by the EU project: FP7-PEOPLE-2010-IRSES 269243 DWBQS.

Journal of Physics: Conference Series **583** (2015) 012013 doi:10.1088/1742-6596/583/1/012013

References

[1] Schlachter A S *et al.* 1981 *Phys. Rev. A* **23**, 2331

[2] DuBois R D, Toburen L H, Middendorf M E and Jagutzki O 1994 *Phys. Rev. A* **49** 350

[3] Santos A C F and DuBois R D 2004 *Phys. Rev. A* **69** 042709

[4] DuBois R D, Santos A C F, Sigaud G M and Montenegro E C 2011 *Phys. Rev. A* **84** 022702

[5] Montenegro E C, Sigaud G M, and DuBois R D 2013 *Phys. Rev. A* **87** 012706

[6] Bernardi G, Suárez S, Fregenal D, Focke P and Meckbach W 1996 *Rev. Sci. Instrum.* **67** 1761

[7] Fregenal D, Monti J M, Fiol J, Fainstein P D, Rivarola R D, Bernardi G, and Suárez S 2014 *J. Phys. B: At. Mol. Opt. Phys.* **47** 155204

[8] Monti J M, Fiol J, Fregenal D, Fainstein P D, Rivarola R D, Wolff W, Horsdal E, Bernardi G, and Suárez S 2014 *J. Phys.: Conf. Series* **488** 012041

[9] Monti J M, Rivarola R D and Fainstein P D 2011 *J. Phys. B: At. Mol. Opt. Phys.* **44** 195206

[10] Green A E S, Sellin D L, and Zachor A S 1969 *Phys. Rev.* **1** 184

[11] Bruch R, Rødbro M, Bisgaard P, and Dahl P 1977 *Phys. Rev. Lett.* **39** 801

Journal of Physics: Conference Series **583** (2015) 012014 doi:10.1088/1742-6596/583/1/012014

Atomic Physics with Accelerators: Projectile Electron Spectroscopy (APAPES)*

I Madesis[1,2], A Dimitriou[1,2], A Laoutaris[3], A Lagoyannis[2], M Axiotis[2], T Mertzimekis[2,4], M Andrianis[2], S Harissopulos[2], E P Benis[5], B Sulik[6], I Valastyán[6] and T J M Zouros[1,2]

[1]Department of Physics, University of Crete, P.O Box 2208, GR 71003 Heraklion, Greece

[2]Tandem Accelerator Laboratory, INP, NCSR Demokritos, GR 15310 Ag Paraskevi, Greece

[3]Department of Physics, School of Applied Sciences, National Technical University of Athens, GR 15780 Zografou, Athens, Greece

[4]Department of Physics, University of Athens, Zografou Campus, GR 15784 Athens, Greece

[5]Department of Physics, University of Ioannina GR 45110, Ioannina, Greece

[6]Institute for Nuclear Research (MTA ATOMKI), Bem tér 18/c, H 4026 Debrecen, Hungary

E-mail: adimitr@physics.uoc.gr

Abstract. The new research initiative APAPES (http://apapes.physics.uoc.gr/) has already established a new experimental station with a beam line dedicated for atomic collisions physics research, at the 5 MV TANDEM accelerator of the National Research Centre "Demokritos" in Athens, Greece. A complete zero-degree Auger projectile spectroscopy (ZAPS) apparatus has been put together to perform high resolution studies of electrons emitted in ion-atom collisions. A single stage hemispherical spectrometer with a 2-dimensional Position Sensitive Detector (PSD) combined with a doubly-differentially pumped gas target will be used to perform a systematic isoelectronic investigation of K-Auger spectra emitted from collisions of pre-excited and ground state He-like ions with gas targets using novel techniques. Our intention is to provide a more thorough understanding of cascade feeding of the 1s2s2p ^4P metastable states produced by electron capture in collisions of He-like ions with gas targets and further elucidate their role in the non-statistical production of excited three-electron 1s2s2p states by electron capture, recently a field of conflicting interpretations awaiting further resolution. At the moment, the apparatus is being completed and the spectrometer will soon be fully operational. Here we present the project progress and the recent high resolution spectrum obtained in collisions of 12 MeV C^{4+} on a Neon gas target.

Journal of Physics: Conference Series **583** (2015) 012014 doi:10.1088/1742-6596/583/1/012014

1. Introduction

The APAPES initiative establishes in Greece the discipline of Atomic Physics with Accelerators, a field with important contributions to fusion, hot plasmas, astrophysics, accelerator technology and basic atomic physics of ion-atom collision dynamics, structure and technology. This is being accomplished by combining the existing interdisciplinary atomic collisions expertise from three Greek universities, the strong support of distinguished foreign researchers and the high technical ion-beam know-how of the Demokritos Tandem accelerator group into a cohesive initiative.

2. Experimental setup

The new experimental setup includes a single stage hemispherical spectrometer with injection lens and 2-dimensional PSD combined with a doubly-differentially pumped gas target. It has been used in previous work at Kansas State Univ. [1], but recently moved to the Athens Tandem. This high-efficiency, high-resolution ZAPS system is ideally suited for use in the electron spectroscopy of weak ion beams such as the ones called for in this work and due to its PSD is also about 15-20 times more efficient than conventional single channel devices (e.g. two-stage parallel plate electron spectrometers [2]). Additionally, the paracentric entry of the Hemispherical Deflector Analyser (HDA) [3] is a novel feature adding further high resolution capability not available to conventional centric HDAs [4, 5]. In the two following pictures we give a general presentation of the experimental setup. In figure 1 a panoramic view of the beam line is shown and in figure 2 the doubly differentially pumped target gas cell, the HDA, the 4-element input lens and the PSD are shown inside the vacuum chamber.

Figure 1. Panoramic view of the new beam line at the 5 MV Demokritos Tandem.

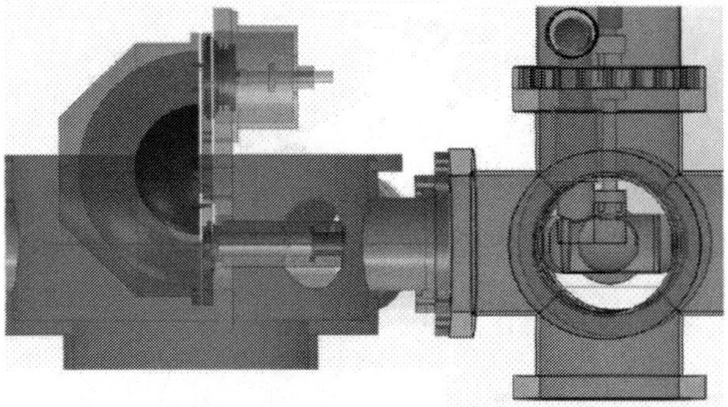

Figure 2. From right to left: The doubly-differentially pumped target gas cell, the 4-element input lens, the HDA and the 2-D PSD. Each colour denotes a different voltage.

Journal of Physics: Conference Series **583** (2015) 012014 doi:10.1088/1742-6596/583/1/012014

3. Electron capture

The various 1s2s2*l* lines observed in the Auger spectra result from the capture of a target electron to one of the possible (1*s*2s2*l*) states. With the use of the well known COWAN Hartree-Fock package [6] the relevant $F^{6+}(1s2snl\ ^{2,4}L_J)$ Li-like energy levels, including dipole and Auger transition rates for principal quantum numbers $2 \leq n \leq 5$ and $l=0$, n-1 were calculated [7]. In figure 3 the energy level scheme along with the corresponding transition rates is shown. Basic quantum mechanics requires the spin coupling of a 2p electron to the 1s2s ^3S state yielding 1s2s2p ^4P quartet and 1s2s2p ^2P doublet states to be in the ratio of 2:1, i.e. $R=\sigma(1s2s2p\ ^4P)/\sigma(1s2s2p\ ^2P) = 2$. However, it has been documented in the literature that the values of R extracted from the spectra are much larger having values of R~6-9 [7-10]. Our intention is to provide a more thorough understanding of cascade feeding of the 1s2s2p ^4P metastable states produced by electron capture in collisions of He-like ions with gas targets and further elucidate their role in the non-statistical production of excited three-electron 1s2s2p states by electron capture, recently a field of conflicting interpretations awaiting further resolution.

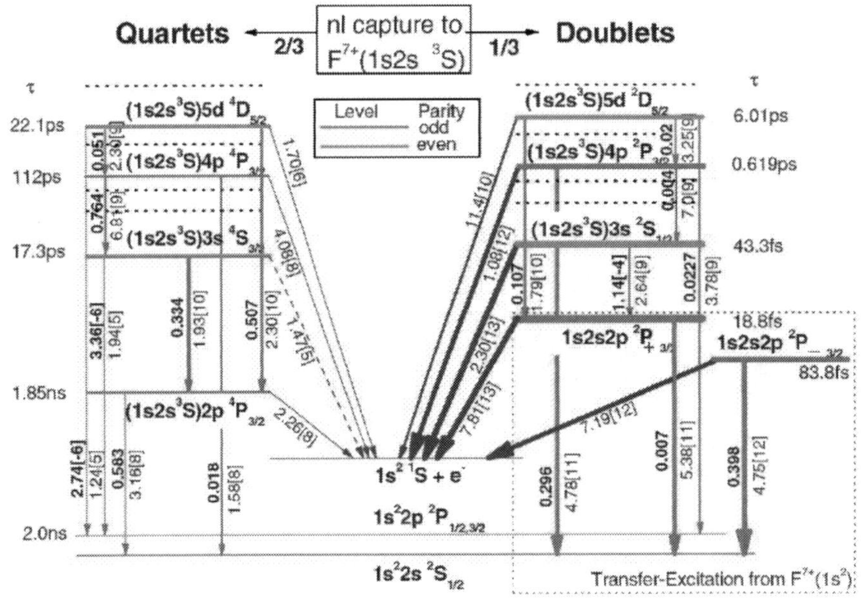

Figure 3. Li-like quartet and doublet F^{6+} 1s2s*nl* energy level scheme (not to scale) resulting from single electron *nl* capture to F^{7+} 1s2s ^3S. Only a few representative levels are indicated for clarity. Arrows represent transitions with widths roughly proportional to their strength radiative *E*1 (dipole) vertical red lines and Auger slanted blue lines. Rates in s^{-1} are given to the right of the arrows the quantity in square brackets indicates power of 10, while radiative transition branching ratios are given in bold to their left. Also indicated are total lifetimes and dashed arrows for Coulomb forbidden transitions (from Ref. [7]).

4. Electron Energy spectrum

Our first high resolution Auger spectrum was obtained recently with our new setup using the 5 MV tandem Van der Graf accelerator and is shown in figure 4. A 12 MeV C^{4+} beam was collimated and directed into the doubly-differentially pumped gas cell that contains gaseous neon target. The gas cell pressure during the measurement was 20 mTorr. The tuning energy W of the analyzer was set at 1525 eV. High resolution was achieved by retarding the electrons using a deceleration factor of F=4. Analysed electrons were counted and normalized to the incident beam current, which was collected at the Faraday cup. The similarity with previous published works is obvious [9].

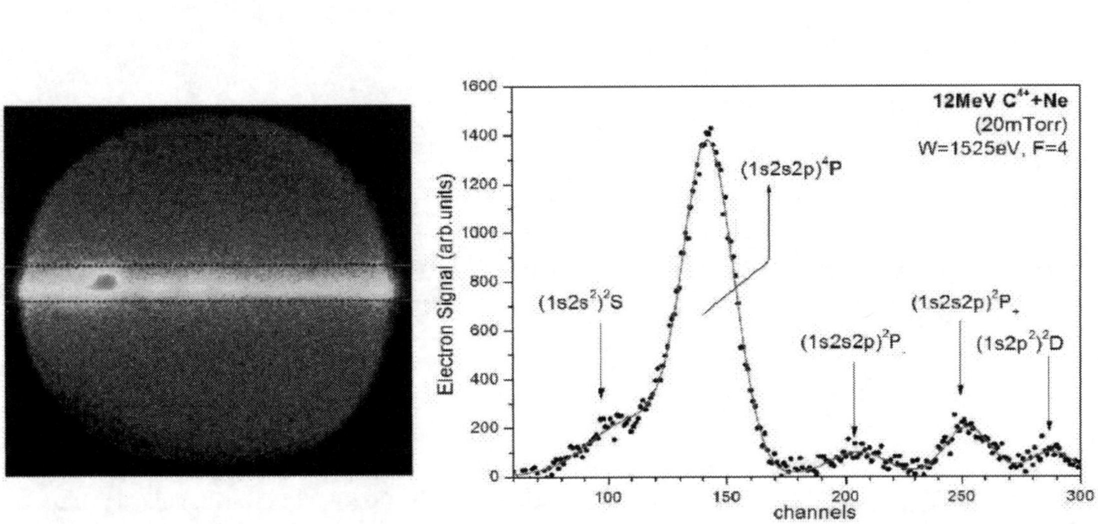

Figure 4. (a) 2-D PSD Image, (b) Projection of the PSD image within the selected region shown in (a). The ^{4}P line is seen to be broadened. Its considerable longer life time makes it decays closer to the spectrometer with increased angular acceptance and therefore kinematic broadening.

5. Future developments

Future developments on the beam line include the upgrade of the Tandem accelerator to also include a recirculating gas stripper in the accelerator terminal, along with a post-stripping stage, for both gas and foil stripping. These, along with the placement of a number of beam profile monitors will allow for better control of the beam both as to its transmission as well as to its composition of the excited states.

6. Acknowledgements

Co-financed by the European Union (European Social Fund—ESF) and Greek national funds through the Operational Program "Education and Lifelong Learning" of the National Strategic Reference Framework (NSRF)—Research Funding Program: THALES. Investing in knowledge society through the European Social Fund (Grant No. MIS 377289).

7. References

[1] Benis E P, Zaharakis K, Voultsidou M M, Zouros T J M, Stockli M, Richard P, Hagmann S 1998 *Nucl. Instrum. & Meth. Phys. Res.* B **146** 120-125.
[2] Benis E P and Zouros T J M and Richard P 1999 *Nucl. Instrum. Meth. Phys. Res.* B **154** 276
[3] Benis E P and Zouros T J M 2000 *Nucl. Instrum. Meth. Phys. Res.* A **440** 462-465.
[4] Sise O, Dogan M, Martinez G and Zouros T J M 2010 *J. Electron Spectroscopy and Related Phenomena* **177** 42
[5] Cowan R D 1981 The theory of Atomic structure and Spectra (*University of California Press, Berkeley CA*)
[6] Dogan M, Ulu M, Gennarakis G G and Zouros T J M 2013 *Rev. Scient. Instrum.* **84** 043105
[7] Zouros T J M, Sulik B, Gulyás L and Tökési K 2008 *Phys. Rev. A* **77** 050701R
[8] Rohrbein D, Kirchner T and Fritzsche S 2010 *Phys. Rev. A* **81** 042701
[9] Strohschein D, Röhrbein D, Kirchner T, Fritzsche S, Baran J and Tanis J A 2008 *Phys. Rev. A* **77** 022706
[10] Tanis J A, Landers A L, Pole D J, Alnaser A S, Hossain S, and Kirchner T 2004 *Phys. Rev. Lett.* **92** 133201

Journal of Physics: Conference Series **583** (2015) 012015 doi:10.1088/1742-6596/583/1/012015

Recoil-ion momentum spectroscopy for He^{2+} + He electron capture reactions

M Alessi[1], P Focke[2] and S Otranto[3]

[1]INVAP SE, Av. L. Piedrabuena 4950, 8403CPV, S. C. de Bariloche, Argentina.
[2]Centro Atómico Bariloche, Av. Bustillo 9500, 8400 S. C. de Bariloche, Argentina.
[3]IFISUR and Departamento de Física, Universidad Nacional del Sur, 8000 Bahía Blanca, Argentina.

[2]focke@cab.cnea.gov.ar

Abstract. Electron capture reactions for $^3He^{2+}$ collisions on He at impact energies in the range 40 keV-300 keV have been studied using the Cold Target Recoil-Ion Momentum Spectroscopy setup which has recently became operational at the Centro Atómico Bariloche. State-selective charge exchange cross sections were obtained and in this work we present recoil-ion transverse momentum distributions. For targets with residual thermal motion, we show that the implementation of a back-projection algorithm based on the transverse momentum distribution component along a direction perpendicular to the jet direction provides results in agreement with those obtained by using previously cooled targets. Present results nicely fit the gaps in the datasets already published by other laboratories and are found to be in good agreement with classical trajectory Monte Carlo simulations.

1. Introduction

During the past 20 years the cold-target recoil-ion momentum spectroscopy (COLTRIMS) method has been implemented in several laboratories worldwide and has led to an important advance in our current understanding of several atomic and molecular collision processes [1-8]. In particular, a COLTRIMS setup has been recently turned operational at the Centro Atómico Bariloche [9-11] and has been married to the Kevatron accelerator that spans a high voltage range from 20 kV up to 300 kV. Initial studies performed with this device were focused on the determination of state-selective cross sections [10]. In contrast to ionization processes, charge exchange reactions lead to discrete values for the longitudinal momentum acquired by the recoil-ion after the collision (p_l). The state-selective charge exchange cross sections, in this sense, can then be obtained from the p_l recorded distribution. Projectile angular scattering, on the other hand, can be inspected via the transverse momentum acquired by the recoil-ion (p_t). Initial studies with a target solely cooled by the adiabatic expansion in the nozzle-skimmer clearly showed a non-isotropic distribution along the plane determined by the $p_{t\text{-}y}$ and $p_{t\text{-}z}$ components. The coordinate system used is as described in ref. [9, 10]. This non-symmetric distribution reflected the residual thermal motion in the target atom which elongated the width of the distribution in the $p_{t\text{-}y}$ direction (in the gas jet direction). Later on, a liquid N_2 cooling stage was implemented nicely leading to p_t distributions which were completely symmetrical in terms of its components $p_{t\text{-}y}$ and $p_{t\text{-}z}$.

In this work, we show that charge exchange p_t distributions obtained for setups with and without pre-cooling stages for the target, provide similar results if a back-projection algorithm for the p_t

Journal of Physics: Conference Series **583** (2015) 012015 doi:10.1088/1742-6596/583/1/012015

component, that is not thermally affected, is considered. In particular, we consider electron capture reactions for $^3He^{2+}$ collisions on He at impact energies in the range 40 – 300 keV proper of our Kevatron accelerator. The present state-selective charge exchange cross sections and p_t distributions are also contrasted to classical trajectory Monte Carlo simulations which incorporate dynamical screening (dCTMC) [12].

2. The back-projection algorithm

In figure 1 we show the transverse momentum distribution at impact energy of 60 keV. The p_y and p_z distributions are shown. p in the abscissa is used for p_y or p_z . All distributions are normalized at the peak value.

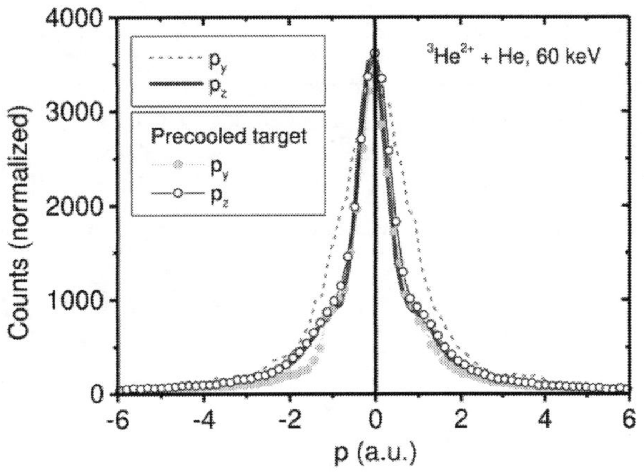

Figure1.Transverse momentum distributions for 60 keV$^3He^{2+}$collisions on He.

From the cylindrical rotation symmetry of the present collision system it is expected that both distribution should have the same shape. For a He gas jet obtained from gas in the reservoir at room temperature the distributions are different. The p_y momentum distribution is broader than the p_z distribution. This shows that the cooling from adiabatic expansion in the nozzle is not enough to reduce the thermal initial momentum of the target along the jet direction. On the contrary, along the z direction, due to geometrical cooling by the skimmer collimation of the gas jet, the residual momentum can be considered irrelevant. When the gas jet originates from a precooled gas both distribution show the same shape, as expected and are also nearly equal as the p_z distribution that results from a non-precooled target.

The transverse momentum p_t for the recoil ion is given by:

$$p_t = \sqrt{p_y^2 + p_z^2}\ . \tag{1}$$

In the case of measurements obtained with a non-precooled target, the p_y distribution is affected by residual thermal broadening, and equation (1) is not applicable. However, a projection of the transverse momentum along the z-axis is not affected by the thermal motion of the target. For the latter the projection $P(p_z)$ is (here $p = p_t$):

$$P(p_z) = 2\int_{p_z}^{\infty} \frac{f_c(p)p}{\sqrt{p^2-p_z^2}}\,dp, \tag{2}$$

here $f_c(p)$ is the transverse momentum distribution resulting from the collision. Equation (2) for the p_z-projection is known as the Abel transform and its inverse reads:

$$f_c(p) = -\frac{1}{\pi}\int_p^{\infty} \frac{d}{dp_z}[P(p_z)]\frac{1}{\sqrt{p_z^2-p^2}}\,dp_z. \quad (3)$$

Equation (3) then provides a way to obtain transverse momentum distributions even when a thermally affected target is being used.

In figure 2 we show the transverse momentum distributions for single capture to the ground and excited states for a precooled target, using equation (1), and a non-precooled target using the back-projection procedure of equation (3).

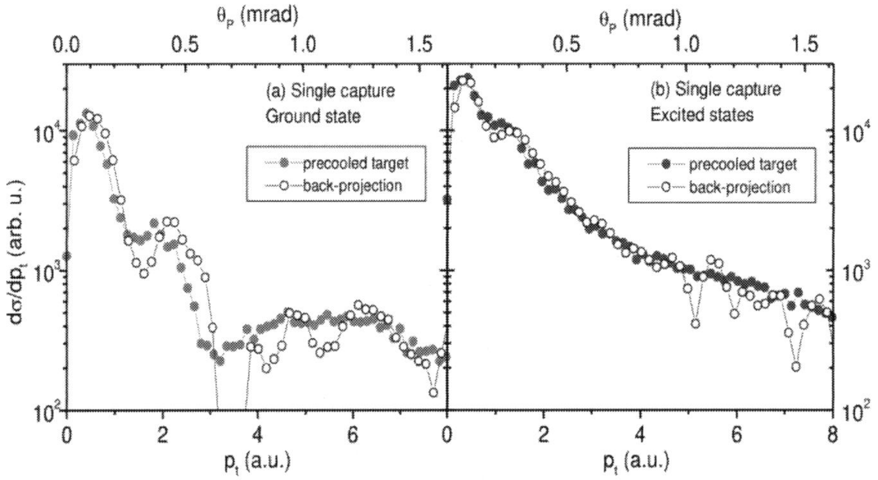

Figure 2. Transverse momentum distributions for 60 keV $^3He^{2+}$ collisions on He. Back-projected results and pre-cooled data are explicitly shown. θ_P is the projectile scattering angle.

The result is satisfactory showing that a back- projection procedure may be applied in the cases in which the target is not cool enough or when the target gas that cannot be precooled, such as Ar or Ne, is used. The procedure has also been checked more extensively with single capture reactions on the H^+ + He system in the range from 25 to 200 keV.

Applying equation (3) to our measurements we obtained the transverse momentum distributions and compared them with results from other laboratories where a precooling stage has been used. In figure 3 we show some of these results. Present experimental data for the transverse momentum distributions are compared to those obtained by Mergel *et al.*[13] at 62.5 keV/amu with a 14 K external cooling system that provides an inner target temperature of 0.1 K. dCTMC results are also shown and found to be in good agreement with the data, with the exception of the deep structure at 1.5 a.u. that cannot be reproduced by the simulation. In figure 4, present experimental charge exchange distributions at 67 keV/amu in terms of the projectile scattering angle θ_P are shown, and compared to those obtained by Schöffler *et al.*[14] at 60 keV/amu. Good agreement is found among both sets of data except for scattering angles larger than 0.4 mrad.

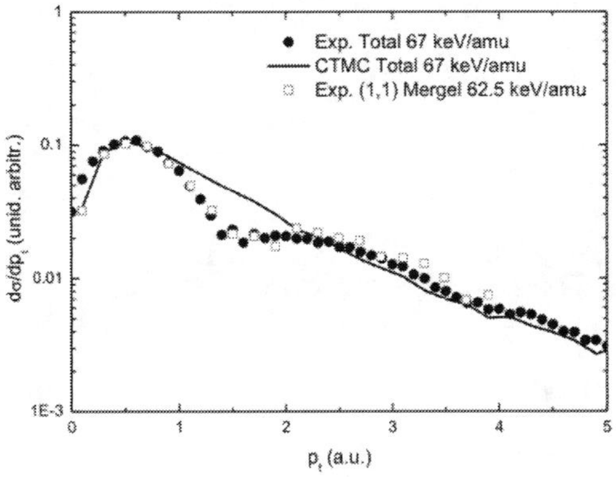

Figure 3. $^3He^{2+}$ collisions on He. Total transverse momentum distributions (experimental and CTMC). Solid circles, present data; data of Mergel *et al* [13].

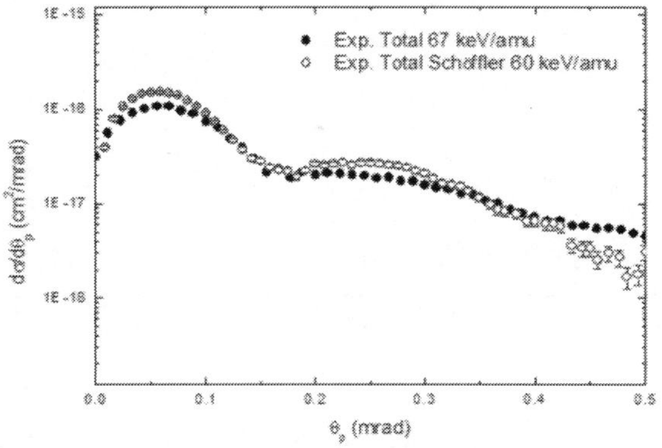

Figure 4. $^3He^{2+}$ collisions on He. Experimental scattering angle distributions. Solid circles, present data; data of Schöffler [14].

3. Conclusions

Electron capture reactions for $^3He^{2+}$ collisions on He have been studied at impact energies in the range 40 - 300 keV using the COLTRIMS technique. Recoil-ion transverse momentum distributions were presented for state-selective charge exchange collisions. A back-projection algorithm based on the transverse momentum distribution component along a direction perpendicular to the jet direction provides results in agreement with those obtained by using previously cooled targets. Present results nicely fit the gaps in the datasets already published by other groups.

Acknowledgments
Work at UNS supported by Grants No. PGI 24/F059 and No. PIP 112-201101-00749 of CONICET (Argentina).

References
[1] Kambara T *et al.*1995, *J. Phys. B: At. Mol. Opt. Phys.* **28** 4593
[2] Moshammer R *et al.* 1996, *Nucl. Instrum. Meth. Phys. Res.* B **108** 425
[3] Cassimi A *et al.*1996, *Phys. Rev. Lett.* **76** 3679
[4] Ullrich J *et al.* 1997, *J. Phys. B: At. Mol. Opt. Phys.* **30** 2917
[5] Abdallah M *et al.* 1997, *Appl. Accel. Res. Ind. AIP* **392** 209
[6] Dörner R *et al.* 2000, *Phys. Rep.* **330** 95
[7] Cocke C L 2004, *Phys. Scr.* T **110** 9
[8] Zhu X L *et al.* 2006, *Chin. Phys. Lett.* **23** 587
[9] Alessi M, Fregenal D and Focke P 2011, *Nucl. Instrum. Meth. Phys. Res.* B **269** 484
[10] Alessi M, Otranto S and Focke P 2011, *Phys. Rev.* A **83** 014701
[11] Alessi M, Cariatore N, Focke P and Otranto S 2012, *Phys. Rev.* A **85** 042704
[12] Montemayor V J and Schiwietz G 1989, *Phys. Rev.* A **40** 6223
[13] Mergel V *et al.* 1995, *Phys. Rev. Lett.* **74** 2200
[14] Schöffler M 2006, PhD Thesis J W Goethe Univ. Frankfurt/M

Journal of Physics: Conference Series **583** (2015) 012016 doi:10.1088/1742-6596/583/1/012016

Charge-sharing in fragmentation of nitrogen molecules in collision with highly charged ions

J Matsumoto, H Tezuka and H Shiromaru
Department of Chemistry, Tokyo Metropolitan University, Hachioji, Tokyo 192-0397, Japan.

E-mail: junmatsu@tmu.ac.jp

Abstract. An apparatus for low-energy collision of highly charged ions with molecules, comprising a position sensitive time-of-flight measurement devise for recoil ions and a charge state analyser for the scattered projectile, was newly constructed at a beam line of an electron cyclotron resonance ion source. Collision experiments of 120 keV Ar^{8+} with the target of nitrogen molecules were conducted as a test run of this apparatus. Focusing on the dissociation channels with asymmetric sharing of the charges by fragmentation, correlation between Auger electron emission of the projectile and Coulomb explosion of the target is discussed.

1. Introduction

When homonuclear diatomic molecules are highly ionized by low-energy collision of highly charged ions (HCI), total charges of the transient molecular ions are eventually shared by the dissociating fragments mainly in the way that these are distributed equally, or near equally with a single charge difference [1]. This is simply understandable in terms of energetics that such sharing minimizes the total ionization energy. In covalent molecules, valence electrons are delocalized, and charge-rearrangement on the way of fragmentation tends to reduce the charge inequality. Nevertheless, highly-asymmetric channels have been observed to some extent, and have attracted attention since the information at the instant of collision may be preserved in such events. It is natural to expect that the asymmetric channels are sensitive to the electron capture conditions, orientation of the target and the impact parameter. Aside from the experimental difficulties, the orientation effect and the impact parameter are measurable. In fact, a sharp orientation preference has been found for rare-gas dimer target [2], although it is an exceptional case in which asymmetric sharing is a major channel [3], most likely due to low mobilities of electrons in the dimer. The question is whether it can be observed for covalent molecules.

To observe the orientation effect, difficulty arises in detection efficiency of lower charged fragments in addition to that caused by small branching ratios of highly charge-asymmetric channels. Generally speaking, fragment ions with large velocity components perpendicular to the TOF axis tend to escape from the detection area. For (near) symmetric channels, the charge states of the energetic fragments are high since these are produced from highly charged parent ions, for example, by the reaction $(N_2)^{5+} \rightarrow N^{3+} + N^{2+}$. In such cases, TOF of the ions are small and radial drift length is relatively small. On the other hand, lower charge fragment in highly asymmetric channel, for example N^{1+} formed in $(N_2)^{5+} \rightarrow N^{4+} + N^{1+}$, has a large TOF with large coulombic energies, and the radial drift length will be larger. Thus, we need a large detector or sophisticated ion optics for this purpose. As for the impact parameter, number of Auger electrons emitted from the scattered projectile can be a rough measure. For example, triple capture collisions may lead to single and double Auger emission processes, in which the latter is considered to be the processes with larger impact parameter.

Recently, we reconstructed a new beam line of the electron cyclotron resonance (ECR) ion source at Tokyo Metropolitan University. A cross-beam apparatus was installed for the study of electron capture dynamics of HCI with a molecular target. In this paper, we describe the details of the test experiments and discuss the preliminary results.

2. Ion-molecule collision apparatus

A schematic drawing of the collision apparatus is shown in figure 1. The apparatus is basically combined from the two existed ones comprising a position-sensitive time-of-flight (PSTOF) device with a large position-sensitive detector (120 mmϕ) [4] and a charge analyzer for scattered projectile ions [5]. It allows us to observe highly charge-asymmetric dissociation channels correlated with different number of Auger electrons. The HCIs produced in a 14.25 GHz ECR ion source are extracted and introduced to the apparatus typically with the energy of 15q keV. The ion beam is trimmed by two 1 mm-pinholes and controlled by two sets of deflectors. The target molecules are first introduced to a beam source chamber evacuated by a 1500 L/s turbo molecular pump, and then trimmed by a skimmer and directed to the collision chamber evacuated by a 400 L/s turbo molecular pump. For a condensable target, a liquid nitrogen trap assists the evacuation. At the collision center, the HCI beam was crossed with the target molecular beam at the right angle.

After the collision, the scattered projectiles are dispersed according to the charge states by a set of deflectors, and detected by an MCP with a position-sensitive anode. To select desired charge state, first the positional image is acquired. A typical image obtained for Ar^{8+} collision with N_2 is shown in the inset of the figure 1, in which the spots of the scattered ions with different charge states are reasonably separated. Then, the center of the desired spot is selected by the four-jaw slit to avoid contamination of ions with the neighboring charge. Detection of the charge-selected scattered ions is used as a TOF trigger for the recoil ion measurement. The fragment ions from highly-charged molecular ions are extracted by a homogeneous electronic field. The TOF spectrometer consists of a stack of 30 electrodes carefully designed so as to minimize the inhomogeneity of the electric field [4]. From the event-by-event multi-hit PSTOF data stored in the list mode, we extract relevant coincidence events to determine the velocity components of all the fragment ions.

In the present study, 120 keV Ar^{8+} ions were used as projectiles. The velocity of the projectile ions is about 0.40 a.u., where multiple electron capture occurs dominantly rather than ionization does. The typical operating pressure of the nitrogen molecular beam was 8×10^{-4} Pa, while the base pressure was 5×10^{-7} Pa, in the collision chamber. The electric field applied to the TOF tube was 15.1 V/mm. To calculate the velocity vectors of the fragments, the momentum conservation rule was imposed on each fragment pair while a momentum transferred by the projectile was neglected. The distance from the collision point to the detector calculated in this way ranges from 199 to 209 mm, agreeing with the

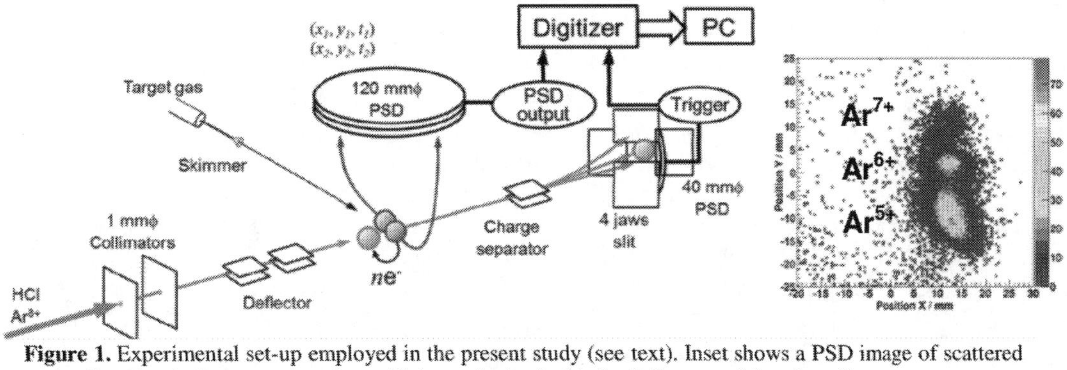

Figure 1. Experimental set-up employed in the present study (see text). Inset shows a PSD image of scattered projectiles after 1- 3-charge changing collisions, obtained with the fully-opened four-jaw slit.

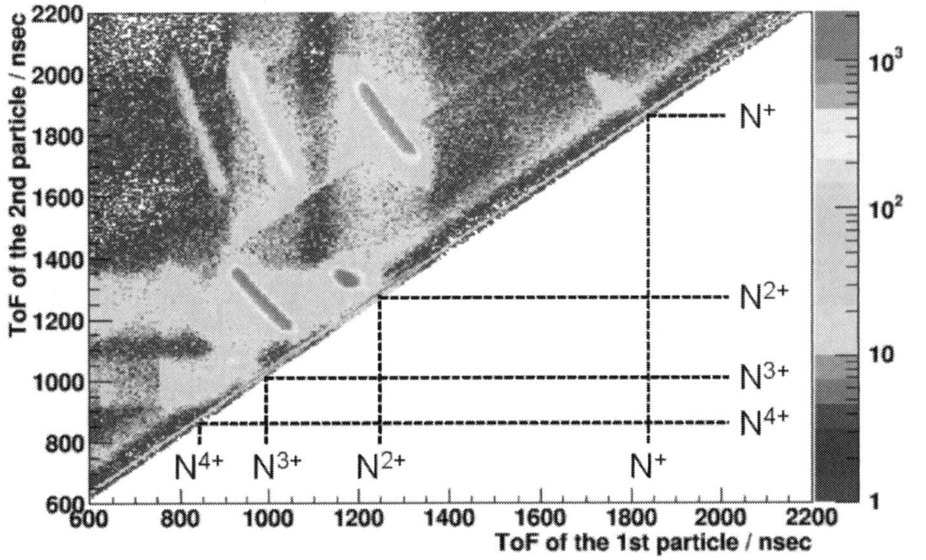

Figure 2. Coincidence map of fragmentation of multiply-charged nitrogen molecules with the trigger of scattered Ar^{6+} ions. To emphasize the minor channels, the color indicator is presented in a log scale. Assignment is given by the lines

mechanical configuration. Then, kinetic energy releases (KER) of the ion-pair fragmentation channels were obtained.

3. Results and discussion

The coincidence map of the nitrogen molecules with the trigger of scattered Ar^{6+} is shown in figure 2, by which the fragmentation channels N^{p+}-N^{q+}: (p, q) = (1, 1), (1, 2), (1, 3), (1, 4), (2, 2), (2, 3), (2, 4), (3, 3) are identified. Since the events are correlated with scattered Ar^{6+}, the initial projectile Ar^{8+} ions capture two- to six-electrons, and then the transient ions de-excite accompanying zero- to four-Auger electron emission, while the highly-charged nitrogen molecules dissociate:

$$Ar^{8+} + N_2 \rightarrow Ar^{(8-n)+**} + N_2^{n+} \rightarrow Ar^{6+} + (n-2)e^- + N_2^{n+}\ (n = 2 - 6),\ N_2^{n+} \rightarrow N^{p+} + N^{q+}.$$

For Ar^{5+} (the figure is not shown here), the channels (p, q) = (1, 2), (1, 3), (1, 4), (2, 2), (2, 3), (2, 4), (3, 3), (3, 4) are also identified with emission of from zero- to four-Auger electrons. As can be seen in the figure, some spots are partly hidden by the degenerate TOFs, and some spots overlap with each other. In such cases, we have to extract the events within a specific range of orientation with respect to the TOF axis. In the present study, we analyze the charge asymmetric dissociation channels with no such problems.

For (1, 2), (1, 3) and (1, 4) channels, the KER distributions were derived from the double-hit PSTOF data, triggered by Ar^{6+} and Ar^{5+}. The numbers of emitted Auger electrons are therefore one to three for Ar^{6+} and zero to two for Ar^{5+}. The results are summarized in figure 3. A striking difference among the channels appears in the peak energies, that is, irrespective of the number of Auger electrons, the peak positions of the KER distributions for higher-asymmetric fragmentation channels (1, 3) and (1, 4) are much larger than those estimated from the Coulomb explosion (CE) model, while the positions for near-symmetric (1, 2) ones are around the CE value. In general, higher peak values and broader distributions of the KER are considered to be outcomes of production of highly excited electronic states at the time of the electron removal. Thus, the results suggest that the asymmetric charge sharing have more chance for survival if these are in highly excited states. It should be noted that the asymmetry in (1, 2) channel is preserved in any case.

Journal of Physics: Conference Series **583** (2015) 012016 doi:10.1088/1742-6596/583/1/012016

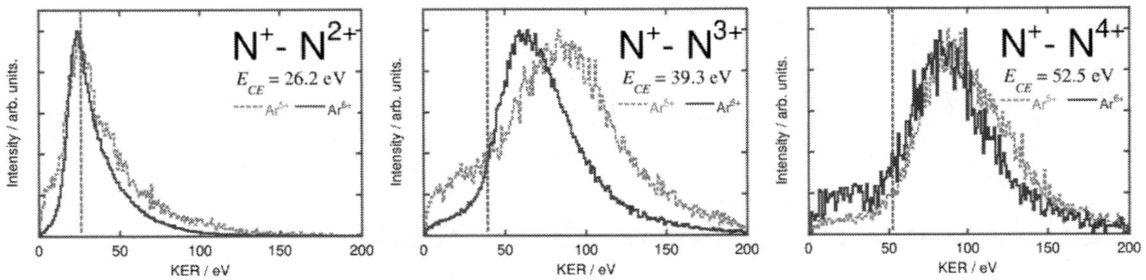

Figure 3. KER distributions of the N^+-N^{q+} (q = 2 - 4) channels. Solid lines (blue) : Ar^{6+} scattered ions, dotted lines (red) : Ar^{5+} ones. The vertical line in each figure indicates the KER value estimated by the CE model, E_{CE}.

For (1, 2) channel, difference in the KER distribution between zero and one Auger emission seems to be small, while that correlated with zero Auger is slightly broader. On the other hand, a clear difference can be seen in (1, 3) channel between the processes with one and two Auger electrons. This is also the case for (1, 4) channel, although the difference is somewhat less clear because of the lower statistics. This propensity can be interpreted by a simple picture based on the classical over-barrier model (COBM). According to the COBM, electron capture at a larger impact parameter leads to production of higher electronic states of the scattered ion (electron acceptor) and lower electronic states of the recoil ion (electron donor). Then the scattered ions in the higher electronic states tend to emit more electrons, and the recoil ions in the lower electronic states tend to give lower KER. These are indeed observed in the present study.

4. Conclusion

The KER distributions for asymmetric fragmentation channels of multiply ionized N_2 were successfully measured for the selected charge states of the scattered ions. The propensities found for KER distributions are qualitatively consistent with a simple picture based on COBM. To find the reason why the (1, 2) channel does not show a clear dependence on the electron capture process, and to explore the generality of this finding, further measurements are needed. So far, we have observed Auger-dependent coulombic dissociation of tri-atomic molecules [6], and plan to study the fragmentation dynamics for more complex molecules [7].

References

[1] For example, Ben-Itzhak I *et al.*, 1993 *Phys. Rev. A* **47**, 2827
[2] Iskandar W *et al.*, 2014 *Phys. Rev. Lett.* **113**, 143201
[3] Matsumoto J *et al.*, 2010 *Phys. Rev. Lett.* **105**, 263202
[4] Matsumoto J *et al.*, 2012 *J. Phys: Conf. Ser.* **388** 102058
[5] Hayakawa K *et al.*, 2011 *J. Phys. B: At. Mol. Opt. Phys.* **44**, 075207
[6] Tezuka H *et al.*, to be published
[7] De S *et al.*, A-T2-4 HCI 2014, Veshapidze G *et al.*, A-T2-29 HCI 2014

Journal of Physics: Conference Series **583** (2015) 012017 doi:10.1088/1742-6596/583/1/012017

Modeling ion-molecule collision using *Monte Carlo* simulation

P M Mishra and U Kadhane
Indian Institute of Space Science and Technology, Thiruvananthapuram 695547, Kerala, India
E-mail: mishra.preetimanjari@gmail.com

Abstract. *Monte Carlo* simulation is performed under local density approximation for electronic energy loss distribution by intermediate velocity protons for several examples of polycyclic aromatic hydrocarbon (PAH) molecules as well as for some of naphthalene derivatives. The energy loss distribution was found to peak in the range of 50-70 eV for all the molecules except C_{60}. Apart from this, a strong orientation effect with respect to projectile for electronic stopping magnitude is observed for all molecules except C_{60} due to its icosahedral symmetry. This work helps in estimation of the ionization and fragmentation cross sections using electronic stopping.

1. Introduction

Intermediate velocity collision experiments where the capture and ionization cross sections have competing contributions, highlight the complexity and importance of theoretical modeling for these processes. For the cases like PAH molecules, it is not practical to perform experiments on each and every member of the family to obtain such cross sections. Therefore electron capture processes are often calculated using coulomb over the barrier model to a very good agreement with experimental measurements. Similarly for ionization as well as fragmentation various approaches are followed. On the other hand the complexity in theoretical calculations is enhanced for the case of large molecules like PAHs where the potential energy surface cannot be calculated numerically to a good accuracy. Moreover in such collision processes the energy loss is not a discrete quantity and is governed by statistical processes [1].

The present work helps in estimating fragmentation cross section for ion-molecule collision from mean energy loss calculations with the help of a simple local density approximation (LDA) and *Monte Carlo* simulation. Using various thresholds and ionization potentials, these values can be converted to ionization or fragmentation cross sections. LDA is a useful approach for large systems where the statistical processes dominate. The validity of the simulation is established for a PAH molecule namely naphthalene for which the experimental fragmentation yields agree well with the simulated geometric cross section [2] and for nucleobases as well [3].

The molecules considered for the present investigation are PAHs and their derivatives interacting with protons. PAHs are a family of hydrocarbons consisting hexagon or pentagon rings of carbon (C) atoms with hydrogen (H) at periphery of the ring. The inherent stability for PAHs and aromatic character is due to the existence of delocalized electrons because of sp^2 hybridized C atoms. The studies of high energy radiation interaction with PAHs are important in interstellar physics [4] as well as on earth [5]. PAHs, their cations and their hydrogenated derivatives are considered to be the possible carrier of diffuse interstellar bands [6, 7] and their spectroscopic observations match with the unidentified infrared (UIR) band emission at 3.3, 6.2, 7.7, and 11.3 μm from different galactic sources. The interstellar heating is also attributed to the PAHs [8]. PAHs also interact with cosmic ray radiation which consist of 99% of protons [4]. On earth PAHs are formed due to incomplete combustion of C rich compounds like fuel, coal, crude oil and they remain in human body on inhalation of polluted air or intake of improperly treated food. Then upon exposure to the solar radiation PAHs become mutagenic and carcinogenic inside human body [5, 9]. Proton therapy is a well known technique in treatment of cancer where the knowledge of energy deposition into the biomolecules inside the body is very crucial to avoid undesirable radiation effect during the treatment [10]. Hence the mean energy loss calculations are helpful in biological, medical as well as physics applications. The simulations are performed for 80 and 100 keV proton beam due to its significance in therapy known as Bragg peak energy which causes the highest dose deposition in radiation therapy [11, 12]. So far the application of LDA to PAHs (except fullerene and anthracene) has not been studied and the estimation of the fragmentation cross section from electronic stopping distribution is for the first time reported as part of this work for naphthalene. Hence this work will help in validating the future experiments at intermediate velocity collision for any ion-molecule combination.

2. Mean electronic energy loss calculation

For intermediate velocity projectile collision, the energy deposited into the molecule due to a light ion like proton is dominated by electronic processes. The electronic stopping power is much larger than the nuclear one with an assumption that removal of electron is faster than the nuclear motion. The projectile can transfer varied amount of its kinetic energy to the target depending on the nature of interaction whether direct ionization or electron capture or transfer ionization. The fragmentation dynamics depends upon the coulombic interaction between projectile and target, and hence on the energy transfer into the molecule. In the energy range under consideration here, the recoil energy of the projectile is negligible compared to the net energy deposited in the molecule [13]. To estimate the energy transferred to the target molecule from the projectile ion we have performed *Monte Carlo* simulations for random trajectories of projectile ions using the LDA model developed by Lindhard *et al.* [14, 15] .

2.1. Local density approximation

The LDA is a well known approach in stopping power for solid targets [14, 15]. This model has also been used for ion-atom [16] and ion-molecule collisions [17] and is applicable for a range from several keV to several MeV energy per atomic mass unit (amu) where collision time is shorter ($10^{-16} - 10^{-17}$ s) than the vibrational or rotational time scales (10^{-13} s or longer) [17]. The detailed formalism of the calculation adapted here is reported in literature [14, 17]. The LDA approximates the target as a non uniform distribution of electron density and the energy loss distribution due to the interaction of projectile (considered as a point charge) is thus calculated at each volume element of the total electron density, which finally provides the mean energy loss.

2.2. Monte Carlo simulation details

The 10^6 randomly generated trajectories are made to pass through a non-uniform electron density distribution representing the target. Using LDA method, energy loss is calculated for each volume element of the electron density and the total energy loss is calculated as the sum of all these contributions. In the simulation, the deceleration of projectile ions along their trajectories as well as the coulomb deflection of these ions in the field of the target nuclei is neglected. Total electron density is considered as the input for LDA model as the magnitude of the projectile velocity is close to that of the velocity of the inner most electron of the target molecule. The target electron density was obtained quantum mechanically using the non-local hybrid Becke three-parameter Lee-Yang-Parr functional (B3LYP) basis at 6-311G (2d, p) with Density functional theory (DFT) method using GAUSSIAN09 [18]. The simulation has been carried out for a fixed orientation of the molecule as well as for fixed inter-nuclear coordinates, thus making it suitable for interpretation of the experimental results. In the *Monte Carlo* simulations, two methods have been used in order to generate the random trajectories of the projectile: i) From all directions, to compare and understand the experimental results, and ii) Plane-wise, to understand the behavior of electronic stopping according to the elemental composition and geometry of the molecule. The first case is reported here showing the electronic stopping distribution. The geometric cross sections (inset of figure 1) are calculated by integrating the area under the curve (see figure 1) of the energy loss distribution (with first ionization potential as the lower limit of area integral) and then normalizing it with the total number of trajectories.

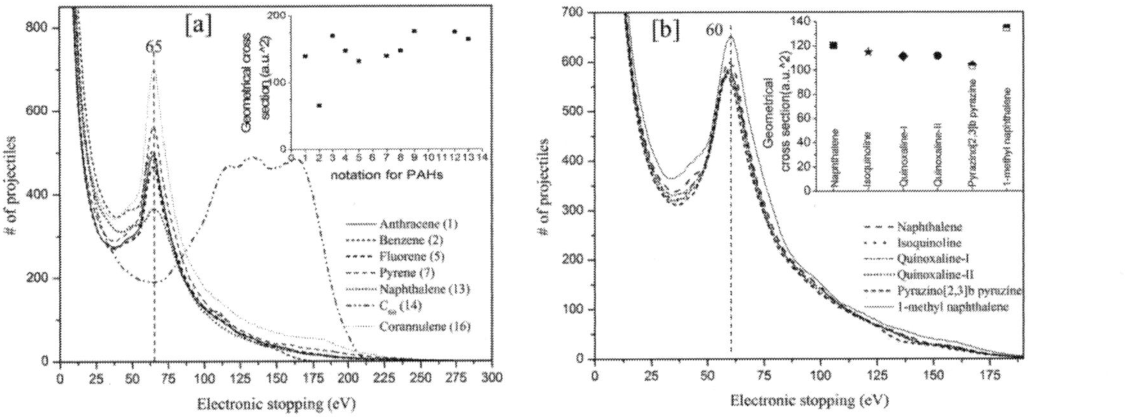

Figure 1. Mean energy loss calculation using *Monte Carlo* simulation for interaction of randomly generated proton projectiles of [a] 100 keV with PAH targets and [b] 80 keV with naphthalene derivative targets. Insets showing the geometrical cross section obtained for different targets: anthracene(1), benzene(2), benzo[c]phenanthrene(3), fluoranthene(4), fluorene(5), phenanthrene(6), pyrene(7), tetracene(8), tetrahydrochrysene(9), tetrahydropyrene(10), tetraphene(11), triphenylene(12), naphthalene(13), C_{60}(14), coronene(15), Corannulene(16) using *Monte Carlo* simulation results with proton projectile.

3. *Monte Carlo* simulation results for electronic stopping

Figure 1 shows the electronic stopping distribution curve for randomly generated 10^6 linear trajectories of the proton projectile ions interacting with PAHs as well as naphthalene derivative target. The peak of mean energy loss remains same for all the molecule except C_{60} due to the difference in symmetry and electron density distribution. However, the probability of higher energy loss decreases rapidly after a pronounced peak. This peak is a result of the trajectories which pass through the molecular volume almost perpendicular to the molecular plane. This is due to the fact that this orientation offers the largest geometrical cross section to the incoming projectile beam. But at the same time this orientation causes much less energy loss compared to the in plane collision and this is the reason of the peak position in the energy loss distribution being independent of the PAH species. The peak height is decided by the total number of C atoms in the molecule. It is also seen from simulation results (not shown here) that contour plot of electronic stopping for perpendicular impact of proton beam with respect to the plane of molecule for planar PAHs demonstrates the structure very well and is in agreement with the distribution of electron density over the molecular plane [2].

In our previously reported work [2], the geometric cross section obtained from this simulation for proton-naphthalene system successfully reproduced the experimental fragmentation cross section. Here the geometric cross section obtained from simulation of proton-PAHs and proton-naphthalene derivative interaction using the first ionization potential as the threshold for fragmentation is shown in inset of figure 1. All the PAHs (excluding benzene) show a large (nearly 30%) variation due to size and composition. In figure 1[a], only few examples out of all molecules is shown (see the caption of figure 1[a] for nomenclature of molecules). As can be seen from inset of figure 1[a], anthracene and phenanthrene show almost same cross section. Similarly tetracene, tetrahydropyrene and tetra hydrochrysene show similar cross sections. In figure 1[b] inset, all the naphthalene and its derivatives show relatively low (about 15%) variation in the cross sections except 1-methyl naphthalene where the additional methyl group introduces excess electron density causing larger cross section.

4. Conclusions

The *Monte Carlo* simulation of intermediate velocity protons interacting with PAHs and their derivative are performed. The electronic energy loss process is found to deposit about 50-70 eV energy with highest probability. This is because the planar orientation of target offers largest cross section for the interaction. The geometric cross sections were found to be dependent on the number of carbon atoms present. Using the first ionization potential as the cut off, the geometric cross sections can be calculated. Similarly for other processes like fragmentation also the cross sections can be estimated using the known or measured threshold values.

5. References

[1] Russek A and Meli J 1970 *Physica* **46** 222–243
[2] Mishra P, Rajput J, Safvan C, Vig S and Kadhane U 2014 *J. Phys. B: At. Mol. Opt. Phys.* **47** 085202
[3] Mishra P and Kadhane U 2014 *Nucl. Instrum. Meth. B* **336** 12–18
[4] Tielens A G 2008 *Annu. Rev. Astron. Astrophys.* **46** 289–337
[5] Yu H 2002 *J. Environ. Sci. Heal. C* **20** 149–83
[6] Salama F, Bakes E, Allamandola L and Tielens A 1996 *Astrophys. J.* **458** 621–36
[7] Dabestani R and Ivanov I N 1999 *Photochem. Photobiol.* **70** 10–34
[8] Tielens A G 2005 *The physics and chemistry of the interstellar medium* (Cambridge University Press)
[9] Dong S, Hwang H M, Harrison C, Holloway L, Shi X and Yu H 2000 *B. Environ. Contam. Tox.* **64** 467–74
[10] Michael B D and O'Neill P 2000 *Science* **287** 1603–1604
[11] Ziegler J F and Biersack J P (Internet address: http://www.srim.org/)
[12] Le Padellec A, Moretto-Capelle P, Richard-Viard M, Champeaux J and Cafarelli P 2008 *J. Phys.: Conf. Ser.* vol 101 (IOP Publishing) p 012007
[13] Cocke C and Olson R E 1991 *Physics reports* **205** 153–219
[14] Lindhard J and Scharff M 1953 *Mat. Fys. Medd. K. Dan. Vidensk. Selsk.* **27**
[15] Lindhard J and Winther A 1964 *Mat. Fys. Medd. K. Dan. Vidensk. Selsk.* **34**
[16] Kabachnik N, Kondratyev V, Roller-Lutz Z and Lutz H 1997 *Phys. Rev. A* **56** 2848
[17] Kabachnik N, Kondratyev V, Roller-Lutz Z and Lutz H 1998 *Phys. Rev. A* **57** 990
[18] Frisch M J, Trucks G W, Schlegel H B, Scuseria G E, Robb M A, Cheeseman J R, Scalmani G, Barone V, Mennucci B and et al G A P 2009 Gaussian 09 Revision D.01 gaussian Inc. Wallingford CT 2009

Journal of Physics: Conference Series **583** (2015) 012018 doi:10.1088/1742-6596/583/1/012018

Energy threshold in multiple ionization by electron or positron impact

C C Montanari and J E Miraglia
Instituto de Astronomía y Física del Espacio, IAFE, CONICET-UBA, casilla de correo 67, sucursal 28, C1428EGA, Buenos Aires, Argentina

Abstract. The energy threshold for the formation of highly charged ions is experimentally well-known and measured. For single ionization, it is the binding energy of the outermost electrons. For multiple ionization, the ionization begins at impact energies much larger than the theoretically expected ones. In this contribution we present a simple expression for the energy threshold for multiple ionization by electron or positron impact. It was obtained as the mean value of the energy transferred to each ionized electron by using Thompson classical approximation. Present results reproduce quite well the experimental thresholds. Moreover, the inclusion of these values in the theoretical multiple ionization cross sections allows describing rather well the experimental data for single up to sextuple ionization of rare gases.

1. Introduction

Electron impact ionization is a basic and active field within the atomic collisions. Most of the multiple ionization experimental values are from electron impact processes [1, 2]. Also in recent years the investigation on ionization by positron impact has received considerable attention (for example, the work of Laricchia and collaborators [3]). In both, electron and positron impact, multiple ionization cross sections depend smoothly on the impact energies in the intermediate to high energy range. But for low energies they fall down drastically, several orders of magnitude, when the energy gets close to certain value. These thresholds, or appearance energies, are clearly observed experimentally [4, 5]. For single ionization the threshold is the energy gap of the outermost sub-shell. However, for N-fold ionization the threshold is much larger than N times this energy. And the difference increases with N, *i.e.* the experimental threshold for quintuple ionization of Kr is around 200 eV $>> 5 \times E_{4p} = 71\ eV$.

In a recent theoretical work on multiple ionization by electron impact [2] not null cross sections were obtained for energies below the experimental threshold. The aim of this contribution is to correct this failure by proposing a value for the appearance energy as the mean energy transferred to the emitted electrons, enough to jump to the continuum and to leave the target with a certain mean kinetic energy.

2. Theoretical model

We calculate the mean energy transfer by resorting to a classical model by Thompson [6]. Briefly, if we consider the scattering problem of two charged particles (charges Z_1 and Z_2, velocities v_1 and v_2, masses m_1 and m_2), the transferred energy is

$$\Delta E = \frac{1}{2}m_2(v_2^2 - v_{2f}^2) = \frac{1}{2}m_2(v_{1f}^2 - v_1^2). \quad (1)$$

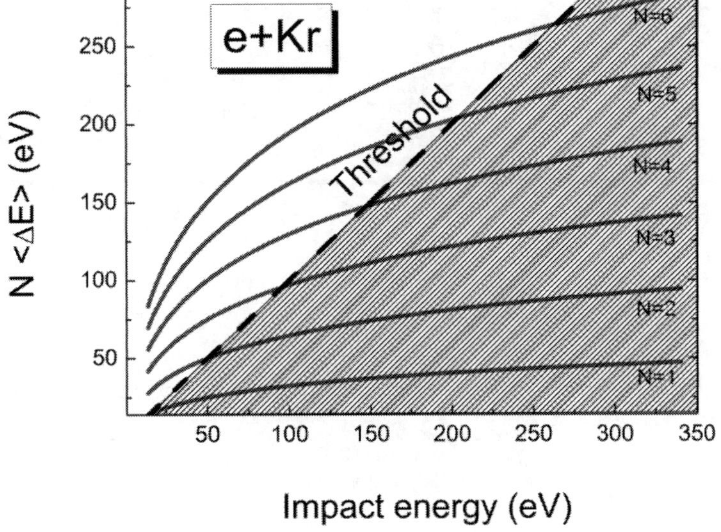

Figure 1. Mean energy transferred in multiple ionization by electron impact, as function of the impact energy (solid-lines). The dashed-line is the impact energy, and the shaded region indicates $E > N\times < \Delta E_{nl} >$.

Considering the momentum and energy conservation and assuming $v_1 << v_2$, the differential cross section in the transferred energy is

$$\frac{d\sigma}{d(\Delta E)} = \frac{2\pi \ (Z_1 Z_2)^2}{m_1 v_2^2 \ (\Delta E)^2}, \tag{2}$$

and the total cross section to emit an electron of the nl sub-shell is

$$\sigma_{nl} = \int_{I_{nl}}^{E_{max}} \frac{d\sigma}{d(\Delta E)} \ d(\Delta E) = \frac{2\pi \ (Z_1 Z_2)^2}{m_1 v_2^2} \ (\frac{1}{I_{nl}} - \frac{1}{E_{max}}), \tag{3}$$

with I_{nl} being the ionization energy of the nl sub-shell, $E_{max} = \frac{4 \ m_1 m_2}{(m_1+m_2)^2} \ E$ being the maximum energy transferred, and E the impact kinetic energy of particle 2 [6]. In the case of electron-electron interaction, replacing charges and masses in (3), the ionization cross section in Thompson approximation is expressed as

$$\sigma_{nl} = \frac{\pi}{E \ I_{nl}} \left(1 - \frac{I_{nl}}{E}\right). \tag{4}$$

Following the same procedure we can calculate the stopping cross section as

$$S_{nl} = \int_{I_{nl}}^{E_{max}} \Delta E \ \frac{d\sigma}{d(\Delta E)} \ d(\Delta E) = \frac{\pi}{E} \ Ln(E/I_{nl}). \tag{5}$$

This logarithmic dependence with the impact energy is similar to Bethe stopping, valid in the high energy region. Then we calculate the mean energy transferred $< \Delta E_{nl} >$ to ionize an electron initially in the nl sub-shell as

$$< \Delta E_{nl} > = \frac{S_{nl}}{\sigma_{nl}} = \frac{E \ I_{nl} \ Ln(E/I_{nl})}{(E - I_{nl})}. \tag{6}$$

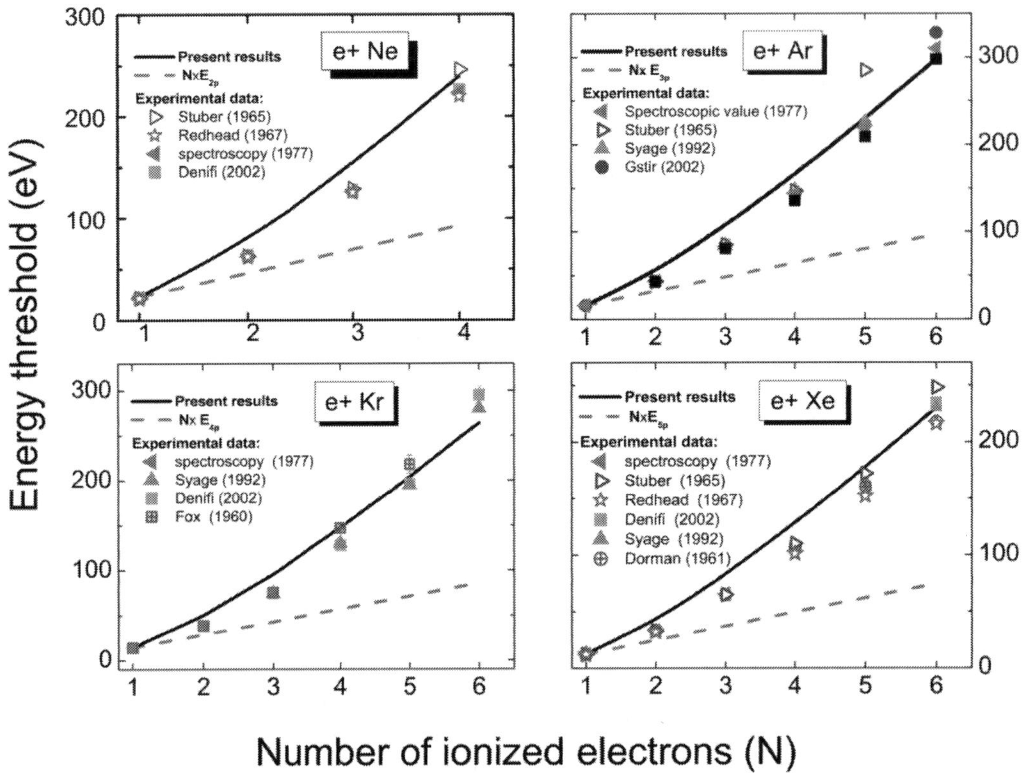

Figure 2. [colour online] Energy threshold for multiple ionization by electron or positron impact. Curves: black-solid, present results; red-dashed line, the binding energy of the outer sub-shell per the number of ionized electrons. Symbols: experimental data, details in the insets and references in [4, 5]; red left-triangles, spectroscopy data [7]

In this model we propose that the minimum impact energy for N-fold ionization of the nl sub-shell is $N\times < \Delta E_{nl} >$. In figure 1 we display the mean energy transferred given by (6) as function of the impact energy E, for ionization of the $4p$ sub-shell of Kr. We also include in this figure the straight line corresponding to the electron impact energy. The energy region where the multiple ionization is allowed is indicated as a shaded area.

3. Results and discussions

We compare the present results for Ne, Ar, Kr and Xe with the experimentally known energy thresholds [4, 5]. In figure 2 we show $N\times < \Delta E_{np} >$ for the outer np electrons of each target. We also include the value $N \times E_{np}$ to emphasize the difference between them. The improvement in the description of the experimental energy thresholds is clear. The advantage is more evident for quadruple to sextuple ionization.

Finally in figure 3 we display the triple ionization cross sections of Kr by electron impact. The solid black curve was obtained using the present energy minimum, while the red-dashed curve is our previous calculation considering $N \times E_{np}$ as energy threshold. A detailed explanation of the calculation of the multiple ionization cross sections and of the experimental data included

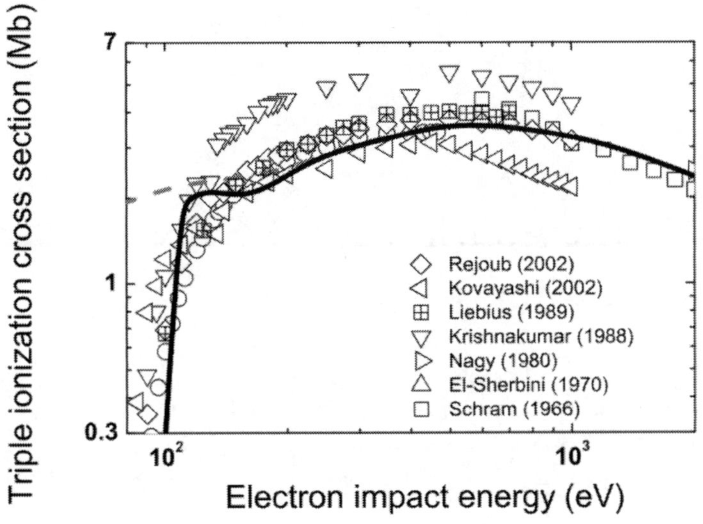

Figure 3. [colour online] Triple ionization of Kr by electron impact. Curves: black-solid line, CDW-EIS results including present expression for the energy threshold; red-dashed line, similar values considering just the binding energy of each shell [2]. Symbols, compilation of experimental data in [2]

in this figure is in [2].

4. Conclusions

We calculate the energy threshold for multiple ionization by electron impact, by using Thompson classical model for the ionization cross section as function of the transfer energy. The results show very good agreement with the experimental thresholds obtained using different technics. We also show that this approximation improves previous multiple ionization cross sections in the region near the threshold.

Acknowledgments

This work was partially supported by the Consejo Nacional de Investigaciones Científicas y Técnicas, the Agencia Nacional de Promoción Científica y Tecnológica, and the University of Buenos Aires, Argentina.

References

[1] Shevelko V and Tawara H 1998, *Atomic Multielectron processes* (Springer-Verlag, Berlin Heidelberg)
[2] Montanari C C and Miraglia J E 2014, *J. Phys. B: At. Mol. Opt. Phys.* **47**, 105203
[3] Laricchia G, Cooke D A, Kover A, Brawley S J 2013 *Experimental Aspects of IonizationStudies by Positron and Positronium Impact* (Cambridge University Press), Vol **56**, pp.116-136
[4] Gstir B, Denifl S, Hanel G, Rümmele M, Fiegele T, Cicman P, M Stano, Matejcik S, Scheier P, Becker K, Stamatovic A and Märk T D 2002, *J. Phys. B: At. Mol. Opt. Phys.* **35** 2993-3007
[5] Denifl S, Gstir B, Hanel G, Feketeova L, Matejcik S, Becker K, Stamatovic A, Scheier P, and Märk T D 2002, *J. Phys. B: At. Mol. Opt. Phys.* **35** 4685-4694
[6] McDowell M R C and Coleman J P 1970, *Introduction to the theory of ion-atom collisions*, (North-Holland Pub. Co.), chapter 3
[7] Rosenstock H M, Draxl K, Steiner B W and Herron J T 1977 *Energetics of gaseous ions*, *J. Phys. Chem. Ref. Data* 6 (suppl. 1)

Journal of Physics: Conference Series **583** (2015) 012019 doi:10.1088/1742-6596/583/1/012019

Intensity ratio of EUV emission lines in Fe XV studied with electron beam ion traps

Erina Shimizu[1], Hiroyuki A. Sakaue[2], Daiji Kato[2,3], Izumi Murakami[2,3], Norimasa Yamamoto[4], Hirohisa Hara[3,5], Tetsuya Watanabe[3,5], and Nobuyuki Nakamura[1]

[1]Institute for Laser Science, The University of Electro-Communications, Tokyo 182-8585, Japan
[2]National Institute for Fusion Science, Gifu 509-5292, Japan
[3]The Graduate University of Advanced Studies (SOKENDAI), Kanagawa 240-0193, Japan
[4]Chubu University, Aichi 487-8501, Japan
[5]National Astronomical Observatory of Japan, Tokyo, 181-8588, Japan

E-mail: `n_nakamu@ils.uec.ac.jp`

Abstract. We present emission spectra of highly charged Fe ions in the extreme ultraviolet range obtained with two electron beam ion traps. The intensity ratio between the $3s3p\ ^3P_2$ - $3s3d\ ^3D_3$ (233.9 Å) and $3s3p\ ^1P_1$ - $3s3d\ ^1D_2$ (243.8 Å) transitions in Fe XV is given for various electron beam parameters. The experimental ratios are compared with theoretical model calculation as well as our previous experiment, where significant discrepancy with theoretical model was found. The present result, which has been obtained with a higher resolution spectrometer to investigate possible line blending, confirms that the discrepancy is not due to line blending.

1. Introduction

The electron density dependence of Fe ion spectra in the extreme ultraviolet (EUV) range is very important for the spectroscopic diagnostics of the solar atmosphere. In our previous study [1], the electron density dependence of the intensity ratio was measured for several density-sensitive emission lines in Fe XIII, XIV, and XV with a compact electron beam ion trap (EBIT) called CoBIT [2]. Although good quantitative agreement was obtained between the experiment and our model calculation for Fe XIII and XIV, a significant discrepancy was found for the intensity ratio between the $3s3p\ ^3P_2$ - $3s3d\ ^3D_3$ (233.9 Å) and $3s3p\ ^1P_1$ - $3s3d\ ^1D_2$ (243.8 Å) transitions in Fe XV. The discrepancy for this line ratio has also been found so far from solar observations (for example, see ref. [3]), and line blending has often been pointed out as the origin of the discrepancy [3, 4, 5]. In this paper, we re-examine the Fe XV spectra with two EBITs in Tokyo. In particular, a spectrometer with a higher resolution is applied to investigate possible line blending.

2. Experiment

In the previous study [1], CoBIT and a flat field grazing incidence spectrometer [6] were used to observe the EUV spectrum of Fe XV. The spectrometer employed a variable line spacing concave grating (Shimadzu 30-002) with an average groove number of 1200 grooves/mm and a

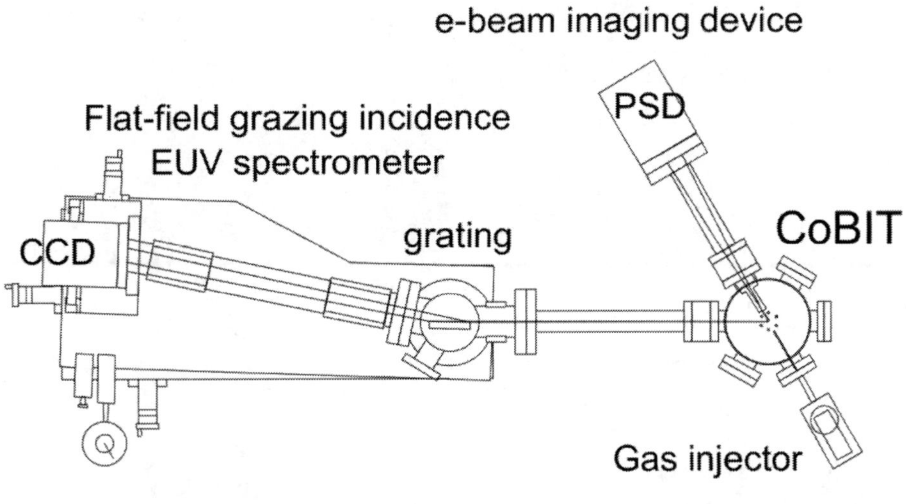

Figure 1. Schematic drawing of the present experimental setup.

radius of curvature of 5649 mm. The distance from the grating to the focal plane was 235 mm and the dispersion was 7.0 Å/mm on the focal plane.

On the other hand, in the present study, a concave grating with a larger radius of curvature (Hitachi 001-0660) was used to obtain higher dispersion. Figure 1 shows the schematic drawing of the present spectrometer installed to CoBIT. Although the average groove number is the same as that of the previous grating, the larger radius of curvature (13450 mm) and the larger distance from the grating to the focal plane (563.2 mm) makes the dispersion on the focal plane higher as 2.6 Å/mm. In addition, in the present study, not only CoBIT but also the Tokyo-EBIT [7, 8] was used. As shown in Fig. 1, the spectrometer is used in the slit-less configuration in which the thin line shaped source in an EBIT is regarded as an entrance slit. The "slit width" is determined by the electron beam width because short lifetime EUV transitions occur immediately after the ion is excited by a beam electron. It is thus possible to obtain higher resolution with the Tokyo-EBIT because higher compression of the electron beam is possible with a magnetic field higher than that in CoBIT. To produce Fe ions, a vapor of ferrocene at a room temperature was introduced from a gas injector for the experiment with CoBIT, whereas a vapor of metal iron heated in an effusion cell [9] was introduced for the experiment with the Tokyo-EBIT. Typical spectra are shown in Fig. 2. The spectral resolution in the present study was 0.4 Å with CoBIT and 0.2 Å with the Tokyo-EBIT, which has been much improved from the previous study where it was 0.8 Å. The improved resolution has a beneficial role in investigating possible line blending.

3. Results and discussion

Figure 3 shows the experimental intensity ratios between the $3s3p\ ^3P_2$ - $3s3d\ ^3D_3$ (233.9 Å) and $3s3p\ ^1P_1$ - $3s3d\ ^1D_2$ (243.8 Å) transitions in Fe XV obtained with various EBIT parameters (electron current, electron energy, central magnetic field, etc.). For the CoBIT experiment, the electron density was experimentally determined with an electron beam imaging device (see Fig. 1) similarly to the previous study [1]. For the Tokyo-EBIT experiment, since no imaging device could be applied, the electron density was determined by applying the experimental ratio between the $3s^23p\ ^2P_{1/2}$ - $3s^23d\ ^2D_{3/2}$ (211.3 Å) and $3s^23p\ ^2P_{3/2}$ - $3s^23d\ ^2D_{5/2}$ (219.1 Å) transitions in Fe XIV to the model calculation, which was confirmed in the previous study to reproduce the experimental ratio well for these transitions. The solid line in the figure is the

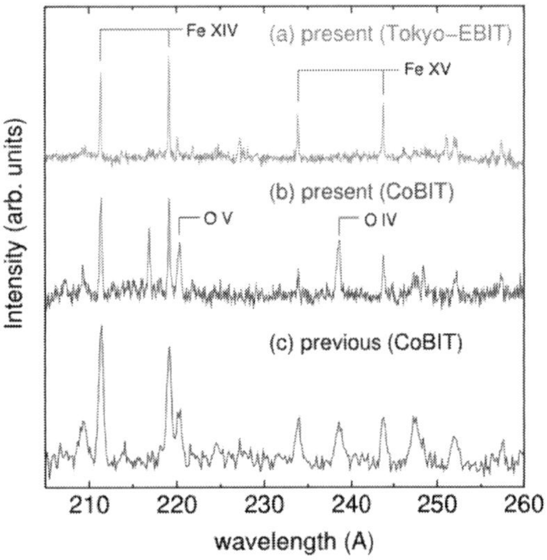

Figure 2. (color online) EUV spectra of Fe ions obtained with (a) the Tokyo-EBIT (610 eV - 12 mA), (b) CoBIT (500 eV - 13 mA). (c) Spectra obtained in previous study [1] with CoBIT (500 eV - 15 mA).

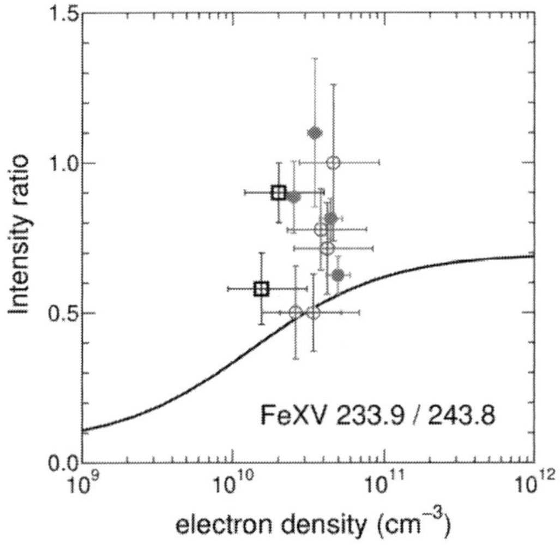

Figure 3. (color online) Intensity ratio between the $3s3p\ ^3P_2$ - $3s3d\ ^3D_3$ (233.9 Å) and $3s3p\ ^1P_1$ - $3s3d\ ^1D_2$ (243.8 Å) transitions in Fe XV. Black open squares: previous results with CoBIT [1]; blue open circles: present results with CoBIT; red closed circles: present results with the Tokyo-EBIT.

theoretical ratio obtained by the model calculation in our previous paper [1].

As seen in the figure, although the scatter of data is found to be relatively large, the general tendency that experimental ratios exceed the model calculation is confirmed. The same tendency in solar observation was already pointed out by Dufton *et al.* [3] in 1990. They analyzed four solar flare spectra obtained with the Naval Research Laboratory's S802A instrument on *Skylab*, and found that the observed ratio exceeded theoretical high density limit (0.59) substantially for three of them, and was almost on the limit value for another one. The origin of this discrepancy has been discussed in their and several subsequent papers [4, 5, 10], and some of them pointed out the influence of possible blending of Ni XVIII at 233.8 Å or Ar XIV at 243.8 Å. However, in the present study with EBITs, the possibility of Ni and Ar is completely excluded. Possible contaminants in EBITs are C, N, O, Ba, and W, but none of them has meaningfully intense line at the Fe XV lines of interest. In particular, there is no evidence of any contaminants in the high resolution spectra obtained with the Tokyo-EBIT (Fig. 2(a)), whereas some lines from O ions were confirmed for the CoBIT spectra. We thus consider that the discrepancy with the theoretical model is not due to line blending, but the accuracy of the atomic data used in the model.

4. Summary and outlook

We have re-examined the previous observation of Fe XV, where a significant discrepancy in intensity ratio with a theoretical model. Possible line blending has been carefully investigated using a higher resolution spectrometer, and we conclude that the discrepancy is not due to line blending.

Since sensitivity correction was not applied in this study, an intensity ratio between lines at close wavelengths was studied. To obtain the ratio between lines at distant wavelengths, sensitivity correction should be applied because the diffraction efficiency of a grating and the quantum efficiency of a CCD depend on wavelength. For this purpose, a manipulation system has been installed to insert a metal wire into the trap region of CoBIT. Sensitivity correction using bremsstrahlung radiation from an inserted wire is in progress.

Acknowledgment

This work was performed under the Research Cooperation Program in the National Institutes of Natural Sciences (NINS).

References

[1] Nakamura N, Watanabe E, Sakaue H A, Kato D, Murakami I, Yamamoto N, Hara H and Watanabe T 2011 *The Astrophysical Journal* **739** 17
[2] Nakamura N, Kikuchi H, Sakaue H A and Watanabe T 2008 *Rev. Sci. Instrum.* **79** 063104
[3] Dufton P L, Kingston A E and Widing K G 1990 *Astrophysical Journal* **353** 323–328
[4] Kastner S O and Bhatia A K 2001 *Astrophysical Journal* **553** 421–428
[5] Keenan F P, Aggarwal K M, Bloomfield D S, Msezane A Z and Widing K G 2006 *Astronomy & Astrophys.* **449** 1203–1208
[6] Sakaue H A, Kato D, Nakamura N, Watanabe E, Yamamoto N, Chen C and Watanabe T 2009 *J.Phys.: Conf.Ser.* **163** 012020
[7] Nakamura N, Asada J, Currell F J, Fukami T, Motohashi K, Nagata T, Nojikawa E, Ohtani S, Okazaki K, Sakurai M, Shiraishi H, Tsurubuchi S and Watanabe H 1997 *Phys. Scr.* **T73** 362
[8] Currell F J, Asada J, Ishii K, Minoh A, Motohashi K, Nakamura N, Nishizawa K, Ohtani S, Okazaki K, Sakurai M, Shiraishi H, Tsurubuchi S and Watanabe H 1996 *J. Phys. Soc. Jpn.* **65** 3186
[9] Yamada C, Nagata K, Nakamura N, Ohtani S, Takahashi S, Tobiyama T, Tona M, Sakurai M, Kavanagh A P and Currell F J 2006 *Rev. Sci. Instrum.* **77** 066110
[10] Keenan F P, Aggarwal K M, Milligan R O, Ryans R S I, Bloomfield D S, Srigengan V, O'Mullane M G, Lawson K D, Msezane A Z, Brosius J W, Davila J M and Thomas R J 2005 *Mon. Not. R. Astron. Soc.* **356** 1592

Journal of Physics: Conference Series **583** (2015) 012020 doi:10.1088/1742-6596/583/1/012020

Single electron ionization of NH_3 and CH_4 by swift proton impact

C A Tachino[1], J M Monti[1], O A Fojón[1], C Champion[2] and R D Rivarola[1]

[1] Laboratorio de Colisiones Atómicas, Instituto de Física Rosario, CONICET and Universidad Nacional de Rosario, Bv. 27 de Febrero 210 bis, 2000 Rosario, Argentina

[2] Université Bordeaux 1, CNRS/IN2P3, Centre d'Etudes Nucléaires de Bordeaux Gradignan, CENBG, 33175 Gradignan, France

E-mail: `tachino@ifir-conicet.gov.ar`

Abstract. Single ionization from ammonia and methane molecules by impact of protons is analysed. Double differential cross sections are calculated, within the post- and prior-versions of the continuum distorted wave-eikonal initial state (CDW-EIS) model, considering two different representations of the initial bound state of the active electron. A comparison between both set of results shows a very little sensitivity of the double differential cross sections to the initial state representation in the case of ammonia molecules. On the other hand, discrepancies are found when CH_4 molecular target is considered, showing that in this case the description of the corresponding molecular orbitals influences the cross section calculations.

1. Introduction

Electron ionization from atoms and molecules due to bare ion impact have an important role in a variety of processes related to different areas of physics and chemistry. For example, methane and ammonia molecules are components of some planetary atmospheres and of the interstellar medium, and their interaction with UV photons and with charged particles may provoke the appearance of highly reactive radicals, as well as ionic and neutral species (for more details see, for example, [1] and references therein). It is necessary then to posses accurate theoretical and experimental cross sections not only for ionization, but also for capture and fragmentation processes in order to modelling, as well as possible, the physical situation of interest.

In this work, we extend our previous theoretical investigations for H_2O targets [2] to other molecules, namely ammonia and methane, whose present a strong monocentric character. Our aim is to analyse the influence of the molecular orbital characterization in the description of the single ionization reaction. To this end, double differential cross sections (DDCS) are computed within the post- and prior-versions of the CDW-EIS approximation. Regarding the molecular target, two different representations of its initial bound states are considered: the MO-LCAO-SCF (molecular orbitals constructed from a linear combination of atomic orbitals in a self-consistent field) approach and the description proposed by Roberto Moccia [3–5]. In both cases, the continuum state of the emitted electron is represented by a double product of a plane wave and continuum factors corresponding to the coulombic interaction with the projectile and the target. The effective nuclear charges used in the formulation of these continuum factors are chosen according to the representation employed to describe the molecular orbitals. Details

Journal of Physics: Conference Series **583** (2015) 012020 doi:10.1088/1742-6596/583/1/012020

concerning the theory used in the present study have been already outlined in [2], thus, only a brief description of the method will be presented here. Atomic units are used throughout unless otherwise stated.

2. Theory

Let us consider the single ionization of a molecule due to the interaction with a bare ion of nuclear charge Z_P and velocity $\vec{v}$ parallel to the z-axis of a laboratory reference frame. Since the energy of the incident particle is high enough, the molecular rotational and vibrational times are much larger than the collision ones and, in consequence, a *frozen-core* picture where the molecular nuclei remain fixed in their initial positions during the ionization process can be considered. Concerning the multielectronic character of the target, the corresponding analysis is reduced to the study of a single-electron system, where the active electron (the one to be ionized) evolves in an effective coulomb field of a residual target consisting of the molecular cores and the remaining passive electrons [6].

As it was mentioned before, our purpose is to analyse the influence of the NH_3 and CH_4 molecular orbital representation in the calculation of DDCS. Double differential cross sections, as a function of the energy ε_k and the angle $\Omega_k = (\varphi_k, \theta_k)$ subtended by the emitted electron, can be obtained using both the post (+) and the prior (-) forms of the scattering matrix element, $\mathcal{R}^{\pm}_{i,f_{MO}}$, through the following expression:

$$\sigma^{(2)}_{MO}(\varepsilon_k, \theta_k) = \frac{d\sigma}{d\varepsilon_k\, d\Omega_k} = \frac{k}{8\pi^2} \int\int d\Omega\, d\vec{\eta}\, |\mathcal{R}^{\pm}_{i,f_{MO}}(\vec{\eta}, \Omega)|^2, \qquad (1)$$

where $k = \varepsilon_k^2/2$ is the momentum of the emitted electron, $\vec{\eta}$ is the transverse momentum transfer, and Ω identifies the spatial orientation of the molecule (see [2] for theoretical details). In order to model the molecular orbitals (MO) of the target, two different approaches were considered. In the MO-LCAO-SCF one, each MO is constructed from a linear combination of atomic orbitals in a self-consistent field, being these atomic states centred on each molecular nuclei [7, 8]. In order to simplify the computation of DDCS, we employed the procedure proposed by Senger *et al.* [9–11], where is assumed that the DDCS for a given molecular orbital can be reduced to a weighted sum of atomic DDCS corresponding to the atomic constituents of the molecule, the weights of this linear combination being obtained via a population analysis. The atomic double differential cross sections needed in the present calculations were obtained employing Roothaan-Hartree-Fock wavefunctions [12] to describe the electron initial bound state. Finally, all these partial DDCS are added in order to get the DDCS for the complete molecule.

In the description proposed by Roberto Moccia [3–5], each molecular orbital is expressed as a linear combination of Slater-type functions (STF) all referred to a common origin, being this point chosen to coincide with the position of the heaviest nucleus. DDCS for the complete molecule is calculated by adding all the partial contributions of the different molecular orbitals of the target, and multiplying them by the corresponding occupation number.

3. Results and discussion

In order to analyse the influence of the molecular orbital characterization in the description of the single ionization reaction of NH_3 and CH_4 targets, we calculate for each molecule the corresponding double differential cross sections considering either the Moccia representation or the MO-LCAO-SCF one.

In figure 1, a comparison between DDCS for ionization of NH_3 by 250 keV-proton impact, as a function of the electron emission angle and for different emission energies, is shown. DDCS presented in the left panel of figure 1 corresponds to calculations made within the post version of CDW-EIS formalism (post-DDCS), whereas the curves shown in the right panel were obtained within the prior version (prior-DDCS). In general, it seems that differences between results for both initial state descriptions are not significant over all the angular range for post- results as well as for the prior- ones. In the case of post calculations, Moccia's double differential cross sections show in general a small enhancement with respect to the MO-LCAO-SCF ones in the region of backward scattering, and this effect seems to be more pronounced as the emission energy increases. On the other hand, from the plots corresponding to prior-DDCS, it can be seen that both sets of theoretical data are so close that they are almost indistinguishable. The same behaviour was observed for single ionization of water molecules by bare ion impact [2].

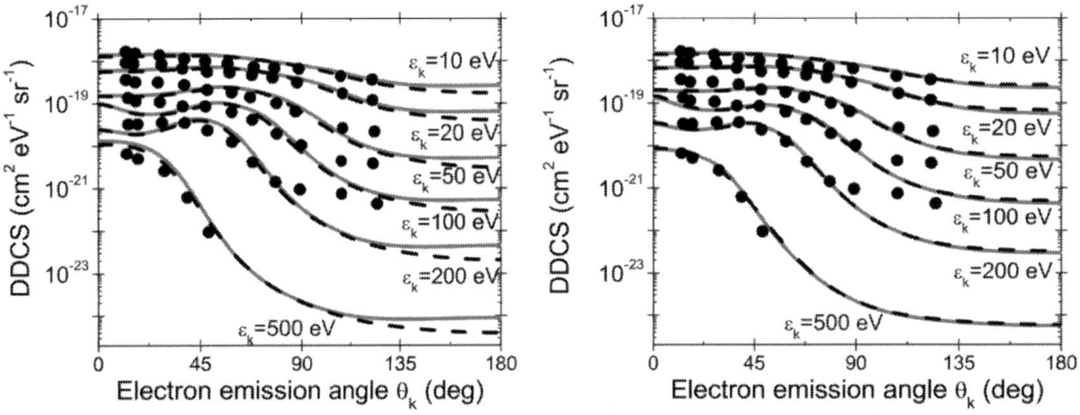

Figure 1. DDCS for ionization of ammonia by 250 keV-H^+. Left panel: post-DDCS. Right panel: prior-DDCS. Theory: solid lines represent results for Moccia description; dashed lines correspond to DDCS obtained with MO-LCAO-SCF approach. Experiments: closed circles [11].

From figure 1, it can be seen that theoretical results, for both post- and prior-versions, underestimate the experiments in the region of backward emission at high enough emission energies. Under these conditions, we may consider that the electron is first impacted by the projectile, suffering consequently a recoil scattering by the target nucleus. Thus, if the effective charge used to represent the target continuum state in the exit channel is replaced by the target nuclear one a better agreement between theoretical results and experiments was found for impact of protons on helium targets [13]. A similar behaviour could be expected for the targets here analysed.

Preliminary post-DDCS for CH_4 targets and for 2 MeV-proton impact, as a function of the electron emission angle and for different emission energies, are presented in figure 2. In the case of the Moccia description, and in order to simplify the computation of DDCS, we have decided to truncate the original basis set, leaving out from the calculation the contribution of Slater-type orbitals corresponding to values of n equal or greater than 7 (see [3] for the complete list of STF contribution to each MO of CH_4), since these states have in general coefficients that are at least one order of magnitude lower than the states with $n \leq 4$ (with the exception of the $2a_1$ MO), so it could be considered that their contribution to a given molecular orbital is negligible.

From figure 2 it can be seen that, unlike the ammonia case, there are some discrepancies between results obtained with both molecular orbital characterizations. These differences seem to be larger as the emission energy decreases. Moreover, Moccia's results seem to give a worst agreement with experiments than the MO-LCAO-SCF ones for all the considered emission energies. These differences could be attributed to the fact that in the computation of these quantities we employed, as a first approximation, a truncated Moccia basis set to describe the molecular orbitals of methane and maybe the contribution of the neglected wavefunctions is more important than we think in a first moment. In a forthcoming work, those CH_4- Moccia states

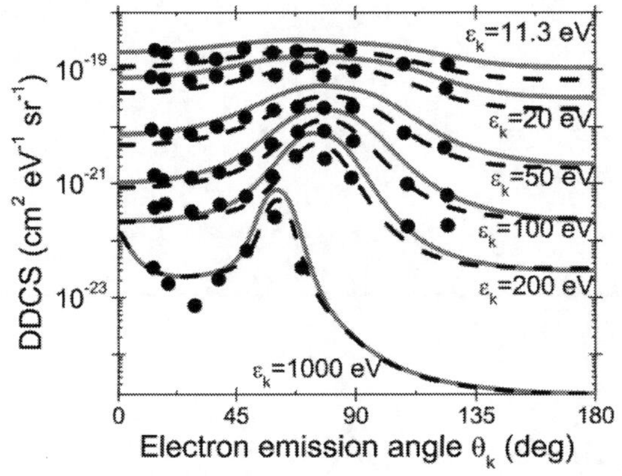

Figure 2. Post-DDCS for ionization of methane by 2 MeV-H^+. Theory: solid lines represent results for Moccia description; dashed lines correspond to DDCS obtained with MO-LCAO-SCF approach. Experiments: close circles [11].

with $n \geq 7$ that were neglected in the present calculation will we taken into account, in order to determine if the differences observed in the presented angular distributions can effectively be attributed to their absence in the preliminary calculations.

4. Conclusions

Single electron ionization of ammonia and methane molecules by swift proton impact was investigated. In the case of ammonia targets, DDCS are found to be not very sensitive to the molecular initial state description. On the other hand, angular spectra for methane molecules, obtained within the post version, showed differences between results obtained with Moccia and with MO-LCAO-SCF representations. These discrepancies may be attributed to the use of a truncated Moccia basis set, where Slater-type function with principal quantum number equal or greater than seven were neglected. Future calculations for methane will be made including these states with $n \geq 7$.

Acknowledgments

CAT, JMM, OAF and RDR acknowledge the CONICET (PIP 11220090101026) and the project PICT 2145 of the Agencia Nacional de Promoción Científica y Tecnológica. This work was partially supported by the ECOS-Sud A09E04, and by the project PICS 5921 (THEOS) of the Centre National de la Recherche Scientifique.

References

[1] Luna H, Cavalcanti E G, Nickles J, Sigaud G M and Montenegro E C 2003 *J. Phys.* B: *At. Mol. Opt. Phys.* **36** 4717
[2] Tachino C A, Monti J M, Fojón O A, Champion C and Rivarola R D 2014 *J. Phys.* B: *At. Mol. Opt. Phys.* **47** 035203
[3] Moccia R 1964 *J. Chem. Phys.* **40**, 2164-2176.
[4] Moccia R 1964 *J. Chem. Phys.* **40**, 2176-2185.
[5] Moccia R 1964 *J. Chem. Phys.* **40**, 2186-2192.
[6] Fainstein P D, Ponce V H and Rivarola R D 1988 *J. Phys.* B: *At. Mol. Opt. Phys.* **21**, 287.
[7] Duncan A B F 1957 *J. Chem. Phys.* **27**, 423-429.
[8] Hoffmann R 1963 *J. Chem. Phys.* **39**, 1397-1412.
[9] Senger B, Wittendorp-Rechenmann E and Rechenmann R V 1982 *Nucl. Instrum. Methods Phys. Res.* B **194**, 437-441.
[10] Senger B and Rechenmann R V 1984 *Nucl. Instrum. Methods Phys. Res.* B **2**, 204-209.
[11] Senger B 1988 *Z. Phys.* D: *At. Mol. and Clusters* **9**, 5631-5638.
[12] Clementi C and Roetti C 1974 *At. Data Nucl. Data Tables* **14**, 177.
[13] Monti J M, Fojón O A, Hanssen J and Rivarola R D 2010 *J. Phys.* B: *At. Mol. Opt. Phys.* **43**, 205203.

Journal of Physics: Conference Series **583** (2015) 012021 doi:10.1088/1742-6596/583/1/012021

Dressed projectile charge state dependence of differential electron emission from Ne atom

S Biswas[1], **J M Monti**[2], **R D Rivarola**[2] **and L C Tribedi**[1]

1 Tata Institute of Fundamental Research, Homi Bhabha Road, Colaba, Mumbai 400 005, India

2 Instituto de Física Rosario (CONICET-UNR), Universidad Nacional de Rosario, 2000 Rosario, Argentina

E-mail: `shubhadeep@tifr.res.in, lokesh@tifr.res.in`

Abstract. We study the projectile charge state dependence of doubly differential electron emission cross section (DDCS) in ionization of Ne under the impact of dressed and bare oxygen ions. Experimental DDCS results measured at different angles are compared with the calculations based on a CDW-EIS approximation using the GSZ model potential to describe projectile active-electron interaction. This prescription gives an overall very good agreement. In general a deviation from the q^2-law was observed in the DDCS. The observations crudely identify the dominance of different projectile electron loss mechanisms at certain electron energy range.

1. Introduction

Collision with dressed ions is considerably different than bare ion impact. In that case, additional phenomena like projectile ionization or excitation can take place due to target electron-projectile electron interaction [1, 2, 3, 4, 5]. Depending on the impact parameter of the collision, it also can enhance or reduce the target ionization cross section [6]. But, the fact is that relatively lesser number of systematic studies are available involving dressed ion projectiles compared to the bare ion impact studies [7, 8, 9, 10], particularly in the double differential cross section (DDCS) level. Especially, there are very few references [3, 11, 12, 13, 14], which investigated in the low ejected electron energy region. As far as theoretical understanding is concerned most of the earlier descriptions approximated the presence of the projectile electrons by a reduced effective projectile nuclear charge. But this prescription did not give very good result. Recently, it has been shown by Monti *et. al.* that along with the asymptotic long range Coulomb interaction, separate interaction term is necessary to incorporate the short range effect of the projectile electrons [15, 16].

In this report, we provide a comparative analysis of both the experimental and the theoretical findings for the electron emission mechanism of Ne atom following 3.75 MeV/u O^{5+}, O^{6+} and O^{7+} ion impact, in the absolute DDCS level. For comparison, the bare ion data are also provided. On the theoretical side, following the above mentioned reference [15, 16], we have carried out calculations using Green-Sellin-Zachor (GSZ) interaction potential [17] in the continuum distorted wave-eikonal initial state (CDW-EIS) approximation framework [18, 19]. We also provide a comparison with the Born approximation, which works well for the bare ion projectile charge state dependence, especially in the total cross section level.

Journal of Physics: Conference Series **583** (2015) 012021 doi:10.1088/1742-6596/583/1/012021

2. Experimental Details

The present experiment was carried out with the 3.75 MeV/u oxygen ions available from the BARC-TIFR 14 MV tandem Pelletron accelerator facility in Mumbai, India. The main scattering chamber was maintained at a base vacuum of about 2×10^{-7} Torr. During the interaction with the target the chamber was flooded with Ne gas at a static pressure of 0.15 mTorr. We used an electrostatic hemispherical analyser of 6% energy resolution [20] for energy analysis of the electrons, and these were detected by the channel electron multiplier (CEM) detector. Further details of experimental setup is available in Ref. [20]. In this case the electrons are detected in the energy range of 10 eV to 300 eV at three different angles, namely 30°, 90° and 150°. The maximum statistical uncertainty in the measurement was estimated to be about 15%. Other than that the main contribution to the error is from gas pressure fluctuation which was about 7%.

3. Results and Discussions

Here we have used the *prior*-version of the CDW-EIS model for single ionization by dressed projectile[15, 16]. It has been shown that for dressed projectiles, it is necessary to incorporate the effect of projectile electron separately, particularly for the short range interactions. Therefore, to serve this purpose, the projectile potential, V_P, is approximated by a parametric GSZ potential [17], which contains a short-range term and a Coulomb long-range term depending on the charge state q of the projectile:

$$V_P(s) = -\frac{1}{s}(Z_P - q)[H(e^{s/d} - 1) + 1]^{-1} - \frac{q}{s} \tag{1}$$

In (1), H and d are the parameters that depend on the nuclear charge Z_P of the projectile and the number of electrons present in it. Also s is the distance between the target active electron and the projectile nucleus.

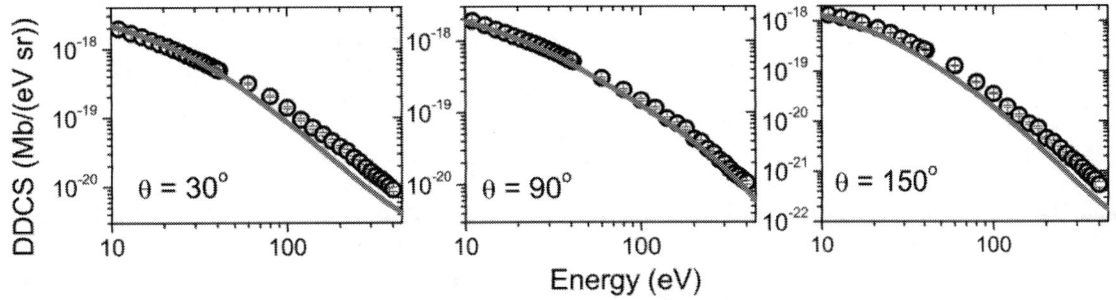

Figure 1. Energy distributions of DDCS for O^{6+}. Solid line denotes the CDW-EIS pridiction.

Fig. 1 shows the typical energy distribution of DDCS for O^{6+} at three different angles. In general, the qualitative features for other charge states are also quite similar. As far as the theoretical agreement is concerned, at 90°, it is very good throughout the entire energy region. For 30° and 150°, in the low energy part, the agreement is quite well. But with increasing energy it gets deteriorated. The theoretical prediction underestimates the DDCS values obtained from the experiment. It is worth mentioning here that for large multi-electronic targets similar kind of deviation was noticed at higher electron energies earlier even for the bare ion impact[21].

In Fig. 2, we have plotted the DDCS as a function of q, at fixed electron energy and at fixed ejection angle. It is evident that the qualitative behaviour of the plots changes a lot with energy and ejection angle. For example, for all angles, at low energy like 11 eV where the target ionization is the dominant process, the DDCS increases gradually with increasing q. This kind of behaviour is expected from the projectile screening argument. As the low energy part of the spectrum corresponds to the large impact parameter collisions, the active electron would be influenced by the projectile as a whole, i.e. by a screened nuclear charge for dressed projectile. Therefore, with increasing q the perturbation strength increases which in turn increases the cross section. For 100 eV plots the behaviour is similar, though the difference between the dressed projectile-DDCS and the bare projectile-DDCS is smaller compared to the earlier case. This can be easily seen for 150°. This decrease can be attributed to the fact of reduction of projectile screening. As the higher energy electrons are emitted in relatively smaller impact parameter collisions, the active electron would see relatively less screened dressed projectile. As a result the cross sections corresponding to the dressed projectiles become comparable to that of the bare ion projectile. Similar discussion of projectile screening effect is available in Ref. [14]. For higher

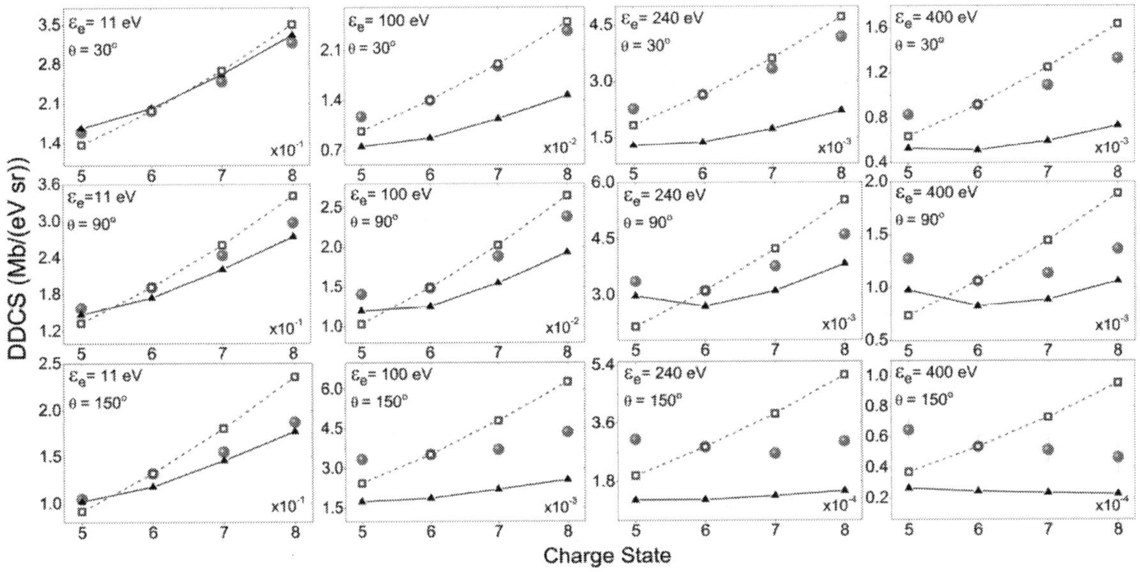

Figure 2. Charge state dependence of DDCS at different observation angle and for different ejected electron energy. The solid spheres represent experimentally measured data. The solid triangles denote the CDW-EIS predictions. The squares denote the q^2-scaling behaviour normalized at the DDCS value corresponding to O^{6+} (i.e. $q = 6$). The vertical scale of each plot is to be multiplied by the factor given at the right hand side corner below of the respective plot. Solid dots represent experimental results.

energies like 240 eV the distribution shapes are considerably different than that of the earlier. The noticeable change is that, for 90°, the DDCS corresponding to O^{5+} is little higher than that of O^{6+}. For 150°, both O^{5+} and O^{6+} DDCSs are larger than that of O^{7+}. Similar behaviour is also observed for 400 eV at 90° and 150°. In fact at 150°, the DDCS distribution shows completely decreasing behaviour with increasing q. This apparently different behaviour can be due to the dominance of other ionization mechanisms. It is well known that in the high energy range the projectile ionization contributes to the electron emission spectra [3]. Other than that the simultaneous ionization also plays an important role. Evidently, for lower q, i.e. with more number of electrons with the projectile nucleus, the probability of the above mentioned processes increases. Therefore, in this higher energy region, there is a mutual competition between the target ionization and the projectile or the simultaneous ionizations. The Born approximation predicted q^2-scaling law (denoted by squares in the figure), to some extent fails to reproduce the experimental findings especially at backward angle and for higher energies. For example, at 240 eV and 150° plot, the predicted behaviour is opposite to the observed behaviour. At lower energies the agreement is relatively better. On the other hand, the CDW-EIS prediction (denoted by solid triangles in the figure) reproduces the qualitative behaviour quite well, more or less for all angles and for entire energy range. The absolute values also match well with the experimental values for 11 eV. For other three energies, the theoretical values underestimate the experimental ones to some extent, particularly for extreme angles.

4. Conclusion

We have measured the projectile q-dependence of DDCS for electron emission in ionization of Ne atom by dressed O-ions of energy 3.75 MeV/u. The CDW-EIS model along with the GSZ model potential, gives an overall good agreement with the experiment. This emphasizes the importance of proper inclusion of the effect due to the projectile electrons in the theoretical models. The Born predicted q^2-scaling does not work so well to describe the DDCS data particularly for the higher energies and at the backward angles.

Acknowledgments

We would like to thank the staffs at BARC-TIFR Pelletron accelerator facility for smooth operation of the machine. JMM and RDR acknowledge support from the Agencia Nacional de Promoción Científica y Tecnológica of Argentina through the project PICT 2011-2145.

References

[1] Bates D R and Griffing G W 1953 *Proc. R. Phys. Soc. (London)* A **66** 961
[2] Bates D R and Griffing G W 1955 *Proc. R. Phys. Soc. (London)* A **68** 90
[3] Stolterfoht N, Schneider D, Burch D, Wieman H and Risley J S 1974 *Phys. Rev. Lett.* **33** 59
[4] Hülskötter H P, Meyerhof W E, Dillard E and Guardala N 1989 *Phys. Rev. Lett.* **63** 1938
[5] Zouros T J M, Lee D H and Richard P 1989 *Phys. Rev. Lett.* **62** 2261
[6] McGuire J H, Stolterfoht N and Simony P R 1981 *Phys. Rev.* A **24** 97
[7] Misra D, Kadhane U, Singh Y P, Tribedi L C, Fainstein P D and Richard R 2004 *Phys. Rev. Lett.* **92** 153201
[8] Misra D, Kadhane U, Singh Y P, Tribedi L C, Fainstein P D and Richard R 2005 *Phys. Rev. Lett.* **95** 079302
[9] Misra D, Kelkar A, Kadhane U, Kumar A, Fainstein P D and Tribedi L C 2006 *Phys. Rev.* A **74** 060701(R)
[10] Nandi S, Agnihotri A N, Kasthurirangan S, Kumar A, Tachino C A, Rivarola R D, Martn F and Tribedi L C 2012 *Phys. Rev.* A **85** 062705
[11] Liao C, Richard P, Grabbe S R, Bhalla C P, Zouros T J M, and Hagmann S 1994 *Phys. Rev.* A **50** 1328
[12] Toburen L H, Stolterfoht N, Ziem P and Schneider D 1981 *Phys. Rev.* A **24** 1741
[13] DuBois R D, Toburen L H, Middendorf M E and Jagutzki O 1994 *Phys. Rev.* A **49** 350
[14] Stolterfoht N, Schneider D, Burch D, Wieman H and Risley J S 1994 *Phys. Rev.* A **49** 5112
[15] Monti J M, Rivarola R D and Fainstein P D 2008 *J. Phys. B: At. Mol. Opt. Phys.* **41** 201001
[16] Monti J M, Rivarola R D and Fainstein P D 2011 *J. Phys. B: At. Mol. Opt. Phys.* **44** 195206
[17] Green A E S, Sellin D L and Zachor A S 1969 *Phys. Rev.* **184** 1
[18] Crothers D S F and McCann J F 1983 *J. Phys. B: At. Mol. Opt. Phys.* **16** 3229
[19] Gulyás L, Fainstein P D and Salin A 1995 *J. Phys. B: At. Mol. Opt. Phys.* **28** 245
[20] Misra D, Thulasiram K V, Fernandes W, Kelkar A H, Kadhane U, Kumar A, Singh Y P, Gulyas L, and Tribedi L C 2009 *Nucl. Instrum. Methods* B **267** 157
[21] Biswas S, Misra D, Monti J M, Tachino C A, Rivarola R D and Tribedi L C (*submitted*)

17th International Conference on the Physics of Highly Charged Ions IOP Publishing
Journal of Physics: Conference Series **583** (2015) 012022 doi:10.1088/1742-6596/583/1/012022

Dielectronic satellite lines of Fe XVII

P Beiersdorfer, G V Brown and A Laska
Lawrence Livermore National Laboratory, Livermore, CA 94550, USA
E-mail: beiersdorfer@llnl.gov

Abstract. We have been measuring the L-shell x-ray emission produced by dielectronic recombination of neonlike Fe^{16+} ions populating doubly excitated levels of the form $1s^2 2l^7 3\ell n'\ell'$ in Fe^{15+}. The measurements are carried out at the Livermore electron beam ion trap facility, where we isolate dielectronic resonances by choice of the electron beam energy. We utilize crystal spectrometers to record the x-ray lines generated in the radiative deexcitation of the $1s^2 2l^7 3\ell n'\ell'$ upper levels. The measurements aim to support the identification of such lines in low-density astrophysical plasmas.

1. Introduction

L-shell dielectronic satellite lines can play an important role in the spectra of stellar coronae because of the potential to use such lines as temperature diagnostics [1]. Dielectronic satellite lines to the well known characteristic Fe XVII L-shell lines are produced by electron capture and subsequent radiative decay:

$$1s^2 2s^2 2p^6 + e^- \rightarrow 1s^2 2s^2 2p^5 n\ell n'\ell' \rightarrow 1s^2 2s^2 2p^6 n'\ell' + h\nu \tag{1}$$

or

$$1s^2 2s^2 2p^6 + e^- \rightarrow 1s^2 2s 2p^6 n\ell n'\ell' \rightarrow 1s^2 2s^2 2p^6 n'\ell' + h\nu'. \tag{2}$$

The strongest such satellite lines $h\nu, h\nu'$ are those with $n = n' = 3$, i.e., those produced by LMM resonances. However, even the strongest resonances do not produce lines that are easily detected in typical astrophysical or laboratory plasmas. Moreover, some of the upper levels produced by dielectronic recombination can also be produced by collisional excitation. In fact, most of the Fe XVI x-ray lines detected so far in low-density coronal plasmas are collisionally excited lines [2]. Similarly, Graf et al. have measured multiple Fe XVI lines using an electron beam ion trap [3]; again, these lines were exclusivly produced by collisional excitation.

Identification of dielectronic satellite lines, i.e. Fe XVI lines that are produced exclusively or nearly exclusively by dielectronic recombination, are now possible because of the low-noise, high-resolution spectra provided by some of the astrophysical x-ray observatories. In particular, Fe XVI lines with a so-called spectator electron $n'\ell$ and $n' = 4$ have recently been identified in the spectrum of Capella observed with the Chandra X-ray Observatory [4]. These lines were used to infer a formation electron temperature of $\log(T) = 6.67 \pm 0.03$ K. There has also been a search for dielectronic satellite lines with $n' = 3$ in the spectrum of Capella based on theoretical predications [5].

Here we report on using an electron beam ion trap to specifically excite and observe the dielectronic satellite lines of Fe XVII in order to guide the search for such lines from low-density astrophysical plasmas.

2. Experiment

Dielectronic recombination is a resonant process, because the energy of the free electron e^- must be such that upon capture into a level $n'\ell'$ it gives off the correct energy to promote a bound electron from an $n = 2$ level to a level $n\ell$. We can use this fact to excite specific resonances by appropriate selection of the energy of the electron beam interacting with the ions trapped in an electron beam ion trap, as illustrated in innumerable such experiments before, e.g., [6, 7, 8].

For the present measurements we employ the Livermore electron beam ion trap facility [9]. It was equipped with two crystal spectrometers [10], which together cover the iron L-shell emission between about 12.0 and 17.5 Å. Individual x rays diffracted by the crystals are detected with thin-window proportional counters [11]. A window-less SiLi detector was used to monitor the trap conditions in real time. Although the SiLi detector has a relatively poor energy resolution of more than 100 eV, it has a much higher count rate than the EBIT x-ray microcalorimeter [12, 13], which has a more than ten times better energy resolution than the SiLI detector. All data from the crystal spectrometers and the SiLi detector have been acquired fully time resolved using an event-mode data acquistion system [14]. Each x-ray event is tagged with the energy of the electron beam and the time.

3. Results and Discussion

By sweeping the electron beam energy from well below the LMM resonance energy, which is located at about 400 eV, to well above the threshold excitation energy of the Fe XVII lines, which varies from about 700 eV for the $3s \rightarrow 2p$ transitions to about 900 eV for the $3p \rightarrow 2s$ transitions, we can map out all $1s^2 2l^7 3\ell n'\ell'$ resonances with $n' \geq 3$. The resultant x-ray emission recorded with the SiLi detector as a function of beam energy is shown in Fig. 1.

As Fig. 1 shows, our measurements clearly resolve the resonances with spectator electrons $n' = 3$ and 4. The resonances with $n' \geq 5$ are not resolved. This is in part because the electron beam energy is not truly mono-energetic; instead, it has an energy spread of about 30 to 50 eV, as demonstrated before [15, 16]. In addition, the resonances of each complex cover a large,

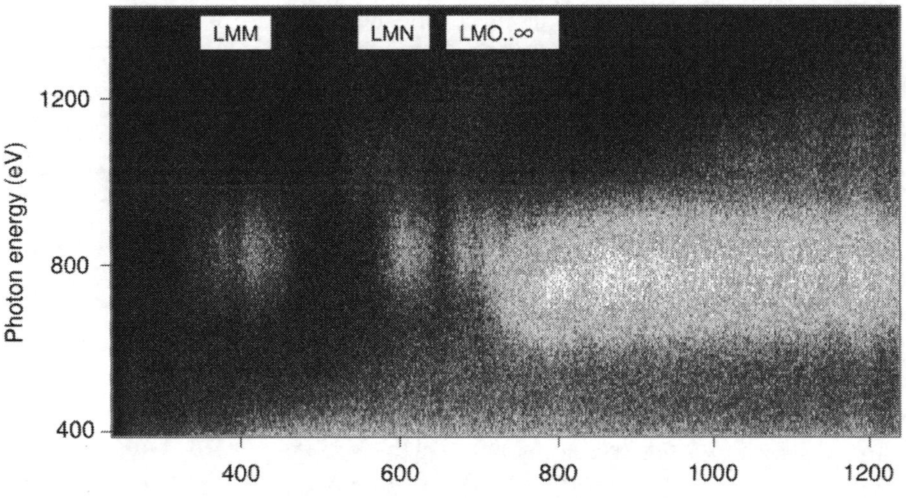

Figure 1. Fe L-shell emission as a function of electron beam energy. The data were recorded with a windowless SiLi detector. The location of the LMM and LMN resonances, as well as those merging with the excitation thresholds, are marked.

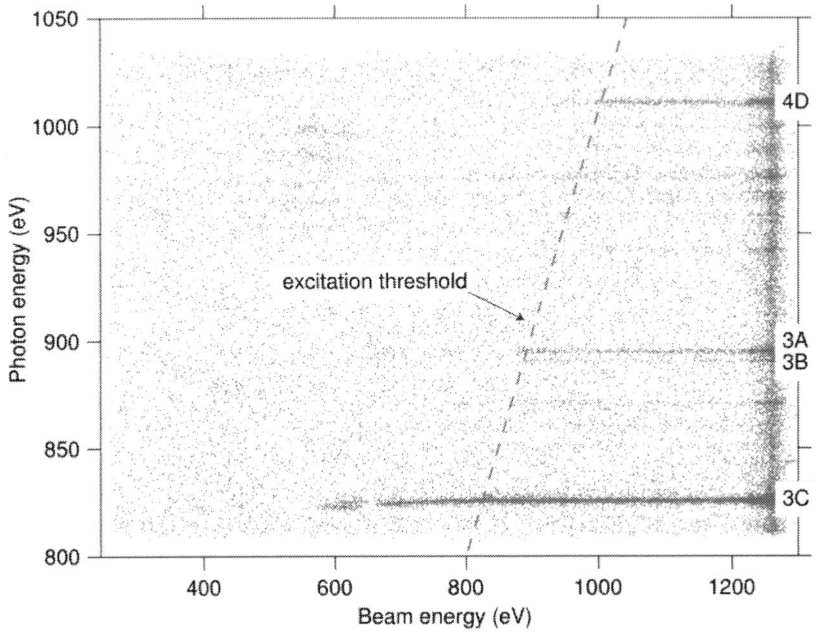

Figure 2. Fe XVII x-ray emission as a function of electron beam energy. The data were recorded with a crystal spectrometer. The emission labeled $3A$, $3B$, $3C$ and $4D$ are transitions in Fe XVII. The dashed line marks the location of the threshold for electron impact-excitation. X rays emitted below this line are from dielectronic recombination.

overlapping range of electron energies.

The x-ray emission observed using one of the two crystal spectrometers as a function of beam energy is shown in Fig. 2. This spectrometer covers the range of the $3p_{1/2,3/2} \rightarrow 2s_{1/2}$ lines $3A$ and $3B$, the $3d_{3/2} \rightarrow 2p_{1/2}$ line $3C$, and the $4d_{5/2} \rightarrow 2p_{3/2}$ line $4D$ in Fe XVII [17]. A lineout of the spectral emission at beam energies above the threshold for electron impact-excitation is shown in Fig. 3(a), while the emission covering the LMN resonances is shown in Fig. 3(b).

An analysis of the full data set is now in progress. From Fig. 2 we can already note that there are no $n' = 3$ dielectronic satellite features in this spectral range. However, we find that there are several dielectronic satellite features with $n' = 4$, as illustrated in Fig. 3(b). Moreover, line $3C$ is associated with a continuous set of high-n dielectronic satellites that form a smooth transition as the electron energy crosses the excitation threshold. This is reminiscent of what has been observed for the K-shell x-ray lines of heliumlike ions [18, 19].

Acknowledgment

This work was performed under the auspices of the U.S. Department of Energy by Lawrence Livermore National Laboratory under Contract DE-AC52-07NA27344 and supported by by NASA Astronomy and Physics Research and Analysis contract NNX12AH84G.

References

[1] Clementson J and Beiersdorfer P 2013 *Astrophys. J.* **763**, 54.
[2] Beiersdorfer P, Díaz F and Ishikawa Y 2012 *Astrophys. J.* **745**, 167.
[3] Graf A, Beiersdorfer P, Brown G V and Gu M F 2009 *Astrophys. J.* **695**, 818.
[4] Beiersdorfer P, Gu M F, Lepson J and Desai P 2011 *in* C Johns-Krull, M. K Browning and A. A West, eds,

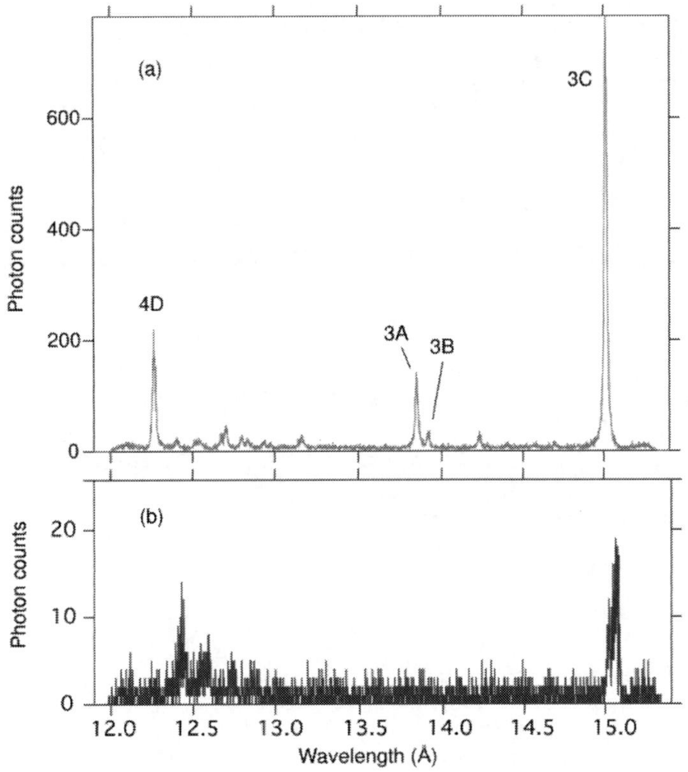

Figure 3. Fe XVII spectral emission in the 12–15 Å range. (a) spectrum formed by direct electron-impact excitation; (b) dielectronic satellite lines from the decay of $1s^22\ell^73\ell'4\ell''$ excited levels.

'16th Cambridge Workshop on Cool Stars, Stellar Systems, and the Sun' Vol. 448 of *Astronomical Society of the Pacific Conference Series* Astronomical Society of the Pacific San Francisco p. 787.
[5] Beiersdorfer P, Bode M P, Díaz F and Ishikawa Y 2014 *Astrophys. J.* in press.
[6] Knapp D A, Marrs R E, Levine M A, Bennett C L, Chen M H, Henderson J R, Schneider M B and Scofield J H 1989 *Phys. Rev. Lett.* **62**, 2104.
[7] P. Beiersdorfer, T. W. Phillips, K. L. Wong, R. E. Marrs, and D. A. Vogel 1992 *Phys. Rev.* A **46**, 3812.
[8] Knapp D A, Beiersdorfer P, Chen M H, Schneider D and Scofield J H 1995 *Phys. Rev. Lett.* **74**, 54.
[9] Beiersdorfer P 2008 *Can. J. Phys.* **86**, 1.
[10] Brown G V, Beiersdorfer P and Widmann K 1999 *Rev. Sci. Instrum.* **70**, 280.
[11] Beiersdorfer P, Brown G V, Goddard R and Wargelin B J 2004 *Rev. Sci. Instrum.* **75**, 3720.
[12] Porter F S, Brown G V, Boyce K R, Kelley R L, Kilbourne C A, Beiersdorfer P, Chen H, Terracol S, Kahn S M and Szymkowiak A E 2004 *Rev. Sci. Instrum.* **75**, 3772.
[13] Porter F S, Beck B R, Beiersdorfer P, Boyce K R, Brown G V, Chen H, Gygax J, Kahn S M, Kelley R L, Kilbourne C A, Magee E and Thorn D B 2008 *Can. J. Phys.* **86**, 231.
[14] Beiersdorfer P, Brown G V, Hildebrandt L, Wong K L and Ali R 2001 *Rev. Sci. Instrum.* **72**, 508.
[15] Beiersdorfer P 1991 *Nucl. Instrum. Methods* **B56/57**, 1144.
[16] Beiersdorfer P, Vogel D A, Reed K J, Decaux V, Scofield J H, Widmann K, Hölzer G, Förster E, Wehrhan O, Savin D W and Schweikhard L 1996 *Phys. Rev.* A **53**, 3974.
[17] Brown G V, Beiersdorfer P, Kahn S M, Liedahl D A and Widmann K 1998 *Astrophys. J.* **502**(2), 1015.
[18] Beiersdorfer P, Chantrenne S, Chen M H, Marrs R, Vogel D, Wong K and Zasadzinski R 1991 *Z. Phys.* D **21**, S209.
[19] Beiersdorfer P, Schneider M B, Bitter M and von Goeler S 1992 *Rev. Sci. Instrum.* **63**, 5029.

Journal of Physics: Conference Series **583** (2015) 012023 doi:10.1088/1742-6596/583/1/012023

Model potentials in liquid water ionization by fast electron impact

M L de Sanctis[1], M-F Politis[2], R Vuilleumier[3], C R Stia[1] and O A Fojón[1]

[1] Instituto de Física Rosario (CONICET-UNR), Av. Pellegrini 250, 2000 Rosario, Argentina.
[2] LAMBE, CNRS UMR 8587, Université d'Evry Val d'Essone, Boulevard François Mitterrand, 91025 Evry, France.
[3] École Normale Supérieure, Dépt. de Chimie, UMR 8640 CNRS-ENS-UPMC, 24, rue Lhomond, 75005 Paris, France.

E-mail: `mdesanctis@ifir-conicet.gov.ar`

Abstract. We study the ionization of water molecules in liquid phase by fast electron impact. We use our previous first-order model within an independent electron approximation that allows the reduction of the multielectronic problem into a monoelectronic one. The initial molecular states of the liquid water are represented in a realistic way through a Wannier orbital formalism. We complete our previous study by taking into account approximately the influence of the passive electrons of the target by means of different model potentials. We compute multiple differential cross sections for the most external orbital $1B_1$ and compare them with other results.

1. Introduction

The ionization of water molecules by electron impact plays an important role in several domains such as radiobiology, radioprotection and medical physics. In fact, this basic reaction is crucial to understand the damage provoked to living tissue by the ionizing radiations. In particular, the production of low energy electrons is of importance to elucidate the mechanisms that lead to cell alteration [1]. In track structure analysis, the liquid water molecules in living cells are often treated as if they were in the gas phase. So, a realistic description of the reaction for molecules in the liquid phase is required to deeply understand the mechanisms in which they are involved. However, a proper description of molecules in the liquid phase is a difficult task that added to the complexities of the collision dynamics makes the problem practically intractable. Hence, several approximations are required in order to obtain the observables of the reaction, i.e., the cross sections.

In a recent work [2], we developed a simple monoelectronic first-order model within the framework of an independent electron approximation to compute multiple differential cross sections (MDCS) for the ionization of the most external $1B_1$ liquid water orbital by fast electron impact. We assume that one of the target electrons (the active one) is ejected in the final channel whereas the other ones (the passive electrons) remain as frozen in their initial states. In this way, the many electron problem is reduced to a monoelectronic one. The orbitals of a single molecule in the liquid phase are represented through a Wannier orbital formalism [3]. The model predicts the presence in the MDCS of the well known binary and recoil peaks similar to the ones of the gas phase [4, 5]. Unfortunately, MDCS for the liquid phase are not available at present. Here, we extend our previous work [2] by considering the influence of the passive electrons on the MDCS by means of model potentials in the entrance channel of the reaction.

Atomic units are used otherwise stated.

2. Theory

As in Ref. [2], the eight-fold differential cross sections (8DCS) for the single ionization of a fixed-in-space liquid water molecule are obtained as,

$$\sigma^{(8)} = \frac{d\sigma}{d\Omega_{mol} d\Omega_e d\Omega_s dE_e} = N(2\pi)^4 \frac{k_e k_s}{k_i} \left| \left\langle \Psi_f^- \mid V_i \mid \psi_i \right\rangle \right|^2 \tag{1}$$

where $\mathbf{k}_i$, $\mathbf{k}_s$ and $\mathbf{k}_e$ are the incident, scattered and ejected electron momenta, respectively. Moreover, $d\Omega_s$ and $d\Omega_e$ denote the solid angles corresponding to $\mathbf{k}_s$ and $\mathbf{k}_e$, respectively. $d\Omega_{mol} = \sin\beta \; d\alpha \; d\beta \; d\gamma$ being α, β, γ the Euler angles of the water molecule. As exchange is not taken into account, $N = 2$ gives the number of electrons in the considered molecular orbital.

The initial and final-state wavefunctions are chosen as,

$$\psi_i = \frac{e^{i\mathbf{k}_i.\mathbf{R}}}{(2\pi)^{3/2}} \Phi_i(\mathbf{r}) \qquad \Psi_f^- \cong \frac{e^{i\mathbf{k}_s.\mathbf{R}}}{(2\pi)^{3/2}} C(\mathbf{k}_e, \mathbf{r}, \nu) \tag{2}$$

respectively, where $\mathbf{R}$ and $\mathbf{r}$ are the position vectors of the incident and *active* electrons, respectively, with respect to the center of mass of the molecule. Φ_i represents the initial molecular orbital of liquid water obtained through a Wannier orbital formalism [3] and the incident electron is described by a plane wave. In the final channel, the ionized electron is described by the Coulomb wavefunction $C(\mathbf{k}, \mathbf{r}, \nu) = \Gamma(1 - i\nu)\,(2\pi)^{-3/2} e^{i\mathbf{k}.\mathbf{r}}\, e^{-\pi\nu/2}\, {}_1F_1[i\nu; 1; -i(kr + \mathbf{k}.\mathbf{r})]$ where ${}_1F_1$ is the confluent hypergeometric function, and $\nu = -1/k$ is the Sommerfeld parameter.

The perturbation in the initial channel is taken as,

$$V_i = \frac{1}{|\mathbf{r} - \mathbf{R}|} + V(\mathbf{R}) \tag{3}$$

where $V(\mathbf{R})$ is a model potential for the projectile-target interaction. We consider no relaxation of the residual target during the reaction. By neglecting the spatial distribution of the nuclei and passive electrons in the water molecule, we obtain an average for the interaction of the projectile with these particles. Firstly, we consider $V(\mathbf{R}) = -1/R$ that corresponds to a complete screening of the net nuclear charge by the passive electrons [2]. In this way, the short range part of the real interaction is neglected. Secondly, to take into account in a simple way the repulsion between the active electron and the passive ones, we include a short range potential given by,

$$V_{short}(R) = -\sum_j N_j \frac{1}{R} \exp(-2Z_j^{eff} R) \left(1 + Z_j^{eff} R\right) \tag{4}$$

where Z_j^{eff} is an effective atomic charge and summation runs over all orbitals j. Here, both Belkić [6] and Slater [7] charges are considered. The corresponding model potentials $V(\mathbf{R})$ are denoted as V_B and V_S, respectively. Furthermore, we test a first order of a Green-Sellin-Zachor type model potential given by $V_G = -\frac{1}{R} - \frac{1}{R}\frac{Z-1}{H}e^{-R/d} - \frac{1}{R}\frac{(H-1)(Z-1)}{H^2}e^{-2R/d}$, where d, H and Z are taken as the ones for the Ne [8] that is isoelectronic to water. Finally, we employ the model potential $V_I = -\frac{8-N}{R} - \frac{N}{R}e^{-2\alpha R} - Ne^{-2\alpha R}$ with the N and α parameter values given in Ref.[9].

Integrating the 8DCS given by Eq. (1) over the Euler angles, five-fold differential cross sections (5DCS) averaged over all possible molecular orientations, are obtained.

Journal of Physics: Conference Series **583** (2015) 012023 doi:10.1088/1742-6596/583/1/012023

3. Results

We compute multiple differential cross sections for single ionization of liquid water molecules from the most external $1B_1$ orbital. Asymmetric kinematic conditions for both scattered and ejected electrons (ejection energies of some eV) are considered. The collision takes place in a coplanar geometry in which the incident, scattered and ejected momenta lie all in the same plane.

First, we present 8DCS for fixed-in-space water molecules as a function of the ejection angle θ_e. Two particular arrangements where the molecule is oriented perpendicularly to the collision plane are considered (see Fig. 1).

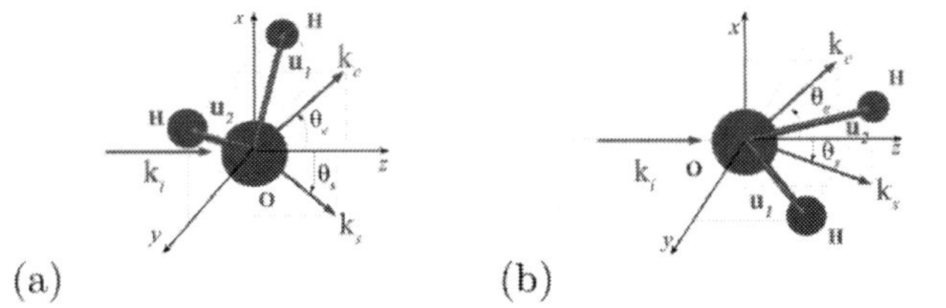

Figure 1. Fixed-in-space molecule configurations considered in this work. Unit vectors $\mathbf{u}_1$ and $\mathbf{u}_2$ define the molecular plane. (a) *Normal I* orientation. (b) *Normal II* orientation.

In Fig. 2, we present results for an incident energy $E_i = 250$ eV, ejected energy $E_e = 5$ eV and scattering angle $\theta_s = 15°$. In this case, the momentum transfer is $q = 1.1$ a.u. and the ionization process is expected to exhibit the typical features of a binary collision regime. This

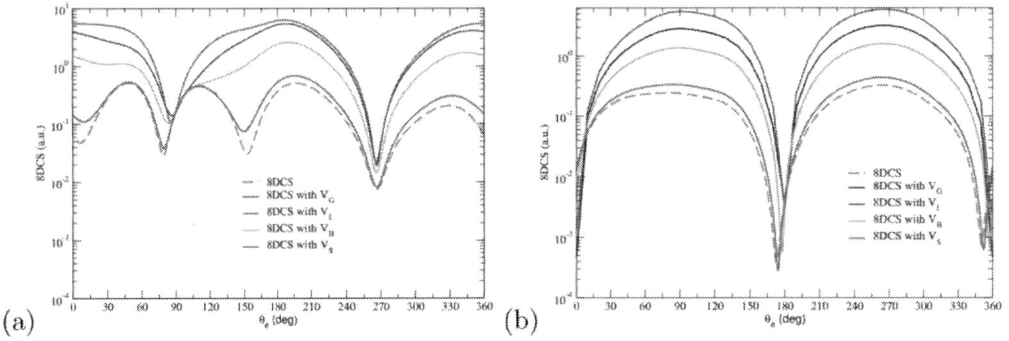

Figure 2. 8DCS per electron for the $1B_1$ orbital as a function of the ejection angle for $E_i = 250$ eV, $E_e = 5$ eV, and $\theta_s = 15°$. (a) *Normal I* and (b) *Normal II* orientations.

is the case for the *Normal I* orientation (Fig. 2(a)) where the $1B_1$ orbital is directed along the incidence direction. The 8DCS computed with $V(\mathbf{R}) = -1/R$ show the characteristic p-orbital two-lobe structure appearing at around the binary peak region ($\theta_e \approx 75°$). On the contrary, for the *Normal II* orientation (Fig. 2(a)) the orbital is oriented perpendicularly to the incidence direction favouring ejection at $\theta_e = 90°$ and $\theta_e = 270°$. The overall behavior observed for both orientations can be explained by the strong p-character of the $1B_1$ orbital and the interplay between both terms of the perturbation given by Eq. (3).

Calculations for the different model potentials are also included in Fig. 2. Slight differences between 8DCS obtained with a Slater potential V_S and the ones for the -1/R contribution are found. It can be seen that the V_S potential preserves the general trend of the angular distribution. On the contrary, for the model potentials V_B, V_G and V_I, strong deviations are observed for both configurations. In particular, for the *Normal I* (Fig. 2(a)) orientation the two-lobe structure is practically erased.

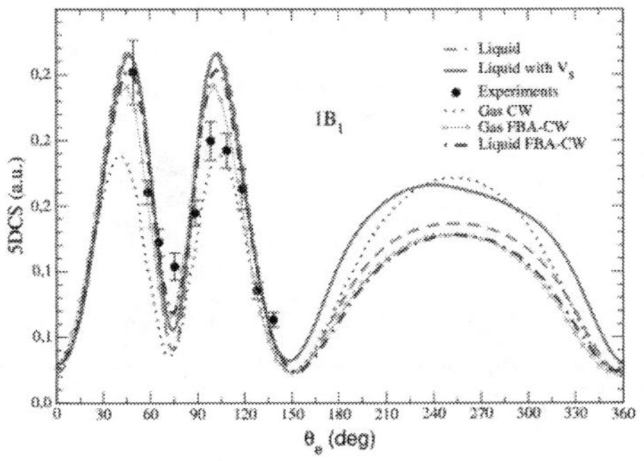

Figure 3. 5DCS for the $1B_1$ orbital as a function of the ejection angle. Incident and ejected energies are $E_i = 250$ eV and $E_e = 10$ eV, respectively, whereas $\theta_s = 15°$.

Second, in Fig. 3, 5DCS are presented for the cases of $V(\mathbf{R}) = -1/R$ and $V(\mathbf{R}) = V_S$ and compare with experiments for the gas [4]. The incident and ejected energies are $E_i = 250$ eV and $E_e = 10$ eV, respectively, whereas $\theta_s = 15°$. Experimental and theoretical results present the same shape at the binary region. In general, qualitative agreement is observed between liquid and gas phases calculations [10, 11]. Moreover, very good accord is observed between our calculations and the previous ones for the liquid [11] obtained also through a first order approximation and a Gaussian monocentric expansion for the water orbitals. Finally, the inclusion of the model potential V_S in our perturbation provokes enhanced cross sections at the recoil region.

4. Conclusions

We have studied the ionization from water molecules in the liquid phase by means of our previous first-order model based on an independent electron approximation [2]. The wavefunction of a single molecule was obtained through the use of Wannier orbitals techniques [3]. The main physical features observed in experiments for vapor (such as binary and recoil peaks) are similar to the ones observed in our predictions (see also Ref.[2]). Moreover, there is a very good agreement with previous predictions for liquid water. From our present results, we estimate that the passive electrons of the reaction may play a role in the recoil region. In particular, the Slater model potential V_S seems to be more suitable to describe the cross sections. It would be interesting to check if this behavior is verified for the other orbitals of the liquid water. Work in this direction is in progress.

Acknowledgments

Authors acknowledge financial support from the Agencia Nacional de Promoción Científica y Tecnológica (PICT No. 01912), and the Consejo Nacional de Investigaciones Científicas y Técnicas de la República Argentina (PIP No. 11220090101026).

References

[1] Boudaiffa B, Cloutier P, Hunting D, Huels M A and Sanche L 2000 *Science* **287** 1658
[2] de Sanctis M L, Politis M F, Vuilleumier R, Stia C R, and Fojón O A 2012 *J. Phys. B* **45** 045206
[3] Hunt P, Sprik M and Vuilleumier R 2003 *Chem. Phys. Lett.* **376** 68
[4] Milne-Brownlie D S *et al* 2004 *Phys. Rev.* A **69** 032701
[5] Kaiser C *et al* 2007 *J. Phys. B: At. Mol. Opt. Phys.* **40** 25632576
[6] Belkić Dz *et al* 1979 *Phys. Rep.* **56** 279
[7] Slater J C 1930 *Phys Rev.* **36** 57
[8] Szydlik P P and Green E S 1974 *Phys. Rev. A* **9** 1885
[9] Illescas C *et al* 2011 *Phys. Rev. A* **83** 052704
[10] Champion C, Dal Cappello C, Houamer S and Mansouri A 2006 *Phys Rev.* A **73** 012717
[11] Champion C 2010 *Phys. Med. Biol.* **55** 11

Double ionization of helium by ion impact: second Born order treatment at the fully differential level

S D López[1], S Otranto[2] and C R Garibotti[3]

[1]Instituto de Astronomía y Física del Espacio, CC67-Suc28, C1428ZAA CABA, Argentina.
[2]IFISUR and Departamento de Física, Universidad Nacional del Sur, 8000 Bahía Blanca, Argentina.
[3]CONICET and Centro Atómico Bariloche, 8400 S. C. de Bariloche, Argentina.

E-mail: sebastlop@gmail.com

Abstract. In this work, a theoretical study of the double ionization of He by ion impact at the fully differential level is presented. Emphasis is made in the role played by the projectile in the double emission process depending on its charge and the amount of momentum transferred to the target. A Born-CDW model including a second-order term in the projectile charge is introduced and evaluated within an on-shell treatment. We find that emission geometries for which the second-order term dominates lead to asymmetric structures around the momentum transfer direction, a typical characteristic of higher order transitions.

1. Introduction

Three collision mechanisms have been identified as responsible of the atomic double ionization (DI) at intermediate to large impact energies. In the two-step-1 (TS1) mechanism, the projectile interacts with only one of the target electrons which subsequently ejects the other via the electron-electron interaction. In the shake-off (SO) mechanism, the second electron relaxes from an ionic bound state to the continuum, following a sudden removal of the primary electron. The third mechanism, usually referred to as two-step-2 (TS2), considers that electrons are sequentially removed by the projectile. In this sense, the TS1 and SO mechanisms involve only one interaction of the projectile with a single electron and can therefore be described as first order terms in a perturbative expansion of the scattering amplitude in the projectile charge Z_P. In contrast, the TS2 mechanism involves two successive projectile-electron interactions and its description requires a second-order term in Z_P. At very high impact energies the SO mechanism dominates the DI process and an accurate description of the initial state correlation is needed, provided the large asymmetry in the electron emission energies. As the impact energy decreases and the scattering event duration increases, emission energies can be less asymmetric, and the electron-electron interaction gains relevance throughout the whole process, and the probability of a double emission mediated by the projectile at all times increases. In this sense, none of these mechanisms is expected to solely dominate the DI at intermediate impact energies. It is clear then that from a theoretical perspective a coherent treatment of them should be proposed at the transition amplitude level.

In this work, we focus on DI processes in ion-helium collisions at an impact energy of 500 keV/amu. At this impact energy the competition among the different mechanisms is expected to be

Journal of Physics: Conference Series **583** (2015) 012024 doi:10.1088/1742-6596/583/1/012024

large. Fully differential cross sections are calculated and analyzed for different projectile charges and momentum transfers. Atomic units are used unless otherwise stated.

2. Theoretical Model

In the present Born-continuum distorted waves (CDW) treatment, the transition amplitude up to second order reads:

$$T_{fi}=T_{fi}^{(1)}+T_{fi}^{(2)}$$

The first order term is given by the First Born Approximation (FBA), in which the projectile is considered as a plane wave in either the initial and final channels:

$$T_{fi}^{FBA}(\vec{k}_1,\vec{k}_2,\vec{\Omega})=\langle\chi_f|V_i|\chi_i\rangle$$

$$V_i=\frac{Z_P Z_T}{R}-\frac{Z_P}{|\vec{R}-\vec{r}_1|}-\frac{Z_P}{|\vec{R}-\vec{r}_2|}$$

We evaluate this FBA term with a distorted wave method in which both target electrons are explicitly considered [1,2]. The final wave function is given by a 3C model with the dynamical screening scheme proposed by Berakdar and Briggs [3]. This model considers infinite collisions between the components of the target. Hence, it accounts for the SO and TS1 mechanisms.

For the second-order term, we consider the on-shell second Born approximation that we have recently introduced [4], and that accounts for the TS2 mechanisms:

$$T_{fi}^{(TS2)}(\vec{k}_1,\vec{k}_2,\Omega)=\int d\vec{\Omega}_n\langle\chi_f|W_f|\xi_1\rangle\langle\xi_1|W_i|\chi_i\rangle$$

The single ionization amplitudes are evaluated within the CDW-EIS approximation and the corresponding perturbations are given by,

$$W_i=\frac{-\nabla_{r_P}^2}{2\mu_P}+\nabla_{r_P}\cdot\nabla_{r_T}$$

$$W_f=-\nabla_{r_P}\cdot\nabla_{r_T}$$

In this study, only one single intermediate state is considered in which one electron is bound to the parent ion while the other is emitted to the continuum with its final momentum.

3. Results

In figure 1 we show the fully differential cross section calculated by means of the FBA at an impact energy of 500 keV for proton impact. The momentum transferred by the projectile Q = 3 a.u.. Electrons are emitted in the scattering plane with equal energies of 10 eV each. Electrons polar angles are measured counter clockwise from the projectile incidence direction. The projectile final momentum is chosen such that the momentum transfer vector lies in the first quadrant. Structures are symmetric with respect to the momentum transfer direction (θ_Q = 74°). Results for other projectiles look identical provided that the FBA transition amplitude only retains the projectile charge information via a Z_P factor. Hence, the only expected change would be that of absolute magnitudes which scale with Z_P^2.

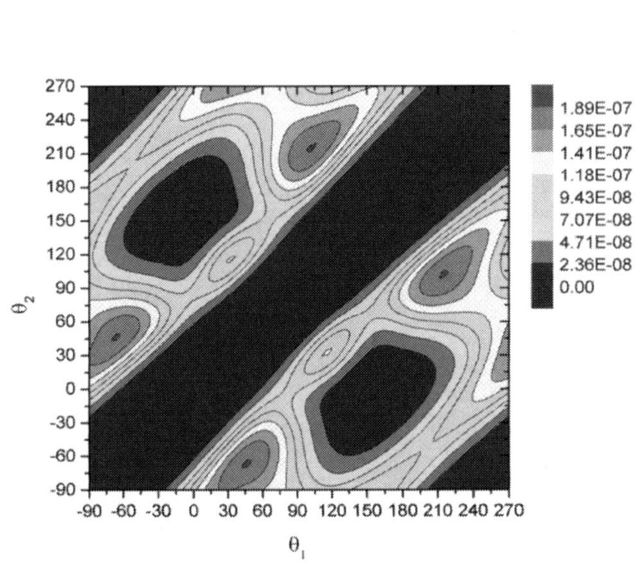

Figure 1: FDCS calculated with FBA of DI of He by proton impact.

In figure 2, we show the fully differential cross sections obtained for the coherent sum of the first and second-order terms of the transition amplitude. In this case, the three collision mechanisms (SO, TS1 and TS2) are incorporated in our theoretical description. The Ward and Macek correlation factor [5] is included in the TS2 term to avoid the unphysical picture of two electrons being emitted hand-in-hand. Three projectiles are considered: proton, He^{2+} and C^{6+}. We now note that the inclusion of the second-order term breaks the symmetry around the momentum transfer direction. This behavior is further highlighted in figure 3 where selected cuts of figures 2 a)-c) are presented. Each cross section has been divided in this case by Z_P^2 to help elucidate how structures deviate from the FBA results. The emission angle of one of the electrons is set as $\theta_1 = 30°$. If the TS1 mechanism were the dominating one, the other electron would be expected to be departing at a relative angle close to 90° which in this case would correspond to $\theta_2 = 120°$ or -60°. The former case can be related to a passive parent ion which plays a spectator role during the second electron removal, while the latter would indicate a recoiling parent ion that strongly mediates the double emission process.

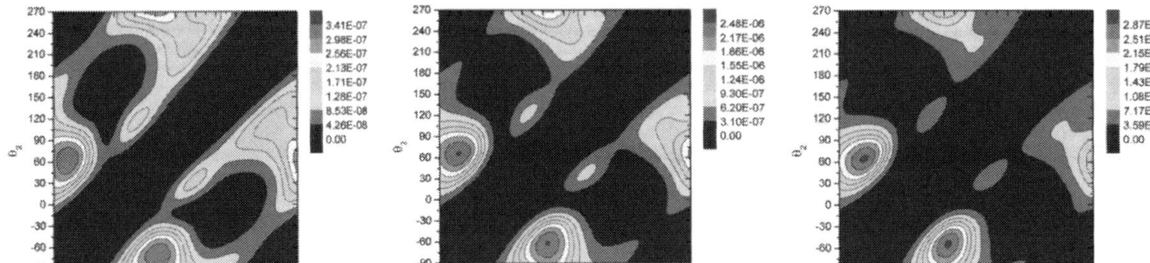

Figure 2: Coherent sum of first and second order for DI of He by proton (a), He^{2+}(b) and C^{6+} (c) impact. The considered impact energy are of 500 keV/amu, the momentum transferred is $Q = 3$ a.u., and the electrons are emitted with 10 eV.

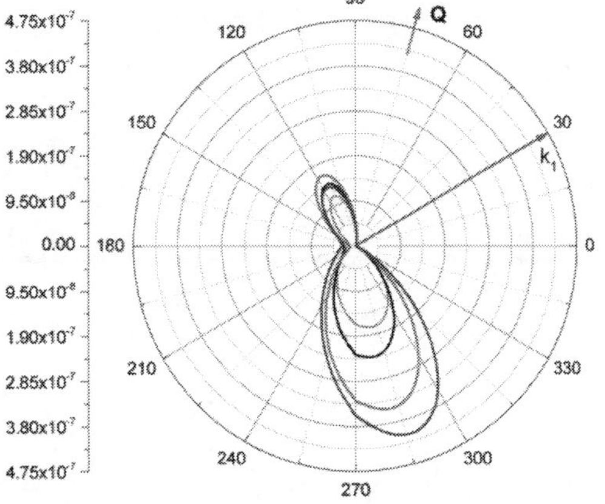

Figure 3: Selected cut of figure 3 for one electron emitted at 30 degrees. DI of He by proton (black line), He^{2+} (red line) and C^{6+} (blue line) impact. FBA (gray line) The considered momentum transferred is Q = 3 a.u.

From figure 3, we note that the ratio among both lobes changes according to the projectile charge. The lobe located at about -75° drastically increases with projectile charge clearly indicating that the electrons tend to be emitted in the direction of the receding projectile. Moreover, a more subtle shift of this structure towards the projectile direction can also be noticed.

4. Conclusions

Fully differential cross sections for the double ionization of He by ion impact have been theoretically studied by means of a distorted wave model that incorporates the SO, TS1 and TS2 mechanisms. A second-order Born term evaluated on-shell has been introduced to coherently add the TS2 mechanism to the first order term which already contains information on the SO and TS1 mechanisms. From our results we infer that the most visible effect introduced by the second-order term is the rupture of symmetry of the FDCS structures around the momentum transfer direction.

Acknowledgment

Work supported by Grants No. PGI24/F059 of UNS and No. PIP 112-201101-00749 of CONICET (Argentina).

References

[1] López S D, Garibotti C R and Otranto S 2011 *Phys Rev.* A **83** 062702.
[2] López S D, Otranto S and Garibotti C R 2013 *Phys Rev.* A **87** 022705.
[3] Berakdar J and Briggs J S 1994 *Phys. Rev. Lett.* **72** 3799.
[4] López S D, Otranto S and Garibotti C R 2013 *Phys Rev.* A **89** 062709.
[5] Ward S and Macek J H 1994 *Phys Rev.* A **49** 1049.

Journal of Physics: Conference Series **583** (2015) 012025 doi:10.1088/1742-6596/583/1/012025

A hybrid distorted wave model: a (almost) *complete-post* version of the CDW model for single ionization

J M Monti[†], O A Fojón[†], J Hanssen[†] and R D Rivarola[†]

[†] Laboratorio de Colisiones Atómicas, Facultad de Ciencias Exactas, Ingeniería y Agrimensura and Instituto de Física Rosario, CONICET-UNR, Av. Pellegrini 250, 2000 Rosario, Argentina.

E-mail: monti@ifir-conicet.gov.ar

Abstract. A hybrid CDW–CDW-EIS model is developed in order to include the dynamic screening in the *post* CDW approximation for single ionization. Doubly differential cross sections for single ionization of water molecules are calculated and comparison with available experimental data and other theories are shown.

1. Introduction and Theory

The distorted wave models are perturbative methods that were introduced in order to accelerate the convergence of the Born series and to avoid the presence of disconnected diagrams that may originate divergent Born sub-series. Among the distorted wave models, there are two that merit to be cited: the Continuum Distorted Wave (CDW) [1] and the Continuum Distorted Wave-Eikonal Initial State (CDW-EIS) [2]. These models were initially formulated to describe single ionization of monoelectronic targets by bare ion impact. Later on,they were extended to study single ionization of multielectron targets [3]. Since then, they have been widely used to calculate cross sections for single ionization involving a large collection of projectiles, such as heavy bare ions, antiprotons, and more recently highly-charged dressed ions, and also a variety of targets as simple atomic systems to large biological molecules (see [4, 5, 6, 7]).

Recently, an extension of these two distorted wave models was developed by Monti *et al.* for the case of single ionization of atomic targets by swift dressed-ion impact [8].

Here, we study the case of bare projectiles and multielectronic targets within an independent electron model following the procedure given in [4]. By assuming only one active electron, the multielectronic Hamiltonian can be reduced to:

$$H_{el} = -\frac{1}{2}\nabla^2 + V_T(\boldsymbol{x}) + V_P(\boldsymbol{s}) + V_s(\boldsymbol{R}) \tag{1}$$

where $\boldsymbol{x}$ and $\boldsymbol{s}$ give the active electron position in the target and projectile reference frames, respectively. $V_T(\boldsymbol{x})$ is a potential which takes into account the interaction of this electron with the rest of the target, $V_P(\boldsymbol{s}) = -Z_P/s$ is the interaction between the bare projectile and the active electron, and $V_s(\boldsymbol{R})$ the interaction the projectile with the target nucleus and the passive electrons. This potential depends only on the internuclear coordinate $\boldsymbol{R}$ and thus, within the straight-line version of the impact-parameter approximation, gives place to a phase factor which, in our calculations, does not affect the electron dynamics. Within a distorted wave formalism, the initial and final distorted-waves are given by:

$$\chi_i^+(\boldsymbol{x},t) = \Phi_i(\boldsymbol{x},t)\,\mathcal{L}_i^+(\boldsymbol{s}) \tag{2}$$
$$\chi_f^-(\boldsymbol{x},t) = \Phi_f(\boldsymbol{x},t)\,\mathcal{L}_f^-(\boldsymbol{s})\,. \tag{3}$$

Here $\Phi_i(\boldsymbol{x},t) = \phi_i(\boldsymbol{x})\exp(\mathrm{i}\varepsilon_i t)$ and $\Phi_f(\boldsymbol{x},t) = \phi_f(\boldsymbol{x})\exp(\mathrm{i}\varepsilon_f t)$ are the initial-bound and final-continuum states solution of the time-dependent Schrödinger equation. We considered a RHF initial wavefunction and an effective Coulomb target potential $V_T(\boldsymbol{x})$ in the final state. ε_i and ε_f are the electron energies in the initial and final states, respectively.
In particular, considering the CDW approximation, the initial distortion in (2) is proposed as:

$$\mathcal{L}_i^{+CDW}(\boldsymbol{s}) = N(\nu)\;{}_1F_1(\mathrm{i}\nu;1;\mathrm{i}vs+\mathrm{i}\boldsymbol{v}\cdot\boldsymbol{s}) \tag{4}$$

whereas the final distortion in (3) is chosen as:

$$\mathcal{L}_f^{-CDW}(\boldsymbol{s}) = N^*(\zeta)\,{}_1F_1(-\mathrm{i}\zeta;1;-\mathrm{i}ps-\mathrm{i}\boldsymbol{p}\cdot\boldsymbol{s}) \tag{5}$$

where $\boldsymbol{v}$ is the projectile velocity, $\nu = Z_P/v$, $\zeta = Z_P/p$, $\boldsymbol{p} = \boldsymbol{k} - \boldsymbol{v}$ is the ejected electron momentum in the projectile reference frame, $\boldsymbol{k}$ is the ejected electron momentum in the target reference frame, and ${}_1F_1$ is the confluent hypergeometric function, and N its normalization factor.
It is well known that within the distorted wave framework, the transition matrix may be written either in its *prior-* or *post-*version (or *complete-post*, see [9]), depending on whether the perturbation operators act over the initial of final distorted waves, respectively. It was shown [10] that the *prior*-version of the CDW transition amplitude for single ionization has an intrinsic logarithmic divergence near the binary encounter peak. This prevents us from using this form of the transition amplitude to calculate ionization cross sections. Therefore the *post*-version of the CDW approximation, which does not contain such divergence, was used to calculate ionization cross sections.
Recently the *post*-version of the CDW-EIS model was revisited by Monti *et al.* [9] showing that approximating the target potential V_T by an effective-Coulomb one neglects a part of the *dynamic-screening*, leading to the omission of a term in the transition amplitude, causing the well known post-prior discrepancies [11, 12]. Properly including the term containing the dynamic-screening in the *post*-version of the CDW-EIS model, therefore obtaining a *complete*-post-version, almost vanishes the post-prior discrepancies. The same approximation to the V_T potential is made in the CDW approximation. Hence the usual *post*-version of the CDW approximation also lacks the dynamic screening contribution.
The *complete-post*-version of the CDW transition amplitude consists of two terms:

$$\mathcal{A}^{+CDW} = \mathcal{A}^{+CDW(a)} + \mathcal{A}^{+CDW(b)} \tag{6}$$

where the firs term is the usual *post*-version (see [1, 5, 6]) and the second one is the term related to the dynamic screening:

$$\mathcal{A}^{+CDW(b)}(\rho) = -i\int_{-\infty}^{+\infty} dt\; e^{i\Delta\varepsilon\, t}\int d\boldsymbol{x}\; V_T(\boldsymbol{x})\phi_f^*(\boldsymbol{x})\mathcal{L}^*{}_f^{-CDW}(\boldsymbol{s})\phi_i(\boldsymbol{x})\mathcal{L}_i^{+CDW}(\boldsymbol{s}) \tag{7}$$

Unfortunately, this term also contains the logarithmic divergences that are found in the *prior*-version. Therefore, in order to include this term avoiding any divergences, we replace, only in the dynamic-screening term, the $\mathcal{L}_i^{+CDW}$ distortion by its asymptotic approximation, I.E. an eikonal phase,

$$\mathcal{L}_i^{+CDW}(\boldsymbol{s}) \approx \mathcal{L}_i^{+EIS}(\boldsymbol{s}) = \exp\left(-\mathrm{i}\nu\ln(vs+\boldsymbol{v}\cdot\boldsymbol{s})\right), \tag{8}$$

Journal of Physics: Conference Series **583** (2015) 012025 doi:10.1088/1742-6596/583/1/012025

obtaining

$$\mathcal{A}^{+EIS(b)}(\rho) = -i \int_{-\infty}^{+\infty} dt\; e^{i\Delta\varepsilon t} \int d\boldsymbol{x}\; V_T(\boldsymbol{x})\phi_f^*(\boldsymbol{x})\mathcal{L}^{*}{}_f^{-CDW}(\boldsymbol{s})\phi_i(\boldsymbol{x})\mathcal{L}_i^{+EIS}(\boldsymbol{s}) \qquad (9)$$

A hybrid CDW–CDW-EIS model is then obtained, in which the transition matrix is given by the sum of the usual CDW *post*-version and the dynamic screening term as described in the CDW-EIS approximation:

$$\mathcal{A}^{+CDW} = \mathcal{A}^{+CDW(a)} + \mathcal{A}^{+EIS(b)}. \qquad (10)$$

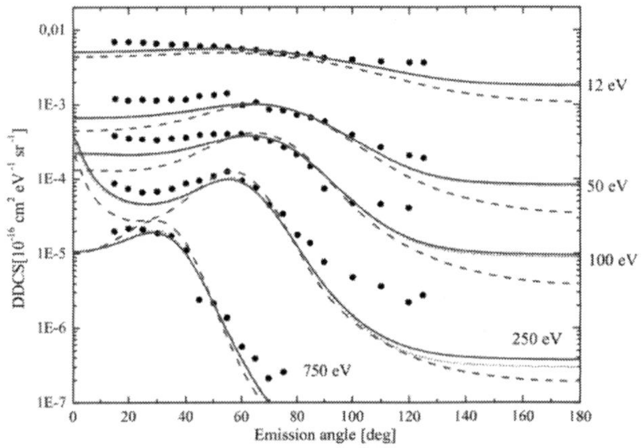

Figure 1. DDCS for single ionization of water molecules by 500 keV proton impact as a function of the ejection angle for 12 eV, 50 eV, 100 eV, 250 eV and 750 eV fixed emission energies. —— Present hybrid model calculations; - - - -*post* CDW calculations; ⋯⋯ *prior* CDW-EIS calculations. ● experimental results from [13].

2. Results

In figures 1 and 2, we show doubly differential cross sections (DDCS) for single ionization of a single isolated water molecule by 500 keV H^+ and 6 MeV/u C^{6+} impact, respectively, as a function of the ejection angle for several fixed emission energies. Calculations performed with the *post*-version CDW model and the hybrid model are shown and compared with *prior*-version CDW-EIS calculations and experimental results. The contribution of the dynamic screening can be directly observed by comparing the *post* CDW calculations and the *hybrid* model. Its inclusion leads to a better agreement with the experimental results for forward and backward emission angles, though underestimation of the experiments is still found. Also, for the 500 keV H^+ impact and 250 keV and 750 keV emission energies, it can be seen that considering the dynamic screening leads to a better description of the binary-encounter peak. In the 6 MeV/u case, a large overestimation in the binary encounter region is found for the 192 ev and 384 eV emission energies case for all theories.

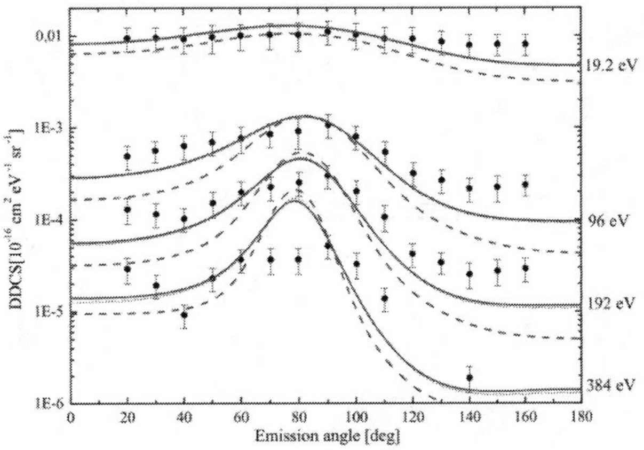

Figure 2. DDCS for single ionization of water molecules by 6 MeV/u C^{6+} impact as a function of the ejection angle for 19.2 eV, 96 eV, 192 eV, and 384 eV fixedemission energies. ——Present hybrid model calculations; - - - -*post* CDW calculations; ······*prior* CDW-EIS calculations. ● experimental results from [14].

3. Conclusions

In order to avoid the logarithmic divergences present in the *prior* CDW model and in its *post*-version dynamic-screening term, a hybrid CDW–CDW-EIS model is presented in which the dynamic screening is considered through an eikonal approximation. It is shown that its inclusion leads to a major improvement of the previous *post* CDW calculations and a better agreement with experimental results.

Acknowledgments

The authors acknowledge the Agencia Nacional de Promoción Científica y Tecnológica de la República Argentina for finalcial support through PICT 2011-2145.

References

[1] Belkić Dž, *J. Phys. B: At. Mol. Phys.* **11** 3529 (1978)
[2] Crothers D S F and McCann J F, *J. Phys. B: At. Mol. Phys.* **16** 3229 (1983)
[3] Fainstein P D, Ponce V H and Rivarola R D, *J. Phys. B: At. Mol. Opt. Phys.* **21** 287 (1988)
[4] Fainstein P D, Ponce V H and Rivarola R D 1991 *J. Phys. B: At. Mol. Opt. Phys.* **24** 3091
[5] Gulyás L, Fainstein P D and Salin A 1995 *J. Phys. B: At. Mol. Opt. Phys.* **28** 245
[6] Gulyás L and Fainstein P D 1998 *J. Phys. B: At. Mol. Opt. Phys.* **31** 3297
[7] Tachino C A, Monti J M, Fojón O A, Champion C and Rivarola R D, *J. Phys.: Conf. Series* **488** 102029 (2014)
[8] Monti J M, Rivarola R D and Fainstein P D 2011 *J. Phys. B: At. Mol. Opt. Phys.* **44** 195206
[9] Monti J M, Fojón O A, Hanssen J and Rivarola R D 2013 *J. Phys. B: At. Mol. Opt. Phys.* **46** 145201
[10] Brauner M and Macek J H 1992 *Phys. Rev. A* **46** 2519
[11] Ciappina M F, Cravero W R and Garibotti C R 2003 *J. Phys. B: At. Mol. Opt. Phys.* **36** 3775
[12] Ciappina M F and Cravero W R 2006 *J. Phys. B: At. Mol. Opt. Phys.* **39** 1091
[13] Toburen H and Wilson W E 1977 *J. Chem. Phys.* **66** 5202
[14] Dal Capello C, Champion C, Boudrioua O, Lekadir H, Sato Y, Ohsawa D 2009 *Nucl. Instrum. Methods Phys. Res. B* **267** 781

Journal of Physics: Conference Series **583** (2015) 012026 doi:10.1088/1742-6596/583/1/012026

When vortices and cusps meet

F Navarrete[1,2], M Feole[1,2], R O Barrachina[1,2] and Á Kövér[3]

[1] Centro Atómico Bariloche ((Comisión Nacional de Energía Atómica), R8402AGP S. C. de Bariloche, Río Negro, Argentina

[2] Instituto Balseiro (Comisión Nacional de Energía Atómica y Universidad Nacional de Cuyo), R8402AGP S. C. de Bariloche, Río Negro, Argentina

[2]3 Institute for Nuclear Research, Hungarian Academy of Sciences (Atomki), P.O. Box 51, H-4001 Debrecen, Hungary

E-mail: `navarrete@cab.cnea.gov.ar`

Abstract. After being overlooked for decades, the presence of quantum vortices in atomic ionization processes was recently uncovered both theoretically and experimentally. On the other hand, the electron capture to the continuum cusp is one of the most conspicuous and well-studied features of the multiple differential cross section in the ionization of atoms by the impact of positively charged projectiles. Here we analyze the conditions for these two structures to approach each other in the configuration space of the transition matrix element, and the effects that this encounter might produce.

1. Introduction

The well-known electron-capture-to-the-continuum (ECC) cusp was first observed more than four decades ago in ion-atom ionization collisions [1], and more recently in positron impact collisions [2]. On the other hand, and in spite of some early evidences [3, 4, 5, 6, 7], quantum vortices in atomic and molecular processes were uncovered only some few years ago. Until now, they were experimentally observed in the ionization of atoms by the impact of electrons [10] and ions [11], and theoretically analyzed for positrons [12] and electric pulses [13].

Vortices and cusps are different in origin and structure. Vortices formed in the wave function during the early stages of the collision [8] might collapse at later times, but some can eventually survive up to the asymptotic regime and manifest themselves as zeros of the ionization matrix element T. The simultaneous conditions $\mathrm{Re}(T) = 0$ and $\mathrm{Im}(T) = 0$ define a manifold $\mathcal{V}$ with co-dimension 2 in the multidimensional configuration space of T. The generalized velocity field $\mathbf{u} = \mathrm{Im}\left[(\nabla_{\mathbf{k}}T)/T\right]$, with $\mathbf{k}$ a two-dimensional momentum orthogonal to $\mathcal{V}$, rotates around it with a 2π quantization that assures the single valuation of T.

On the other hand, the ECC cusps can be explained in terms of a smooth continuation across the ionization limit of capture into highly excited bound states of the projectile [14]. It appears as a $1/k'$ divergence in $|T|^2$ occurring at the threshold of the charge exchange process, where k' is the electron-projectile relative momentum. The condition $\mathbf{k}' = 0$ defines a manifold $\mathcal{C}$ with dimension 3 in the configuration space of T.

While the ECC cusp is a quite ubiquitous feature of ion-atom and positron-atom ionization collisions, and has been extensively observed and studied, the experimental observation of vortices, i.e. deep and confined minima of the electron momentum distribution, is not exempt of difficulties. In particular, they have only been observed in (e,2e) experiments and in 10 keV/amu He^{2+}+He transfer ionization collisions, and in this latter case only at the saddle point region and for large projectile scattering angles [11]. No experimental observation of vortices in positron-atom collisions has been reported so far.

In this article we propose an indirect method for observing a vortex, consisting in the analysis of the strong distortion that it might produce on the ECC cusp for particular experimental conditions. Calculations performed with the Lattice-Time-Dependent Schrödinger Equation method (LTDSE) for the ionization of atomic Hydrogen by 5 keV protons at unit impact parameter have already showed the presence of vortices near the ECC cusp [8]. Similarly, in a previous article we uncovered the presence of a vortex near the ECC cusp for positron impact at 100 and 200 eV.

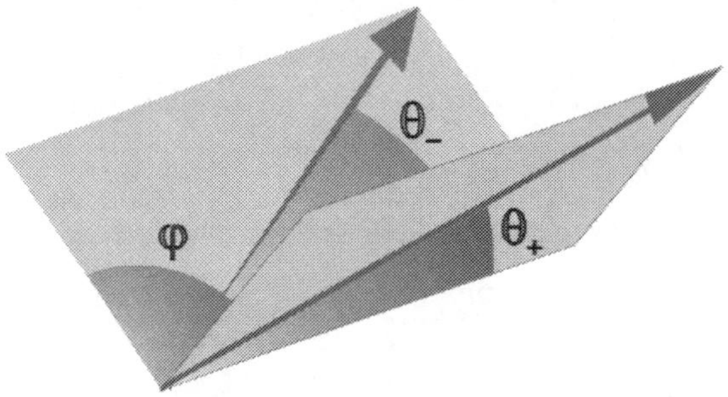

Figure 1. In the energy sharing or collinear geometry employed in this article, the angles formed by the electron θ_- and the projectile θ_+ with respect to the forward direction are equal, $\theta_- = \theta_+$, while their relative azimuthal angle is zero, $\varphi = 0$.

2. Transition Matrix Element

We evaluate the transition matrix element T for the ionization of Hydrogen by the impact of energetic charged projectiles by means of a correlated approximation of the final three-body state. Details of the theoretical approach were described in previous articles [12, 16, 15]. Here, instead of fixing the projectile's emission angle or using the "symmetric geometry" [17] that is standard in studies of (e,2e) collisions, we employ an "energy sharing" or collinear arrangement, where the electron and the projectile move along the same direction in the final state, as it is shown in figure 1.

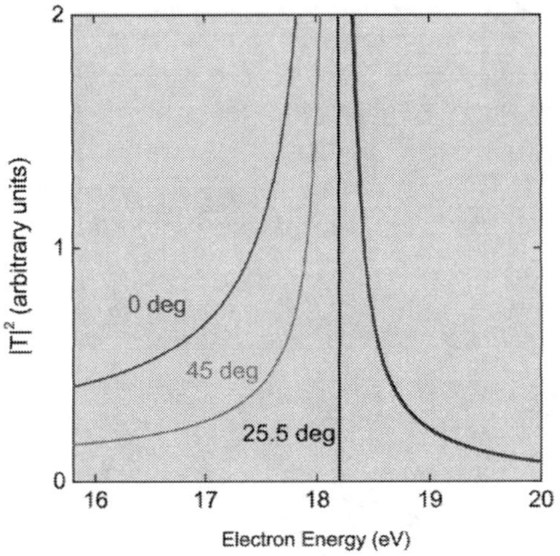

Figure 2. Square modulus of the transition matrix element, $|T|^2$, in a collinear geometry, for the ionization of hydrogen atoms by the impact of a 50 eV positrons, as a function of the electron energy for different emission angles $\theta_- = 0°$ (blue line), 25.5° (black line) and 45° (red line). The curves have been renormalized so as to coincide on the right side of the ECC cusp.

Journal of Physics: Conference Series **583** (2015) 012026 doi:10.1088/1742-6596/583/1/012026

In figure 2 we display the square modulus of the matrix element, $|T|^2$ for a 50 eV $e^+ + H$ ionization collision, as a function of the electron energy for different emission angles θ. For $\theta = 0°$ the ECC cusp shows the well-known asymmetry towards lower energies, an effect common to ion and positron impact that has been extensively studied in the literature. But we now observe that a sudden modification of the the cusp's shape occurs at $\theta = 25.5°$, where the lower energy side of the ECC cusp is strongly suppressed. Up to our best knowledge this effect has neither been observed nor predicted before. Finally, the standard shape of the ECC is recovered for $\theta = 45°$.

We claim that this effect is due to the presence of an isolated zero in the transition matrix element, as shown in figure 3. In fact, this zero corresponds to a vortex [12] at its very emergence, which remarkably occurs near the *ECC* cusp. It can be further demonstrated that its angular position is fairly insensitive to the impact energy. Thus, it produces the strong distortion of the lower-energy side of the ECC cusp.

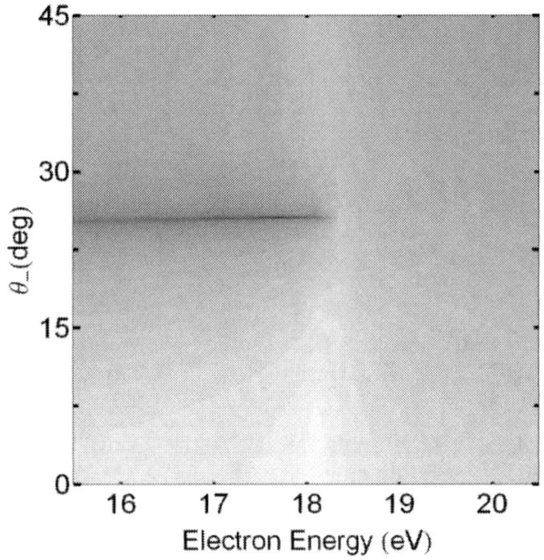

Figure 3. Square modulus of the transition matrix element for the ionization of a Hydrogen atom by the impact of a positron of 50 eV. Conditions are set to a collinear geometry configuration (as explained in Fig 1). The logarithmic scale in arbitrary units sets the lowest and highest values in dark red and light yellow, respectively.

This kind of indirect evidence of a vortex might show up in different characteristics of the Multiple Differential Cross Section (MDCS) in prospective experiments. For instance, it might be observable as a sudden drop of the full width at half maximum (FWHM) or as a shift towards higher energies of the ECC cusp position at intermediate emission angles. However these rather subtle effects might be difficult to resolve experimentally. Bearing this last limitation in mind, we calculated the left-side yield of the ECC cusp. As it is shown in figure 4 this quantity, that should be more feasible to be measured, displays a sharp minimum at precisely the angular position of the vortex.

3. Conclusions

Vortices are very well-known features of many-body systems. They are routinely observed in gases, liquids and plasmas, and in connection with quantum effects as superconductivity, superfluity, and Bose - Einstein condensation. The description of these many body systems customarily resorts to the inclusion of ad-hoc potentials or nonlinear terms. Here, on the other hand, we have investigated their appearance in an extremely simple three-body quantum system in the continuum.

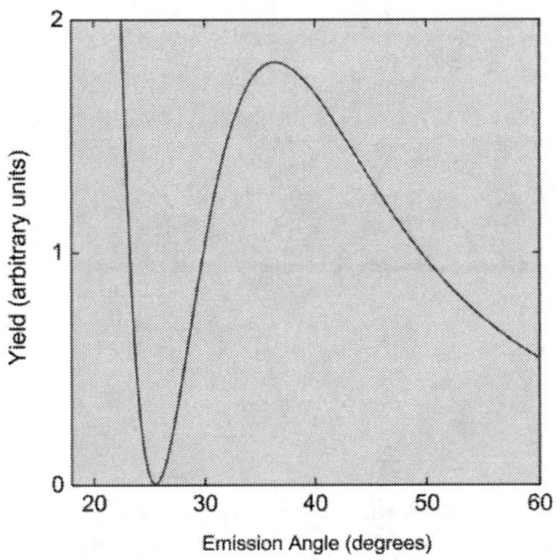

Figure 4. Yield of the square modulus of T in an energy range between 15.7 and 18.2 eV (the latter being the energy corresponding to the ECC cusp) as a function of the emission angle.

However, since the only fingerprint of a vortex in the MDCS is a deep and confined minimum, its experimental observation is hindered by difficulties of resolution and low intensity. Fortunately, as we showed in this article, the emergence of one of them in the proximity of such a conspicuous feature of the MDCS, as the ECC cusp, could help in its experimental determination. We also presented the calculation of the yield, which we hope that could make its experimental observation easier.

Acknowledgments
This work was supported by the Hungarian - Argentinean MINCYT-NIO Cooperation Programme in Science and Technology (grant no HU/10/07), by Comisión Nacional de Energía Atómica, CNEA and Universidad Nacional de Cuyo (Grant 06/C416) and by the Hungarian Scientific Research Foundation (OTKA K104409). FN and ROB are also members of the Consejo Nacional de Investigaciones Científicas y Técnicas (CONICET), Argentina.

References

[1] Crooks G B and Rudd 1970 *Phys. Rev. Lett.* **25** 5309
[2] Kövér A and Laricchia G 1998 *Phys. Rev. Lett.* **80** 5309
[3] Brauner M and Briggs J S 1991 *J. Phys. B: At. Mol. Opt. Phys.* **24** 2227
[4] Botero J and Macek J H 1992 *Phys. Rev.* A **45** 15
[5] Murray A J and Read F H 1993 *J. Phys. B: At. Mol. Opt. Phys.* **26** L359
[6] Berakdar J and Briggs J S 1994 *Phys. Rev. Lett.* **72** 3799
[7] Berakdar J and Briggs J S 1994 *J. Phys. B: At. Mol. Opt. Phys.* **27** 4271
[8] Macek J H, Sternberg J B, Ovchinnikov S Y, Lee T-G, and Schltz D R 2009 *Phys. Rev. Lett.* **102** 143201
[9] Ovchinnikov S Y, Macek J H, Schmidt L PH H and Schultz D R 2011 *Phys. Rev.* A **83** 060701
[10] Macek J H, Sternberg J B, Ovchinnikov S Yu and Briggs J S 2010 *Phys. Rev. Lett.* **104** 033201
[11] Schmidt L Ph H, Goihl C, Metz D,Schmidt-Böcking H, Dörner R, Ovchinnikov S Yu, Macek J H and Schultz D R 2014 *Phys. Rev. Lett.* **112** 083201
[12] Navarrete F, Della Picca R, Fiol J and Barrachina R O 2013 *J. Phys. B: At. Mol. Opt. Phys.* **46** 115203
[13] Ovchinnikov S Yu, Sternberg J B, Macek J H, Lee T-G and Schultz D R 2010 *Phys. Rev. Lett.* **105** 203005
[14] Macek J 1970 *Phys. Rev.* A **1** 235
[15] Fiol J, Rodríguez V D and Barrachina R O 2001 *J. Phys. B: At. Mol. Opt. Phys.* **34** 933
[16] Della Picca R, Fiol J and Barrachina R O 2005 *Nucl. Instrum. Methods* B **247** 52
[17] Gottschalk B, Shlaer W J and Wang K H 1965 *Phys. Lett.* **16** 294

Journal of Physics: Conference Series **583** (2015) 012027 doi:10.1088/1742-6596/583/1/012027

Quantum interference of fast atoms scattered off crystal surfaces

M S Gravielle

Abstract. The striking observation of interference structures produced by grazing impact of fast atoms on crystal surfaces reported a few years ago [1,2] has given rise to the development of a powerful surface analysis technique. This article gives a brief account of the main features of the process, using the Surface Eikonal (SE) approximation as a theoretical tool to analyze the different mechanisms responsible for the quantum interference. The SE approach is a semiclassical method based on the use of the eikonal wave function, which takes into account the coherent superposition of transition amplitudes for different axially channeled trajectories. It has proved to provide a quite good description of experimental diffraction patterns for different collision systems.

Instituto de Astronomía y Física del Espacio (IAFE, CONICET-UBA). Casilla de correo 67, sucursal 28 (C1428EGA) Buenos Aires, Argentina.

E-mail: msilvia@iafe.uba.ar

1. Introduction

In the field of particle-surface dynamic interactions, one of the most interesting findings of recent years corresponds to the observation of quantum interference effects produced by fast atoms after grazingly colliding with crystal surfaces [1, 2]. The diffraction of particles from crystal surfaces is a well understood phenomenon from the beginning of quantum mechanics, having been key to the confirmation of one of the most fundamental concepts of quantum mechanics - the wave-particle duality. Hitherto the presence of interference effects in atom-surface scattering was believed to be confined to de Broglie wavelengths comparable or larger than the interatomic distances in the crystal, beyond which the collision process was supposed to be governed by Newton´s laws. However, a few years ago two different groups reported simultaneously [1, 2] experimental evidences of diffraction effects from insulator surfaces produced by grazing scattering of swift atoms, with energies in the range of keVs, whose de Broglie wavelengths are several orders of magnitude smaller than the shortest interatomic distance in the crystal. This unexpected phenomenon [3, 4], now known as grazing-incidence fast atom diffraction (GIFAD or FAD), is the focus of the present contribution.

Since the first measurements of FAD [1, 2] the process has attracted remarkable attention, being the aim of extensive experimental and theoretical research [5, 6, 7, 8, 9, 10, 11, 12, 13, 14, 15, 16, 17]. Such an interest is motivated not only by the fast velocity of incident projectiles that made the observation of interference effects unforeseen, but also by the exceptional sensitivity of the diffraction patterns to the projectile-surface interaction [7, 8, 10, 18, 19, 20], which is paving the way for the development of a powerful surface analysis technique. The essential feature of FAD is the geometry of grazing collision, which makes somehow possible to decouple the movement of the projectile along the channeling direction, too fast to allow for diffraction, from its much slower motion within the plane perpendicular to it. Hence, diffraction patterns are originated by this transverse motion.

From the theoretical point of view, different methods have been employed to simulate experimental FAD patterns. They range from fully quantum treatments in terms of a wave packet propagation [2, 7, 21] to semiclassical approximations [5, 8] based on the use of classical projectile trajectories. Among these last theories we can mention the Surface Eikonal (SE) approximation [8, 22], which is a distorted-wave method that makes use of the eikonal wave function to represent the elastic collision with the surface, while the motion of the swift projectile is classically described by considering axially channeled trajectories for different initial positions. The SE approach includes a clear description of the main mechanisms of the FAD process, being simpler to evaluate than a full quantum calculation [7, 21]. It has been successfully applied to investigate FAD patterns for different collision systems [18, 22, 23, 24].

This work presents a general overview of the FAD phenomenon based on results derived within the SE approach. Concerning experimental research on FAD, a complete review, including basic features and recent experimental developments, has been recently published in Ref. [25]. Parts of this article have been separately published in Refs. [22, 23, 24]. Atomic units ($e^2 = \hbar = m_e = 1$) are used unless otherwise stated.

2. Surface Eikonal approach

Within the SE approximation the transition matrix per unit surface area $\mathcal{A}$ reads [8, 22]

$$T_{if}^{(SE)} = \frac{1}{\mathcal{A}} \int_{\mathcal{A}} d\vec{R}_{os}\; a_{if}(\vec{R}_{os}), \tag{1}$$

where

$$a_{if}(\vec{R}_{os}) = \int_{-\infty}^{+\infty} dt\; \frac{\left|v_z(\vec{\mathcal{R}}_t)\right|}{(2\pi)^3} \exp[-i\vec{Q}.\vec{\mathcal{R}}_t - i\eta(\vec{\mathcal{R}}_t) - i\phi_M]\; V_{SP}(\vec{\mathcal{R}}_t) \tag{2}$$

represents the transition amplitude associated with the trajectory $\vec{\mathcal{R}}_t$, with $\vec{\mathcal{R}}_t \equiv \vec{\mathcal{R}}_t(\vec{R}_{os})$ the projectile position at a given time t obtained by considering a classical path with initial momentum $\vec{K}_i$ and starting position on the surface plane $\vec{R}_{os}$. In Eq. (2), $v_z(\vec{\mathcal{R}}_t)$ denotes the component of the projectile velocity perpendicular to the surface plane and $\vec{Q} = \vec{K}_f - \vec{K}_i$ is the projectile momentum transfer, with $\vec{K}_f$ the final projectile momentum satisfying the energy conservation, i.e. $K_f = K_i$. The phase η is the eikonal phase, which is defined along the projectile path as

$$\eta(\vec{\mathcal{R}}_t) = \int_{-\infty}^{t} dt'\; V_{SP}(\vec{\mathcal{R}}_{t'}), \tag{3}$$

where V_{SP} is the projectile-surface interaction, and $\phi_M = \nu\pi/2$ is the Maslov correction phase, with ν the Maslov index as defined in Ref. [26].

The SE differential probability, per unit of surface area, of elastic scattering with final momentum $\vec{K}_f$ in the direction of the solid angle $\Omega_f \equiv (\theta_f, \varphi_f)$ is obtained from Eq. (1) as $dP/d\Omega_f = (2\pi)^4 m_P^2 \left|T_{if}^{(SE)}\right|^2$, where θ_f and φ_f are the final polar and azimuthal angles, respectively, φ_f is measured with respect to the incidence direction on the surface plane, and m_P is the projectile mass. A schematic picture of the FAD process and the angular coordinates is displayed in Fig. 1. With regard to the applicability of the method, the SE approach is valid for small de Broglie wavelengths, smaller than the characteristic distance of variation of the surface potential, which makes it especially suitable for the problem under study.

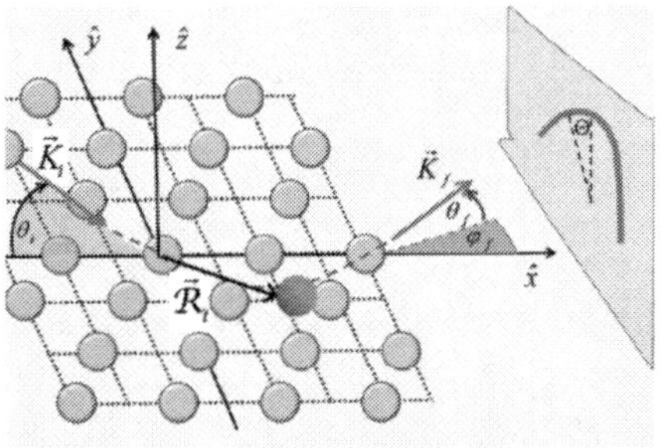

Figure 1. (Color online) Sketch of the angular coordinates for the FAD process.

3. Energy dependence of FAD patterns

When swift atoms grazingly impact on a well ordered surface along a low indexed direction, the projectile distribution shows the typical banana shape characteristic of axial surface channeling [27]. But in the FAD regime, such a distribution presents additionally clear interference structures [28], like the ones observed in Fig. 2 (a) for He atoms scattered off a LiF(001) surface [22]. In the cases of Fig. 2 (a), even though the de Broglie wavelengths of incident atoms are almost three orders of magnitude smaller than the shortest interatomic distance in the crystal, the experimental spectra show well-defined interference patterns, with maxima and minima located symmetrically with respect to the direction of incidence, which corresponds to the azimuthal angle $\varphi_f = 0$. Similar interference structures are also obtained with the SE approximation, which predicts projectile distributions with pronounced maxima whose positions, displayed with black dots in the figure, are in good agreement with the experimental ones. Notice that as a result of the energy conservation, final dispersion angles of classically scattered projectiles satisfy the relation $\theta_f^2 + \varphi_f^2 \simeq \theta_i^2$, with θ_i the incidence angle, which is approximately verified by the experimental distribution as well.

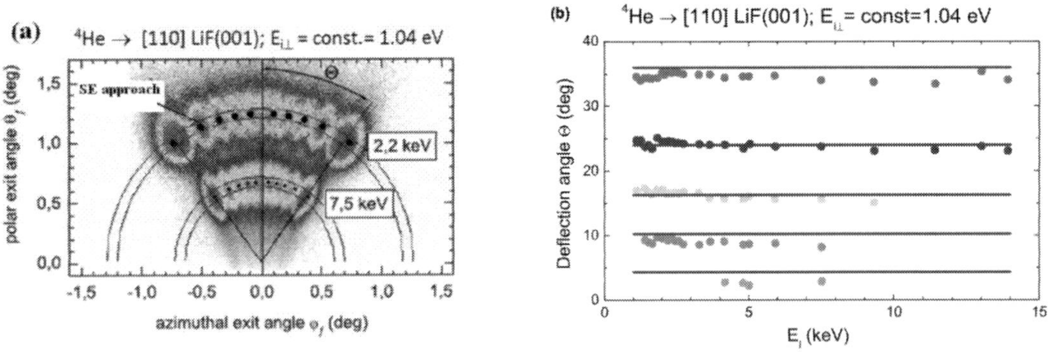

Figure 2. (Color online)FAD for He atoms scattered from LiF(001) [22]. (a) Experimental two-dimensional intensity distributions for two incidence energies, E_i=2.2 keV and 7.5 keV, but with the same perpendicular energy. Black dots, positions of SE maxima. (b) Deflection angles Θ corresponding to maxima of angular distributions, as a function of E_i, for $E_{i\perp}$ =1.04 eV. Full circles, experimental data; curves, SE results.

In Fig. 2 (a) the incidence angles for the two impact energies - E_i = 2.2 keV and 7.5 keV - were set to obtain the same perpendicular energy - $E_{i\perp}$ =1.04 eV - where $E_{i\perp} = E_i \sin^2\theta_i$ is the energy associated with the initial motion perpendicular to the axial channel. This particular setting reveals that the number and Θ-positions of diffraction maxima, with $\Theta = \arctan(\varphi_f\, /\theta_f)$ the deflection angle, are completely governed by the normal energy, being independent of E_i

Journal of Physics: Conference Series **583** (2015) 012027 doi:10.1088/1742-6596/583/1/012027

at the same $E_{i\perp}$ [22]. This behavior is more evident from Fig. 2 (b) where deflection angles corresponding to the experimental and SE maxima are plotted as a function of the total energy E_i while keeping $E_{i\perp}$ constant.

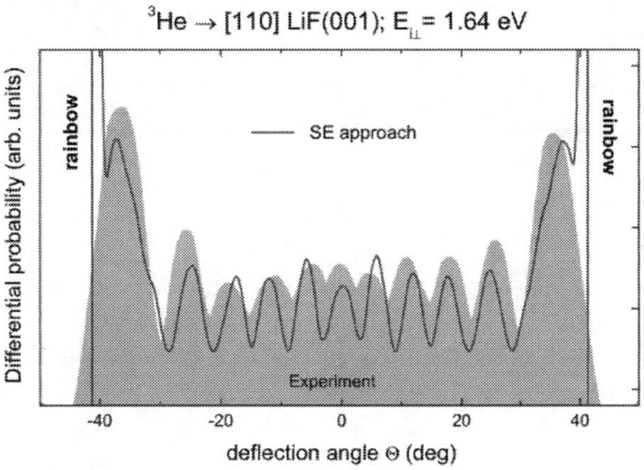

Figure 3. (Color online) Projectile distribution, as a function of the deflection angle Θ, for 3.5 keV He atoms scattered from LiF(001) along the [110] direction [22]. Solid blue line, SE results (for a reduced unit cell); shadow gray line, experimental data.

Experimental and theoretical projectile distributions, as a function of Θ, are plotted in Fig. 3 for He atoms impinging on LiF(001) with $E_{i\perp}$ =1.64 eV. The positions and relative intensities of the experimental peaks are fairly well described by the SE model. However, the concordance deteriorates at the largest deflection angles, which coincide with the classical rainbow ones, where the SE probability presents cusped peaks. The sharp increase of the SE probability at rainbow angles is a characteristic of several semi-classical approximations, widely studied in atom-surface scattering [29], which originates from the accumulation of classical trajectories (caustics) at this point. In quantum mechanics, instead, the sharp rainbow peaks are replaced by smooth maxima that display an exponentially decaying behavior beyond classical rainbow angles, just on the dark side, i.e. in the region of classically forbidden transitions [30]. A semi-quantum approach to solve the rainbow problem of the SE approach has been recently introduced in Ref. [31].

4. Mechanisms of FAD: unit-cell and Bragg diffractions

As it happens for most of diffraction phenomena from periodic structures, FAD patterns have two different origins: unit-cell and Bragg diffractions [5]. Both mechanisms are included in the SE description and can be separately analyzed as follows.

Within the SE approximation, by considering the area $\mathcal{A}$ involved in Eq. (1) as composed by n identical reduced unit cells centered on different sites $\vec{X}_{sj}$ of the crystal surface, the transition matrix can be factorized as

$$T_{if,n}^{(SE)} = T_{if,1}^{(SE)}\ S_n(\vec{Q}_s), \qquad (4)$$

where $T_{if,1}^{(SE)}$ is derived from Eq. (1) by evaluating the $\vec{R}_{os}$-integral over only *one* reduced unit cell, while $S_n(\vec{Q}_s) = \sum_{j=1}^{n} \exp\left[-i\vec{Q}_s.\vec{X}_{sj}\right]/n$ is an oscillatory function that takes into account the crystallographic structure of the surface, with $\vec{Q}_s$ the component of $\vec{Q}$ parallel to the surface plane.

Journal of Physics: Conference Series **583** (2015) 012027 doi:10.1088/1742-6596/583/1/012027

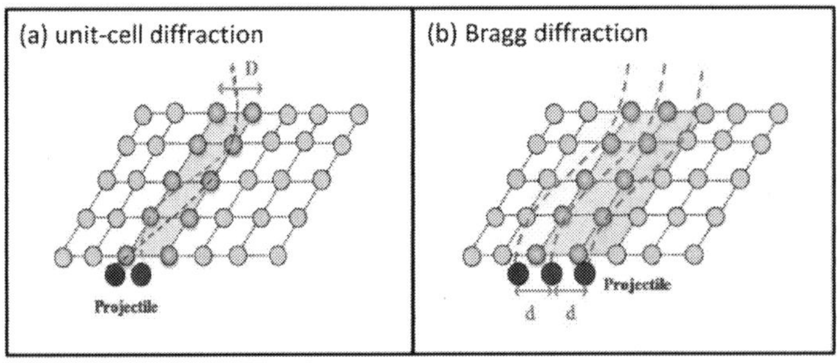

Figure 4. (Color online) Schematic depiction of the different mechanisms involved in the FAD process: (a) unit-cell diffraction, and (b) Bragg diffraction.

Each factor of Eq. (4) describes a different mechanism of FAD. The first factor, $T_{if,1}^{(SE)}$, is a form factor associated with unit-cell diffraction, being produced by the interference of trajectories whose initial positions $\vec{R}_{os}$ differ by a distance smaller than D, where D is the width of the reduced unit cell (Fig. 4 (a)). This unit-cell factor carries information on the shape of the interaction potential across the incidence direction and it is related to supernumerary rainbows [32]. The second factor, $S_n(\vec{Q}_s)$, is a crystallographic factor associated with Bragg diffraction, which originates from the interference of identical trajectories whose initial positions $\vec{R}_{os}$ are separated by a distance d equal to the spacial periodicity of the channel (Fig. 4 (b)). Under typical incidence conditions for FAD, it is possible to approximate $S_n(\vec{Q}_s) \approx S_{n_{tr}}(Q_{tr})$, where n_{tr} is the number of reduced unit cells in the direction transversal to the incidence channel and $Q_{tr} = K_f \cos\theta_f \sin\varphi_f \simeq K_{i\perp} \sin\Theta$ is the component of $\vec{Q}_s$ along such a direction, with $K_{i\perp} = K_i \sin\theta_i$ the perpendicular initial momentum. For periodic surfaces the function $S_{n_{tr}}(Q_{tr})$ gives rise to equal intensity Bragg peaks placed at

$$Q_{tr}d = m2\pi, \tag{5}$$

with m an integer that represents the Bragg order. The width of these Bragg peaks is determined by the number of reduced unit cells reached by the incident wave packet, narrowing as n_{tr} increases.

As illustrative example, in Fig. 5 we display the SE differential probability obtained from Eq. (1) by considering an area $\mathcal{A}$ formed by *three* reduced unit cells, which presents Bragg maxima at angular positions derived from Eq. (5). While the positions of Bragg peaks provide information on the spacing between surface atoms only, their intensities are modulated by the unit-cell form factor $T_{if,1}^{(SE)}$, which acts as an oscillatory envelope function that can reduce or even suppress the contribution of a given Bragg order, giving detailed information about the surface potential. Also from Eq. (5) it is straightforwardly deduced that when $K_{i\perp}$ increases, the angular separation between consecutive Bragg peaks decreases, and consequently, the experimental observation of Bragg diffraction becomes limited by the spatial resolution of the detector. This is the main reason why only interference structures due to unit-cell diffraction are observed in the experimental projectile distributions of Figs. 2 and 3, which correspond to perpendicular energies $E_{i\perp} \gtrsim 1$ eV.

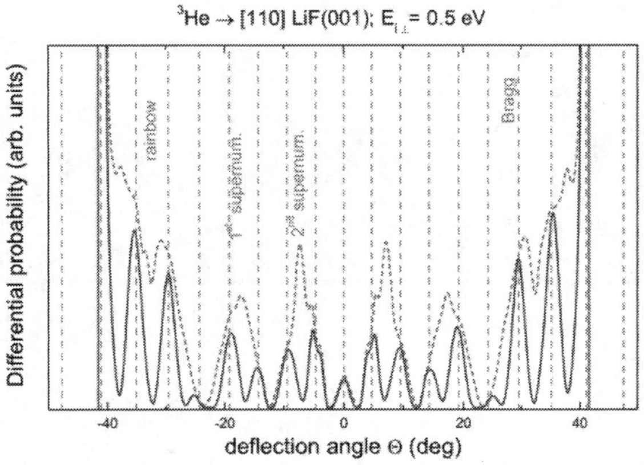

Figure 5. (Color online) SE projectile distribution for He atoms scattered from LiF(001) [22]. Dashed red line, SE probability for unit-cell diffraction; solid blue line, similar by using an extended integration area, as explained in the text. Dotted vertical lines, Bragg peak positions.

5. Sensitivity with the surface interaction

For most of the studied systems, not only the Θ positions but also the relative intensities of diffraction maxima are independent of E_i at the same perpendicular energy [22, 28]. This fact supports the assumption that in the typical FAD regime, even though the complete corrugation of V_{SP} on the surface plane is taken into account within the SE approximation, the Q_{tr} distribution is practically unaffected by the modulation of the potential along the channel, turning the three-dimensional surface potential into an effective two-dimensional (2D) one [21, 33]. Therefore, an overall understanding of the FAD scenario for a given collision system can be obtained by displaying the so-called *diffraction chart*, which shows the intensity of the projectile distribution as a function of both, $E_{i\perp}$ and $Q_{\rm tr}$. In Fig. 6 we compare experimental and SE diffraction charts for He atoms impinging on a silver surface [24], which present reach interference structures, specially as the perpendicular energy increases. Taking into account that different values of $E_{i\perp}$ allow one to probe potential contours across the incidence direction for different distances to the surface, the general good agreement between the experimental and simulated diffraction charts of Fig. 6 is a signature of the quality of the potential model used in the calculations [24].

One of the most attractive characteristic of FAD is its extraordinary sensitivity to the morphological and electronic characteristics of the crystal surface. FAD patterns are strongly dependent on the corrugation of the projectile-surface potential across the incidence direction and subtle changes in the surface interaction introduce substantial modifications of the interference structures [7, 18, 25]. In Fig. 7, experimental momentum spectra for 0.5 keV He atoms impinging on a silver surface are compared with SE distributions considering three different perpendicular energies. The figure clearly shows that the intensity of the central peak, corresponding to the Bragg maximum of order $m = 0$, is extremely sensitive to the corrugation amplitude of the surface potential and a small variation in the $E_{i\perp}$ value strongly modifies the relative intensity of the zeroth order Bragg peak, which switches from being an absolute maximum for $E_{i\perp} = 0.20$ eV to an almost absolute minimum for $E_{i\perp} = 0.28$ eV [24].

6. Conclusions

In this paper we have used the SE approximation to present a brief overview of the FAD phenomenon, focusing on the main aspects of the quantum interference process. The SE

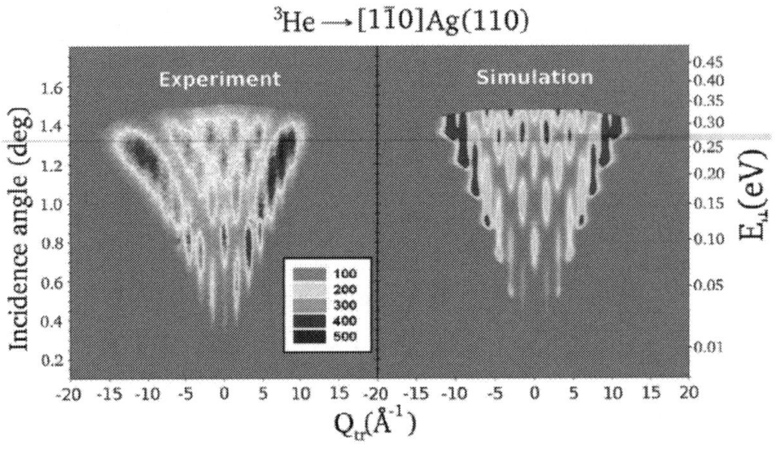

Figure 6. (Color online) Diffraction charts displaying (a) experimental intensities and (b) SE probabilities, as a function of the normal energy $E_{i\perp}$ and the transverse momentum $Q_{\rm tr}$, for 0.5 keV He atoms scattered off Ag(110) [24].

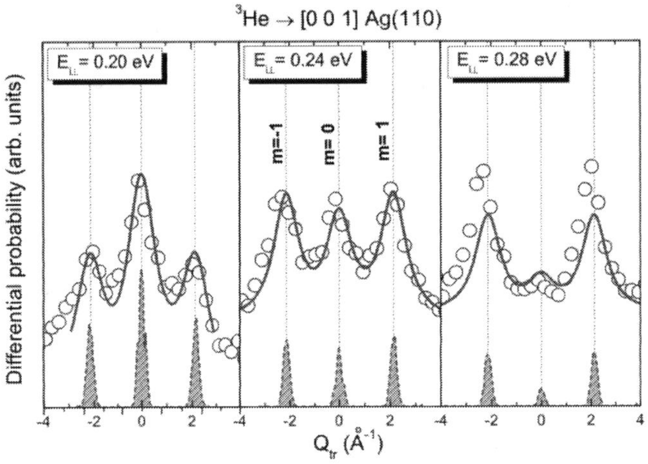

Figure 7. (Color online) Momentum distributions, as a function of $Q_{\rm tr}$, for 0.5 keV He atoms scattered off Ag(110) [24]. The normal energy $E_{i\perp}$ is: (a) 0.20 eV, (b) 0.24 eV, and (c) 0.28 eV. Circles, experimental data; solid and dashed blue lines, SE probabilities convoluted to include inherent uncertainties and without convolution, respectively. The vertical dashed lines indicate Bragg peak positions.

approximation is a semi-classical method that takes into account the quantum interference in terms of the coherent superposition of transition amplitudes for different projectile paths with the same deflection angle. This method has proved to give a fairly good representation of experimental diffraction patterns for different collision systems [8, 18, 22, 23, 24], providing a simple way of analyzing the different mechanism of FAD.

Even though we have here considered only two illustrative surfaces - LiF(001) and Ag(110), which correspond to an insulator and a metal, respectively - the FAD technique has been successfully applied to investigate a wide range of materials, including semi-conductors [13, 36] and thin films [15] and superstructures adsorbed on metal surfaces [10, 17]. In relation to projectiles, FAD patterns were observed for He and Ne impact [5, 23], both inert atoms, as well as for incidence of atomic and molecular hydrogen [1, 2]. This last case, corresponding to fast molecular diffraction, represents a challenge for the theoretical methods and will be the focus of our future research.

Acknowledgments

Financial support from CONICET, UBA, and ANPCyT of Argentina is acknowledged.

[1] Schüller A, Wethekam S and Winter H 2007 *Phys. Rev. Lett.* **98** 016103
[2] Rousseau P, Khemliche H, Borisov A G and Roncin P 2007 *Phys. Rev. Lett.* **98** 016104
[3] Interference structures of FAD had been predicted, before its experimental observation [1, 2], in a widely unknown article by Andreev [4].
[4] Andreev E A 2002 *Russ. J. Phys. Chem.* **76** S164
[5] Schüller A and Winter H 2008 *Phys. Rev. Lett.* **100** 097602
[6] Manson J R, Khemliche H and Roncin P 2008 *Phys. Rev. B* **78** 155408
[7] Aigner F, Simonović N, Solleder B, Wirtz L and Burgdörfer J 2008 *Phys. Rev. Lett.* **101** 253201
[8] Gravielle M S and Miraglia J E 2008 *Phys. Rev. A.* **78** 022901
[9] Bundaleski N, Khemliche H, Soulisse P and Roncin P 2008 *Phys. Rev. Lett.* **101** 177601
[10] Schüller A, Busch M, Seifert J, Wethekam S, Winter H and Gärtner K 2009 *Phys. Rev. B* **79** 235425
[11] Ruiz A, Palao J P and Heller E J 2009 *Phys. Rev. A* **79** 052901
[12] Díaz C, Rivière P and Martín F 2009 *Phys. Rev. Lett.* **103** 013201
[13] Khemliche H, Rousseau P, Roncin P, Etgens V H and Finocchi F 2009 *Appl. Phys. Lett.* **95** 151901
[14] Schüller A, Busch M, Wethekam S and Winter H 2009 *Phys. Rev. Lett.* **102** 017602
[15] Seifert J, Schüller A, Winter H, Wlodarczyk R, Sauer J and Sierka M 2010 *Phys. Rev. B* **82** 035436
[16] Seifert J and Winter H 2012 *Phys. Rev. Lett.* **108** 065503
[17] Seifert J, Busch M, Meyer E and Winter H 2013 *Phys. Rev. Lett.* **111** 137601
[18] Gravielle M S and Miraglia J E 2009 *Nucl. Instrum. Meth. Phys. Res. B* **267** 610
[19] Schüller A, Wethekam S, Blauth D, Winter H, Aigner F, Simonović N, Solleder B, Burgdörfer J and Wirtz L 2010 *Phys. Rev. A* **82** 062902
[20] Schüller A, Blauth D, Seifert J, Busch M, Winter H, Gärtner K, Wlodarczyk R, Sauer J and Sierka M 2012 *Surf. Sci.* **606** 161
[21] Zugarramurdi A and Borisov A G 2012 *Phys. Rev. A* **86** 062903
[22] Schüller A, Winter H, Gravielle M S, Pruneda J M and Miraglia J E 2009 *Phys Rev. A* **80** 062903
[23] Gravielle M S, Schüller A, Winter H and Miraglia J E 2011 *Nucl. Instrum. Meth. Phys. Res. B* **269** 1208
[24] Ríos Rubiano C A, Bocan G A, Gravielle M S, Bundaleski N, Khemliche H and Roncin P 2013 *Phys. Rev. A* **87** 012903
[25] Winter H and Schüller A 2011 *Prog. Surf. Sci.* **86** 169 and references therein
[26] Avrin W F and Merrill R P 1994 *Surf. Sci.* **311** 269
[27] Meyer F W, Folkerts L and Schippers S 1995 *Nucl. Instrum. Meth. Phys. Res. B* **100** 366
[28] Schüller A and Winter H 2009 *Nucl. Instrum. Meth. Phys. Res. B* **267** 628
[29] Miret-Artés S and Pollak E 2012 *Surf. Sci. Rep.* **67** 161 and references therein.
[30] Berry M V and Mount K E 1972 *Rep. Prog. Phys.* **35** 315
[31] Gravielle M S and Miraglia J E 2014 *arXiv:1407.5669 [physics.atom-ph]*
[32] Garibaldi U, Levi A C, Spadacini R and Tommei G E 1975 *Surf. Sci.* **48** 649
[33] The 2D approach of the surface potential breaks down for low parallel velocities [21], for surfaces with strong corrugation [34], or for long distances of periodicity along the axial channel [35].
[34] Gravielle M S, Bocan G A and Díez Muiño R 2010 *Phys. Rev. A* **82** 052904
[35] Zugarramurdi A and Borisov A G 2013 *Nucl. Instrum. Meth. Phys. Res. B* **317** 83
[36] Atkinson P, Eddrief M, Etgens V H, Khemliche H, Debiossac M, Momeni A, Mulier M, Lalmi B and Roncin P 2014 *Appl. Phys. Lett.* **105** 021602

Journal of Physics: Conference Series **583** (2015) 012028 doi:10.1088/1742-6596/583/1/012028

Asymmetric behavior in the excitation of plasmons by ions entering or leaving a material in an oblique trajectory

J L Gervasoni[1,2], S Segui[1], R O Barrachina[1,2]

[1] Centro Atómico Bariloche (Comisión Nacional de Energía Atómica), R8402AGP S. C. de Bariloche, Río Negro, Argentina

[2] Instituto Balseiro (Comisión Nacional de Energía Atómica y Universidad Nacional de Cuyo), R8402AGP S. C. de Bariloche, Río Negro, Argentina

E-mail: gervason@cab.cnea.gov.ar

Abstract. In this work we analyze the production of bulk and surface plasmons by fast ions entering or leaving a material at different incidence angles. In the framework of a semi classical dielectric formulation, we study the different contributions to the induced potential, mainly in the vicinity of the surface, analyzing the differences and similarities that arise when in and out trajectories are compared. In particular, we find some novel oscillatory structures for both perpendicular and oblique trajectories.

1. Introduction

The excitation of plasmons in a material bombarded by swift ions is relevant in different techniques for the spectroscopic analysis of surfaces, thin-foils and nanostructures [1, 2]. On the theoretical ground, these collective excitations are usually studied by means of second quantization or dielectric formulations, as complementary approaches for the analysis of quantum and semiclassical aspects of the problem [3]. For instance, we have recently analyzed the excitation of plasmons by a swift ion traversing a metal-vacuum planar interface in a normal trajectory and studied the differences and similarities that arise between incoming and outgoing trajectories [4]. The aim of the present communication is to extend this analysis to oblique trajectories. This seemingly simple objective is not without pitfalls. For instance, the standard method of cutting off the plasma response function only along the direction parallel to the interface, which has been customarily employed since its introduction by G. D. Mahan [5] four decades ago, is sound for perpendicular trajectories, but fails at other incidence angles. The reason for this failure is that it introduces a spurious spatial asymmetry that might lead to incorrect results, as for instance an unrealistic dependence of the energy loss on the angle at asymptotic distances from the surface. In the present article we avoid this approximation on the cutoff, a generalization that differentiates the present results from previous analyses.

In the following two sections we evaluate the induced potential and the energy dissipation rates due to bulk plasmon excitations. Finally, in section 4, we analyze the dependence of the in-out asymmetry with respect to the incidence angle θ.

Journal of Physics: Conference Series **583** (2015) 012028 doi:10.1088/1742-6596/583/1/012028

2. Induced Potential

In order to analyze the in-out asymmetry of plasmon production by swift ions in oblique trajectories, we avoid any complication of the problem and consider the simplest possible scenario. For instance, we use a Drude-type approximation of the dielectric function $\epsilon(\mathbf{k},\omega)$, namely $\epsilon(\omega) = 1 - \omega_P^2/\omega(\omega + i\gamma)$ [6]. Here, ω_P and γ are the plasma frequency and damping rate of the metal, respectively (in particular, we use aluminum constants, with $\omega_P = 15.8$ eV). Let us point out that for swift projectiles, the discussion in this article still applies when allowance is made for spatial dispersion, provided that their velocity is large compared with the minimum phase-velocity of the plasmons in the medium [7].

We describe the metal-vacuum interface by means of the specular reflection and extended pseudo-medium models [8, 9, 10, 11]. Furthermore, we assume that the ion's trajectory is undisturbed by the plasmon excitation events, an approximation which holds for sufficiently large kinetic energies, namely $mv^2/2 \gg \hbar\omega_p$, where m and v are the mass and velocity of the projectile, respectively. With these provisions, and following the analysis in our previous articles [3, 4], the surface (S) and bulk (B) contributions to the potential induced by an ion of charge Z and velocity $\mathbf{v}$ should be given by (atomic units are used throughout this article),

$$\phi(\mathbf{r},t) = \phi_S(\mathbf{r},t) + \phi_B(\mathbf{r},t) \quad ,$$

where

$$\begin{aligned}\phi_S(\mathbf{r},t) &= Z\,\omega_S \int_{-\infty}^{t} \mathrm{d}t'\, e^{-\gamma(t-t')/2} \sin[\omega_S(t-t')] \\ &\quad \times \int_0^{k_c} \mathrm{d}q\, J_o\left(q|\vec{\rho} - \mathbf{v}_{||}t'|\right) e^{-q(|z|+|v_\perp t'|)} \qquad (1)\\ \phi_B(\mathbf{r},t) &= Z\,\omega_P\,\Theta(-z) \int_{-\infty}^{t} \mathrm{d}t'\, e^{-\gamma(t-t')/2} \sin[\omega_P(t-t')]\,\Theta(-v_\perp t') \\ &\quad \times \int_0^{k_c} \mathrm{d}q\, J_o\left(q|\vec{\rho} - \mathbf{v}_{||}t'|\right)\left(e^{-q|z-v_\perp t'|} - e^{-q|z+v_\perp t'|}\right) \quad . \qquad (2)\end{aligned}$$

Here $\mathbf{v}_{||}(t)$ and $v_\perp(t)$ are the components of the projectile's velocity parallel and perpendicular to the surface, respectively. Similarly, $\vec{\rho}$ and z are the projections of the position $\mathbf{r}$ in directions perpendicular and parallel to $\hat{\mathbf{z}}$. The interface is located on the plane $z = 0$, with the metal occupying the semi-space $z < 0$. We have also defined the usual cutoff in the plasma response function,

$$k_c = \omega_P \sqrt{\frac{1}{v_F^2} - \frac{1}{v^2}} \quad ,$$

where $v_F = (3\sqrt{\pi}\omega_P/2)^{2/3}$ is the Fermi velocity. Both cases, of a particle leaving ($v_\perp > 0$) or entering ($v_\perp < 0$) the solid are considered.

However, as it was stated in the previous section, these results were obtained on the assumption that the cutoff k_c could be kept only in the components parallel to the surface, while the integration over the perpendicular component could be extended to infinity. This approximation might lead to spurious dependencies on the incidence angle, as it can be seen, for instance, in Fig. 7a of ref. [3]. To avoid this effect, we had to trace back the calculations leading to (1) and (2). For instance, the bulk contribution to the induced potential can now be written as

$$\phi_B(\mathbf{r},t) = \Theta(-z)\left[\tilde{\phi}(\mathbf{r},t) - \tilde{\phi}(\mathbf{r} - 2z\hat{z}, t)\right] \quad , \qquad (3)$$

where $\Theta(x)$ is the Heaviside function and

$$\tilde{\phi}(\mathbf{r},t) = \frac{Z\,\omega_P}{2\pi^2}\,\Theta(-z)\int_{-\infty}^{t}\mathrm{d}t' \int_{k<k_c} \mathrm{d}\mathbf{k}\, e^{-\gamma(t-t')/2} \sin[\omega_P(t-t')]\,\Theta(-v_\perp t')\,\frac{e^{i\mathbf{k}.(\mathbf{r}-\mathbf{v}t')}}{k^2} \quad .$$

Journal of Physics: Conference Series **583** (2015) 012028 doi:10.1088/1742-6596/583/1/012028

Note that the integral in $\mathbf{k}$ can be performed analytically in terms of Sine integral functions.

3. Results and Conclusions

Let us first investigate the particular case of a particle heading towards the interface from inside the material (i.e., $v_\perp > 0$ and $t < 0$). In this case, since $\tilde{\phi}(\mathbf{r}, t)$ decreases exponentially in front of the projectile (i.e. $z > 0$), its reflection $\tilde{\phi}(\mathbf{r} - 2z\hat{z}, t)$, which is moving towards the interface from outside the metal, would be negligible inside it. Thus, if the projectile is still far away from the surface, it would be possible to neglect the reflected contribution and approximate the induced potential $\phi_B(\mathbf{r}, t)$ by its direct term alone. However, once the particle has crossed the surface (i.e. $t > 0$), the potential will start to decrease (due to the reduction of the integration range on t'), but its reflection will start to be relatively important. It would seem as if the wake that has been accompanying the projectile inside the solid almost without any distortion, is reflecting at the surface while begins to fade away (see figure 1, left and central panels).

For the opposite situation of a particle moving towards the surface from outside the solid (i.e., $v_\perp < 0$ and $t < 0$), the Heaviside function vanishes and so does the induced potential $\phi_B(\mathbf{r}, t)$. Only when the particle has crossed the surface (i.e. $t > 0$), the range of integration on t' becomes different from zero (namely, $0 < t' < t$), and the induced potential starts to build up. In this case, the projectile will be moving across the waves of its own reflected wake (see figure 1, right panel).

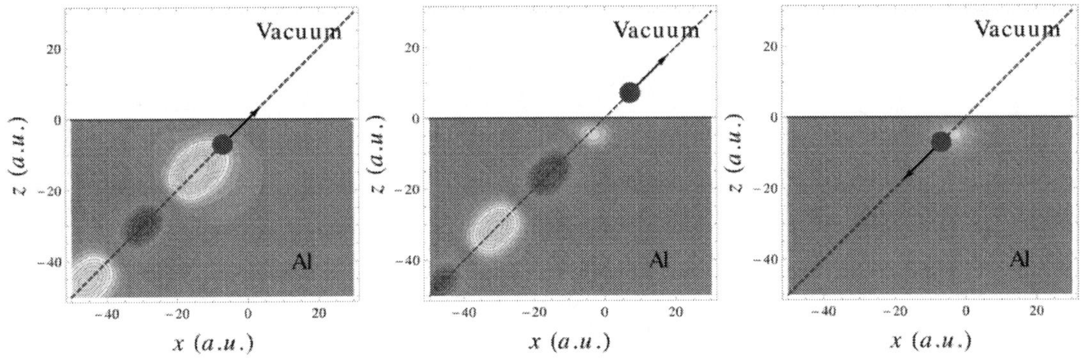

Figure 1. Bulk induced potential in aluminum due to an external ion, normalized to the ion atomic number Z, with $v = 4$ a.u. and incidence angle $\theta = 45$ deg. Left and central panel: outgoing trajectory, particle inside and outside the material, respectively. Right panel: incoming trajectory, particle inside the material (in this case, the bulk induced potential is zero before the particle enters the material).

These differences between the incoming and outgoing trajectories have their effect on the energy lost by the ion per time unit which is evaluated as

$$\frac{\mathrm{d}W}{\mathrm{d}t} = -Z \left. \frac{\partial \phi(\mathbf{r}, t)}{\partial t} \right|_{\mathbf{r}=\mathbf{v}t} ,$$

In figure 2 we show the corresponding bulk contribution

$$\left. \frac{\mathrm{d}W}{\mathrm{d}t} \right|_B = -Z\Theta(-v_\perp t) \left[\left. \frac{\partial \tilde{\phi}(\mathbf{r}, t)}{\partial t} \right|_{\mathbf{r}=(\mathbf{v}_\parallel + \mathbf{v}_\perp)t} - \left. \frac{\partial \tilde{\phi}(\mathbf{r}, t)}{\partial t} \right|_{\mathbf{r}=(\mathbf{v}_\parallel - \mathbf{v}_\perp)t} \right] ,$$

as a function of the distance to the surface $z = v_\perp t$ of the incoming ($v_\perp < 0$) or outgoing ($v_\perp > 0$) ion. Firstly we note that the direct contribution to the energy lost by an outgoing projectile is constant (and equal to that within an infinite medium) up to the crossing of the surface, and becomes zero beyond it. The reflected contribution is practically zero also inside the material, except in the vicinity of the surface. This *begrenzung* effect is due to the interaction of the projectile with the reflection of the front part of the wave accompanying it.

On the other hand, we see in figure 2 that the energy lost by an incoming projectile is very different to the outgoing case. We still see the same direct and reflected terms, but now they show oscillations that were not present for outgoing trajectories. These terms cancel partially, but a remanent oscillation persists that is still clearly visible in the total energy loss. This effect was already studied in a previous article [4] for normal incidence, but it was erroneously attributed to an interference between the direct and reflected contributions to the induced potential acting on the ion. Here we see that the oscillations are already present in both terms and occur as result of the transient construction of the direct wake upon penetration and the surfing of the reflected wake by the projectile. Furthermore, we see that they persist with oblique trajectories, even though their amplitude decrease with increasing angles.

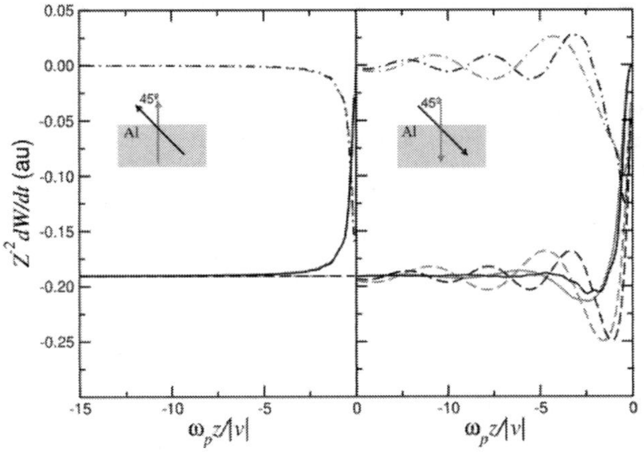

Figure 2. Energy lost per time unit for oblique incoming and outgoing trajectories of the charged particle (black lines), compared with the perpendicular case (red lines). Dashed: direct term; dash-dotted: reflected term; continous: total energy loss rate. Velocity and incidence angle are the same as in previous figures.

Acknowledgments
This work was supported by the Consejo Nacional de Investigaciones Científicas y Técnicas (CONICET) (Grant PIP11220120100374) and Universidad Nacional de Cuyo (Grants no 06 / C416 and 06 / C435).The authors are members of CONICET, Argentina. JLG and ROB are also members of the Comisión Nacional de Energía Atómica (CNEA), Argentina.

References

[1] Raether R 1980 Excitations of plasmons and interband transitions by electrons *Springer tracts in Modern Physics* **88** (Berlin: Springer-Verlag)
[2] Egerton R F 1996 *Electron Energy-Loss Spectroscopy in the Electron Microscope* (Berlin: Springer-Verlag)
[3] Denton C D, Gervasoni J L, Barrachina R O and Arista N R, *Phys. Rev. A* **57** 4498
[4] Gervasoni J L, Barrachina R O, Segui S and Werner W 2013 *Surf. Interface Anal.* **45** 1849
[5] Mahan G D 1978 *Electron and Ion Spectroscopy of Solids* eds. Fiermans, Vennik and Dekeyser (New York: Plenum Press) chapter 1 pp 1-53
[6] Lindhard J 1954 *Kgl. Danske Videnslab. Selskab, Mat-Fys. Medd.* **28** 1
[7] Barrachina R O and Gras-Martí A 1992 *Ultramicroscopy* **45** 345
[8] Ritchie R H and Marusak A L *Surf. Sci.* **4** 234
[9] García Moliner F and Flores F 1978 *Introduction to the Theory of Solid Surfaces* (Cambridge: Cambridge University Press)
[10] Flores F and García Moliner F 1979 *J. Phys. C: Solid State Phys.* **12** 907
[11] Gervasoni J L and Arista N R 1992 *Surf. Sci* **260** 329

Journal of Physics: Conference Series **583** (2015) 012029 doi:10.1088/1742-6596/583/1/012029

Fragmentation of multicharged C_{70}^{q+} prepared in collisions with F^+ ions at 3 keV

S Martin[1], R Brédy[1], C Ortéga[1], M Ji[1], L Chen[1], J Bernard[1], G Montagne[1], X Zhu [2], D Qian[2], and X Ma[2]

[1]Institut Lumière Matière, UMR5306 Université Lyon 1-CNRS, Université de Lyon 69622 Villeurbanne cedex, France

[2] Institute of Modern Physics, Chinese Academy of Sciences, Lanzhou, 730000, Gansu, China

E-mail : smartin@univ-lyon1.fr

Abstract. Ionization and successive evaporation of C_{70}^{r+} is studied as a function of the internal energy using collision induced dissociation under energy control. Multicharged ions C_{70}^{r+} are prepared in F^+ (3 keV) + $C_{70} \rightarrow F^- + C_{70}^{r+}$ + ne collisions. Up to seven successive evaporation of C_2 are observed in an excitation energy range from 40 to 100eV. The dissociation energies of C_{70-2m}^{2+} (m=1-7) are determined using a statistical cascade model to reproduce the excitation energy distribution of C_{70}^{2+} parent ions for each dissociation channel. Results are in good agreement with previous theoretical calculations.

1. Introduction

The fragmentation of mono and multicharged C_{60} has drawn great attention from broad scientific communities during the past decades. The specific scientific interest of the fullerene molecule, considered now as model system, is related to its exceptional symmetric and stable structure[1][2][3][4][5]. It offers the opportunity to easily prepare a high-density mass selected jet in the laboratory. Fragmentations of C_{60} or C_{70} depend sensitively on its charge and on its internal excitation energy[6][7][8]. In collisions between multi-charged ions and C_{70}, multi-charged C_{70} with relatively low internal energy can be prepared via electron capture process.

In this work, we present a study on the fragmentation patterns of C_{70}^{r+} with well-controlled internal excitation energy in $F^+ + C_{70} \rightarrow F^- + C_{70}^{r+}$ + ne collisions using the CIDEC method (collision induced dissociation under energy control) [7]. Very briefly, the mass spectrum of the recoil ions is recorded in coincidence with the kinetic energy loss of the scattered negative ions F^- leading to the measurement of the internal energy distribution of the charged C_{70} parent ions. Using the Arrhenius law we have estimated the dissociation rates for each step in the fragmentation cascade. The internal energy distributions have been reproduced with a cascade statistical model without any assumption on the excitation energy.

2. Experimental set-up

The experimental arrangement has been described in detail in ref. [9] [10]. A relatively pure C_{70} powder around 98% (C_{60} is around 2%) is evaporated from an oven heated at 650°C corresponding to the initial internal energy estimated theoretically to 6 eV. The C_{70} jet crosses the fluorine ion beam delivered by an ECR ion source at 3 keV energy. Downstream from the interaction region, the scattered projectile ions are analyzed by a cylindrical electrostatic analyzer (R=200 mm) and detected with a channeltron. The charged C_{70} or fragments are extracted from the collision region and then analyzed with a Time of Flight (TOF) spectrometer equipped with multichannel plates associated with a multi-anode detector. The electrons ejected from the projectile are accelerated to 20 keV and detected by a silicon surface barrier detector PIPS (Passive Implanted Planar Silicon) in coincidence with the other products in an event-by-event mode. Recoil ion spectra are recorded in coincidence with the detection of F^- outgoing projectile and the number of ejected electrons. The energy loss calibration is done using an Argon gas target. To record a spectrum, the scattered F^- ions are selected according to their kinetic energy loss by scanning the voltage of the electrostatic analyzer. For each voltage, the TOF of recoil ions and the ejected electron number are measured in multi-coincidence and in event-by-event data acquisition mode.

3. Results and discussions

A typical 2D spectrum named in the following EX-RI (Excitation-Recoil Ion) (figure 1(a)) is built by plotting the Recoil Ion TOF along the horizontal axis and the voltage of the projectile analyser along the vertical axis. The vertical scale was converted to the excitation energy of the target in the $F^+ + C_{70}$ -> $F^- + C_{70}^{2+*}$ reaction based on the CIDEC method[7]. Partial projections on horizontal axis for low and high excitation energies parts of the figure 1(a) are shown in figures 1(b) and 1(c) respectively. For the low energy part, C_{70}^{2+}, C_{60}^{2+} and C_{70}^{3+} peaks are observed. C_{60}^{2+} and C_{70}^{2+} peaks correspond to the collision for which two electrons capture process occurs leaving the molecule intact

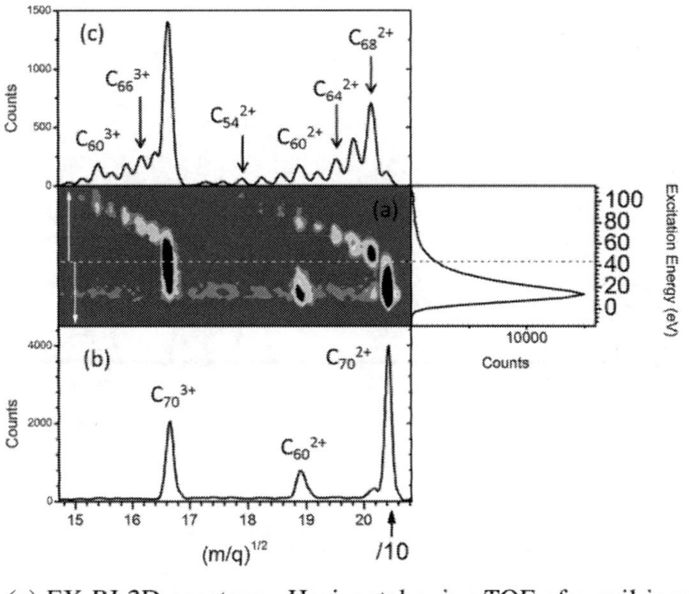

Figure 1. (a) EX-RI 2D spectrum. Horizontal axis : TOF of recoil ions, Vertical axis : kinetic energy loss ΔE converted to excitation energy; (b) partial projection on horizontal axis of figure1(a) for excitation energy of C_{70}^{2+} parent ion from 0 to 45eV; (c) : partial projection on horizontal axis of figure1(a) for excitation energy of C_{70}^{2+} parent ion from 45 to 110 eV.

with an excitation lower than 40 eV. C_{70}^{3+} peak corresponds to the double capture process followed by ionization. As seen on figure 1a, the ionization process starts at relatively low energy, lower than 20 eV. Comparable to the third ionization potential of C_{70} estimated to 13.6 eV, this value shows that the ionization process is probably a direct process analogue to what is observed in interaction with femtosecond laser[11]. The figure 1c shows the high-energy part from 40 to 100eV of figure 1(a). Peaks of C_{70-2m}^{2+} (m=1-9) and C_{70-2m}^{3+} (m=1-7) are attributed to the successive evaporation processes of C_2. Between 40 to 60 eV, the ionization process of C_{70}^{2+} competes with the C_2 evaporation channel. As expected, the dissociation of C_{70}^{2+} and C_{70}^{3+} presents a statistical behavior. Indeed, for the first C_2 emission the average excitation energy of the C_{70}^{2+} parent ion is about 53 eV, which is much larger than the dissociation energy around 9.5 eV. For the second and third evaporations, an extra amount of energy of about 8 and 16 eV is required, respectively. In the time range of the experiment (extraction time is about of 1.7 μs) up to nine C_2 evaporations have been observed leading to C_{52}^{2+} daughter ion. The corresponding internal energy of the C_{70}^{2+} parent ion is about 100 eV. From the projection onto the vertical axis of each spot of the EX-RI spectrum (figure 1a), we obtained the excitation energy distribution of the parent C_{70}^{2+} ions for the evaporative sequence. Each distribution is deconvoluted from the experimental resolution (5.5eV) and fitted with Gaussian distributions N(C_{70-2m}^{2+}, E) (figure 2).

It is noticeable that the distribution corresponding to the C_{60}^{2+} fragment corresponds to a higher stability of fullerene and that the distribution associated to C_{58}^{2+} and C_{56}^{2+} are broadened. The experimental evaporative sequence of C_{70}^{2+} ions leading to the production of C_{70-2m}^{2+} has been simulated using a cascade model (see the paper [6] for details) up to m=7. We have used the Arrhenius law with the heat capacity of C = 0.0138 eV/K. On figure 3, the modeled excitation energy distributions of C_{70-2m}^{2+} are compared to the experimental distributions N (C_{70-2m}^{2+}, E).

It is difficult to adjust at the same time the centers, intensities and widths. While centers of each distribution are well reproduced, slight disagreements are observed for both intensities and widths. In order to obtain better agreement, additional competitive fragmentation channels should probably be

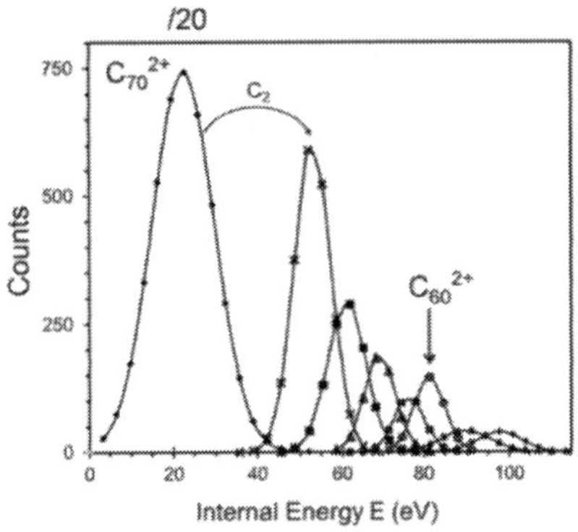

Figure 2. Internal energy distribution of C_{70}^{2+} parent ions for successive C_2 evaporations. The amplitude of the C_{70}^{2+} peak is divided by a factor 20.

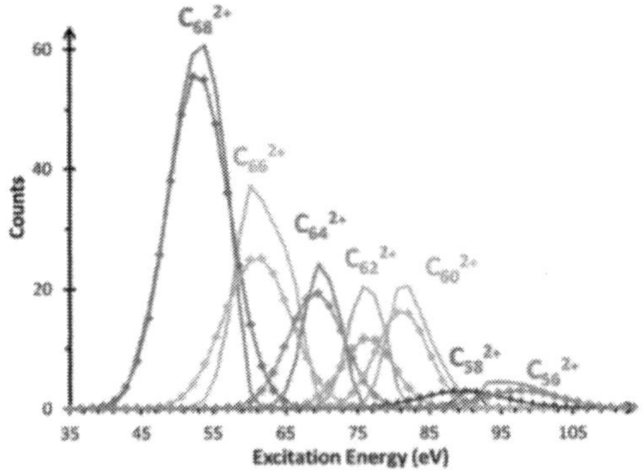

Figure 3. Comparison between experimental and adjusted distributions from C_{68}^{2+} to C_{56}^{2+}.

introduced in the model[12]. Indeed the first emitted fragment could be a C_4 or C_6 followed by successive emission C_2 especially for high excitation energy. Although these fragmentation channels are probably very weak, they could explain the broadening of the experimental distributions especially for C_{58}^{2+} and C_{56}^{2+}. The obtained dissociation energies of C_{70-2m}^{2+} (m=1-7) (table1), treated as free parameters in the model, are found in good agreement with recent calculations[13] . The small and high values for C_{62}^{2+} and C_{60}^{2+} respectively clearly indicate the well-known high stability structure of C_{60} molecule. Comparison with previous measurements on dissociation energies for C_{60-2m}^{2+} (m=1-5) [6] using a C_{60} powder shows also a good agreement.

Table 1. Adjusted pre-exponential factors and comparison between theoretical and experimental dissociation rates obtained with the cascade model from parent ions C_{70}^{2+}.

	A (s^{-1})	D_{exp}(eV)	D_{theo}(eV) [11]
C_{70}^{+2}	4 10^{22}	9.4	9.44
C_{68}^{+2}	1 10^{19}	8.5	8.6
C_{66}^{+2}	5 10^{19}	9.0	8.26
C_{64}^{+2}	1 10^{20}	8.5	8.53
C_{62}^{+2}	5 10^{17}	6.8	6.77
C_{60}^{+2}	1 10^{23}	10.1	10.18
C_{58}^{+2}	1 10^{23}	8.3	8.45
C_{56}^{+2}	1 10^{19}	8.2	8.54

4. Conclusion

As a conclusion, we have applied the CIDEC method to measure the internal excitation energy of multicharged C702+ prepared in the primary two-electron capture process,
F+ + C70→ F-+ C702+*. The internal energy distribution of C702+ for the evaporation channels, C702+ → C70-2m2+ + mC2 has been simulated using a cascade statistical evaporation model. The dissociation energies Dm2+ of doubly charged fullerenes have been derived and compared with success to theoretical values.

Acknowledgments
This work was supported by the French-China grant (CNRS PICS 5722).

References

[1] Campbell E E B, Raz T and Levine R D 1996 *Chem Phys Lett* **7** 253
[2] Cederquist H, Jensen J, Schmidt H T, Zettergren H, Tomita S, Huber B and Manil B 2003 *Phys Rev A* **67** 062719
[3] Rentenier A, Bordenave-Montesquieu A, Moretto-Capelle P and Bordenave-Montesquieu D 2004 *J Phys B Mol Opt Phys* **37** 2455
[4] Manil B, Maunoury L, Huber B A, Jensen J, Schmidt H T, Zettergren H, Cederquist H, Tomita S and Hvelplund P 2003 *Phys Rev Lett* **91** 215504
[5] Lebeault M-A, Baguenard B, Concina B, Calvo F, Climen B, Lépine F and Bordas C 2012 *J. Chem. Phys.* **137** 054312
[6] Chen L, Martin S, Bernard J and Brédy R 2007 *Phys Rev Lett* **98** 193401
[7] Martin S, Chen L, Salmoun A, Li B, Bernard J and Brédy R 2008 *Phys Rev A* **77** 043201
[8] Qian D B, Ma X, Chen Z, Li B, Zhu X L, Zhang S F, Martin S, Bredy R, Bernard J and Chen L 2013 *Phys. Rev. A* **87** 063201
[9] Chen L, Bernard J, Denis A, Martin S and Desesquelles J 1999 *Phys. Scr.* **T80A** 52
[10] Bredy R, Bernard J, Chen L, Montagne G, Li B and Martin S 2009 *J. Chem. Phys.* **130** 114305
[11] Campbell E E B, Hoffmann K, Rottke H and Hertel I V 2001 *J Chem Phys* **114** 1716
[12] Díaz-Tendero S, Alcamí M and Martín F 2005 *Phys. Rev. Lett.* **95** 013401
[13] Díaz-Tendero S, Sánchez G, Alcamí M and Martín F 2006 *Int. J. Mass Spectrom.* **252** 133

Journal of Physics: Conference Series **583** (2015) 012030 doi:10.1088/1742-6596/583/1/012030

Fluctuation of electronic density in the plasmon fields excited by a diatomic molecule

J L Gervasoni[1,2], **S Segui**[1]

[1] Centro Atómico Bariloche (Comisión Nacional de Energía Atómica), R8402AGP S. C. de Bariloche, Río Negro, Argentina

[2] Instituto Balseiro (Comisión Nacional de Energía Atómica y Universidad Nacional de Cuyo), R8402AGP S. C. de Bariloche, Río Negro, Argentina

E-mail: gervason@cab.cnea.gov.ar

Abstract. We study the fluctuations of the electronic density due to plasmon field associated to the wake potential generated by a dimer formed by two identical charged particles (such as two nuclei of the ionized hydrogen molecule) traveling through a semi-infinite dielectric medium. We use coherent states to describe plasmons as wave packets that raise fluctuations in the electronic density of the material. We analyze different configurations of interest, regarding the geometry of the surface and the orientation of the dimer.

1. Introduction

The purpose of this work is to study the effects on the plasmon field of a solid when the incident external charged particles have a simple structure, as a diatomic molecule, using coherent states [1, 3]. This formalism provides a very simple tool to analyze different quantities of interest by means of a hamiltonian formulation of the interaction of the plasmon field with the incident particles. We calculate the wake potential and the fluctuations in the electronic density $\Delta\rho$, and compare them with the results obtained for a single particle. In the present paper, and without loss of generality, rectilinear trajectory is employed. We analize the case of planar and cylindrical geometries of the solid. We also assume that the valence electrons of the bulk and the electrons at the surface interact independently with the external dimer, generating the *bulk* and *surface* plasmons, each with their own characteristic frequencies $\omega_{\mathbf{k}}^{b,s}$.

The structure of the paper is as follows: in section 2 we write the basic formulae used to describe the excitation of a plasmon field due to the passage of a swift dimer formed by two identical charged particles. In section 3 we show calculations of the wake potential and the charge density fluctuations, for bulk and surface excitations in aluminum. Finally, we present some concluding remarks in section 4.

2. Coherent states applied to molecules

In this work, we consider a dimer or molecule as formed by two identical point particles with charge Ze, separated by a given distance d and traveling along a rectilinear trajectory with constant velocity v. We neglect the interaction between the charges, while each charge interacts with the electron gas of the material.

We make use of the coherent states formulation to describe the interaction of the incident particles with the plasmon field. Within this frame, plasmon modes form a set of harmonic oscillators with characteristic frequencies ω_ν determined by the composition and geometry of the studied system [4, 5, 6] and described by an unperturbed hamiltonian H_0 in terms of creation and annihilation operators $a^\dagger_{\mathbf{k}}, a_{\mathbf{k}}$ and with $|n_{\mathbf{k}}\rangle$ spanning the possible states of the plasmon field with n plasmons in the mode $\mathbf{k}$. The interaction of the incident particles with the electron gas is given by

$$H_{int}(t) = \int d^3r \phi(\mathbf{r},t)\rho_{ext}(\mathbf{r},t) \tag{1}$$

where $\rho_{ext}(\mathbf{r},t)$ is the external charge density,

$$\rho_{ext}(\mathbf{r},t) = Ze(\delta(\mathbf{r}-\mathbf{r}_1(t)) + \delta(\mathbf{r}-\mathbf{r}_2(t))),$$

with $\mathbf{r}_i$ the position of the ith particle, and $\phi(\mathbf{r},t)$ is the potential associated to the plasmon field, in terms of $a^\dagger_{\mathbf{k}}$ and $a_{\mathbf{k}}$. The time evolution of the field interacting with the external particles can be obtained within the interaction representation using the Schrödinger equation

$$i\hbar\frac{\partial|\Psi(t)\rangle}{\partial t} = H_{int}|\Psi(t)\rangle \tag{2}$$

where

$$|\Psi(t)\rangle = e^{-i\sum_{\mathbf{k}}(I_{\mathbf{k}}(t)a_{\mathbf{k}}+I^*_{\mathbf{k}}(t)a^\dagger_{\mathbf{k}})}|\Psi(-\infty)\rangle \tag{3}$$

The coefficients $I_{\mathbf{k}}(t)$ are the evolution operators [2] depending on the trajectory of the incident particle, and $|\Psi(-\infty)\rangle$ is the initial state of the field, in terms of the eigenstates $|n_{\mathbf{k}}\rangle$; the states $|\Psi(t)\rangle$ constitute a set of *coherent states*, describing the plasmon field as a wave packet that propagates in a classical electromagnetic way. This formalism allows us to calculate the wake potential as:

$$\phi_W(\mathbf{r},t) = \langle\Psi(t)|\phi|\Psi(t)\rangle, \tag{4}$$

as well as the variation of electronic density $\Delta\rho$ as

$$\Delta\rho = -\frac{1}{4\pi}\langle\Psi(t)|\ \nabla^2\phi|\Psi(t)\rangle \tag{5}$$

and other quantities of interest such as the energy lost by the incident particles dW/dt.

3. Results

3.1. Bulk excitations

First, we consider bulk excitations generated by a dimer of length d traveling along a rectilinear trajectory with constant velocity $\mathbf{v} = v\hat{z}$, with its axis forming an arbitrary angle with the z-axis, which is perpendicular to the surface.

We use the plasmon pole approximation to describe the dielectric properties of the material, with a hydrodynamic dispersion relation, $\omega_{\mathbf{k}} = (\omega_p^2 + \beta k^2 + \hbar^2k^4/2m)^{1/2}$ (where ω_P is the plasma frequency, m is the electron mass and $\beta = v_F/\omega_p$; with v_F the Fermi velocity). The wake potential calculated from equation 4 yields

$$\begin{aligned}\phi^b_W(\mathbf{r},t) &= \frac{Ze\omega_P^2}{2\pi^2}\Re\left[\int\frac{d^3k}{k^2\omega_k}\cos(\mathbf{k}\cdot\mathbf{r}_0)\frac{e^{i\mathbf{k}\cdot(\mathbf{r}-\mathbf{v}t)}e^{\eta t}}{\mathbf{k}\cdot\mathbf{v}-\omega_k t+i\eta}\right]\\ &= 2\Re\left[\int d^3k\cos(\mathbf{k}\cdot\mathbf{r}_0)\widetilde{\phi}^b_w(\mathbf{r},t)\right]\end{aligned} \tag{6}$$

where $\mathbf{r}_0$ is the position of one of the particles constituting the dimer with respect to the center of mass (CM). ϕ_w^b corresponds to the wake potential generated by a single charge Ze following the trajectory of the center of mass, being η the damping parameter.

Figure 1 shows the wake potential generated by the dimer along the z axis for two separations d, being $d = 0$ the limit case of one single particle. The dimer moves with velocity v $= 2$ a.u. perpendicularly to the surface of an homogeneous aluminum medium ($\omega_p = 0.55$ a.u.). The surface is located in $z = 0$.

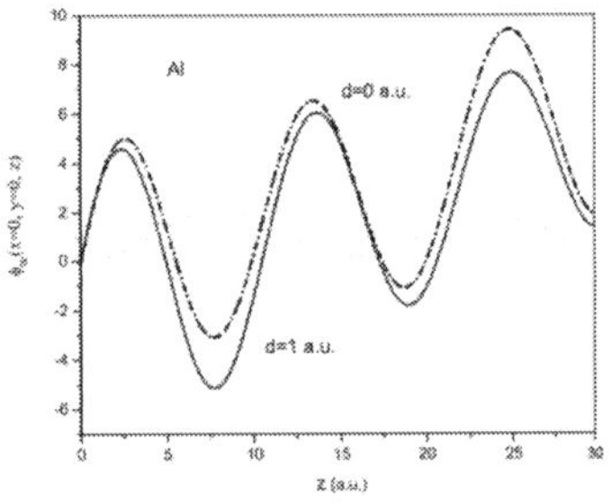

Figure 1. Wake potential for a dimer traveling with velocity $v = 2$ a.u. and different distance d between the particles.

It is important to point that, when the movement of the dimer is parallel to the surface, the interatomic distance for which differences with the case of a point particle start to be significative varies with the distance between the external charged particle and the surface.

3.2. Surface excitations

Now, we turn to the case of a cylindrical geometry, which is important for the study of nanowires and tubes. In previous works we analyzed in detail the interaction of single external charged particle with nanostructures [7, 8]. In this work, we consider only the case of a dimer travelling parallel to the surface of the cylinder, with a fixed distance z_0 from the CM. In this configuration, different orientations can be of interest; as an illustrative example, we calculate the wake potential due to the passage of the dimer with its two particles moving on a plane parallel to the axis of the wire. We solve the equations with a simple dependence for the dispersion relation ($\omega_{\mathbf{k}} = \omega_s = \omega_p/\sqrt{2}$), and obtain the following expression for ϕ_W^s:

$$\phi_W^s(\mathbf{r}, t) = \frac{Z}{4\pi} 2\omega_s e^{-\gamma t} \Re \left\{ i \int \frac{d^2 q}{q} e^{-q z_0} \, 2\cos(\mathbf{q} \cdot \mathbf{r}_0) \frac{e^{i\mathbf{q}\cdot(\mathbf{v}t - \mathbf{r})}}{\mathbf{q}\cdot\mathbf{v} - \omega_s + i\gamma} \right\} \qquad (7)$$

with γ a damping rate which accounts for the finite lifetime of the plasmons. Figure 2 shows the variation of the electronic density due to surface excitations for different interatomic distances, along a line following the trajectory of the dimer on the interface with velocity $v = 2$ au. The distance from the trajectory to the surface is $z_0 = 1$ u.a., with its axis parallel to $\mathbf{v}$.

The induced charge density is obtained from equation 5, which is easily evaluated when combined with the result of equation 7 for the wake potential. $\Delta\rho$ accompanies the wake potential on the surface and has approximately the same dependence with the distance to the surface and orientation of the incident molecule.

4. Conclusions

We have applied the coherent states formalism to describe the bulk and surface plasmon excitation for a diatomic molecule with two identical charges interacting with a semi-infinite solid in different configurations. Note that for a flat surface, the largest contribution to the energy loss

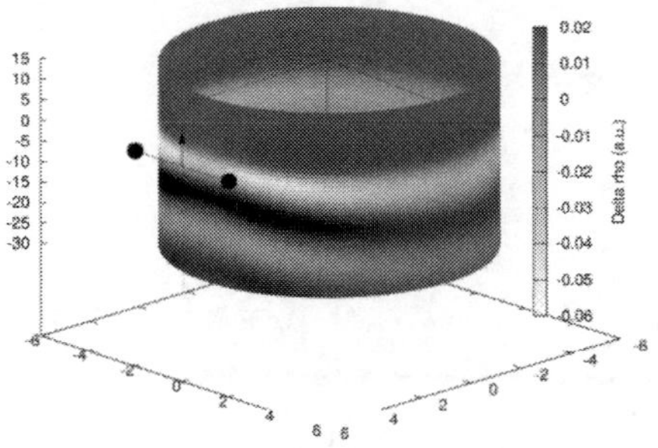

Figure 2. Variation of the electronic density induced by a dimer (2Ze) of velocity $v = 2$ au on the surface of a metallic wire ($w_p = 15.5$ eV) of 5 a.u. radius. The distance between the charges is $d = 4$ a.u. and the distance between the trajectory and the axis of the wire is 6 a.u.

is given by bulk plasmons, but for cylindrical geometry, the relationship is reversed. Because of this fact we have prioritized in this work to show bulk plasmon excitation for flat surface (considering neighborhood effects to surface in the results), and surface plasmons excitations for cylindrical surface. This example is of relevance due to that surface plasmons are the principal contributions to interaction of molecules with nanostructures.

Within a first order of approximation, the calculations are useful to determine critical values of the parameters for which it is important to take into account the structure of the incident particles.

We conclude that the coherent states formalism used to describe the interaction of the plasmon field with the incident charges has proved to be adequate for a straightforward evaluation of different quantities of interest, providing an insight into the contribution of the different excited modes. The present results and their potential applications motivate us to go on with the study of new configurations and geometries of the surface.

Acknowledgments

This work was supported by the Consejo Nacional de Investigaciones Científicas y Técnicas (CONICET) (Grant PIP 112 2012 0100374) and Universidad Nacional de Cuyo (Grants no 06 / C416 and 06 / C435).The authors are members of CONICET, Argentina. JLG is also member of the Comisión Nacional de Energía Atómica (CNEA), Argentina.

References

[1] Merzbacher E 1970 *Quantum Mechanics* (New York: Wiley)
[2] Landau L. D. and Lifshitz E. M. 1976 *Quantum Mechanics* (Pergamon Press, Oxford, 1976)
[3] Echenique P M, Ashley J C, Ritchie R H 1982 *Eur. J. Phys.* **3** 25
[4] Denton C D, Gervasoni J L, Barrachina R O and Arista N R, *Phys. Rev. A* **57** 4498
[5] Arista N R and Fuentes M A 2001 *Phys. Rev. B* **63** 165401
[6] Gervasoni J L and Arista N R 2003 *Phys. Rev. B* **68** 235302
[7] Segui S, Gervasoni J L and Arista N R 2007 *Surf. Sci.* **601** 4169
[8] Mowbray D J, Segui S, Gervasoni J L, Mišković Z L and Arista N R 2010 *Phys. Rev. B* **82** 035405

Journal of Physics: Conference Series **583** (2015) 012031 doi:10.1088/1742-6596/583/1/012031

Electron emission and exciton production of LiF surfaces by grazing impact of hydrogen neutral atoms

J E Miraglia

Instituto de Astronomía y Física del Espacio. Consejo Nacional de Investigaciones Científicas y Técnicas. Departamento de Física. FCEN. Universidad de Buenos Aires. Casilla de Correo 67, Sucursal 28, (C1428EGA) Buenos Aires, Argentina.

E-mail: `miraglia@iafe.uba.ar`

Abstract. We put forward an alternative mechanism to understand the physics underlying the electron emission and exciton formation occurring during grazing collision of neutral hydrogen on LiF(001) surfaces. Instead of visualizing the negative ion formation as the precursor, we propose the formation of excited H(n=2) as a catalyzer to produce excitons and electrons.

1. Preliminary

In a series of articles Roncin, Winter and co-workers [1, 2, 3, 4, 5] introduced a useful model to explain exciton formation and electron emission mechanisms induced by low-energy neutral hydrogen impact. That model considers two steps: first, the formation of a negative ion H^-, and in the second step, the ion H^- gives rise to either an excitation or emission of electrons, namely

$$H(1s) + F^-_@ \rightarrow H^- + F^0_@, \tag{1}$$

$$H^- + F^-_@ \rightarrow \begin{cases} H^0 + F^-_@ + e, \\ H^- + F^{-*}_@, \end{cases} \tag{2}$$

were $F^-_@$ and $Li^+_@$ denote the ions embedded in the Madelung potential [6, 7]. In Refs.[4, 5] the experiments were fitted with a binomial distribution according to the three processes given by Eqs.(1) and (2) which produce a good agreement with the experiments from 600 eV up to 10 keV impact energies.

In this article we investigate another mechanism which can also explain the exciton and electron production, at higher energies, say some keVs. Our proposal is the following: first the projectile electron is excited to H(2s,2p) due to collisions with the active sites of the insulator

$$H(1s) + \begin{pmatrix} F^-_@ \\ Li^+_@ \end{pmatrix} \rightarrow H(2s, 2p) + \begin{pmatrix} F^-_@ \\ Li^+_@ \end{pmatrix}, \tag{3}$$

And second, the resulting excited projectile can also produce a quasi resonance transfer excitation and ionization with the subsequent sites

$$H(2s, 2p) + F^-_@(2p) \rightarrow H(2s, 2p) + \begin{cases} F^0_@ + e, \\ F^-_@(n = 3), \end{cases} , \tag{4}$$

Journal of Physics: Conference Series **583** (2015) 012031 doi:10.1088/1742-6596/583/1/012031

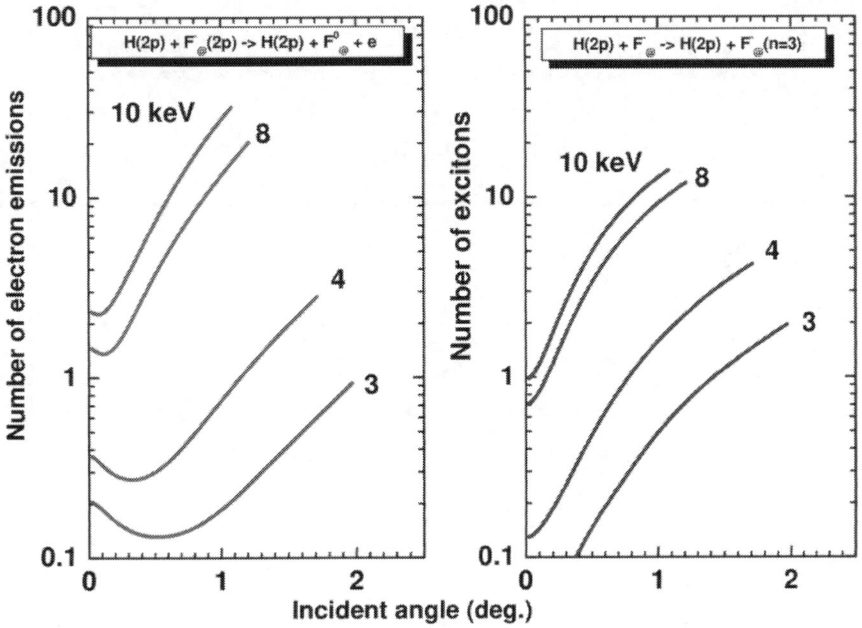

Figure 1. (Color online) Calculated number of electrons and excitons produced by the impact H(2p) on $F^-_@$ as a function of the incident angle θ_i for different impact energies as indicated. To remain in the perturbative regime, the curves were cut off where the corresponding interatomic distances were smaller than 2.5 atomic units.

more efficiently than H(1s) as displayed in Fig. 1. More electron emissions are left to an ionization process of H(n=2) due to other sites

$$\mathrm{H}(n=2) + \begin{pmatrix} \mathrm{F}^-_@ \\ \mathrm{Li}^+_@ \end{pmatrix} \rightarrow \mathrm{H}^+ + e + \begin{pmatrix} \mathrm{F}^-_@ \\ \mathrm{Li}^+_@ \end{pmatrix}. \tag{5}$$

It follows an electron capture to produce H(1s) again and repeat the process several times.

2. Calculations

Based on an onion model developed by our group [6, 7], we have calculated the first perturbative order for the probability of projectile excitation given by reaction (3). The use of the first Born approximation is justified as the projectile rebounds at around 3 atomic units from the surface, warranting the validity of the first perturbative order. Transition probabilities per unit path length as a function of the distance to the surface $W(Z)$ were calculated employing the eikonal approximation. Considering a classical projectile trajectory under planar conditions we can calculate the number of inelastic transitions along the whole projectile path as

$$N = 2 \int_{Z_{\min}}^{\infty} dZ \, \frac{\delta_s v_i \, W(Z)}{\sqrt{(v_i\theta_i)^2 - \frac{2}{M_P} V_i(Z)}}, \tag{6}$$

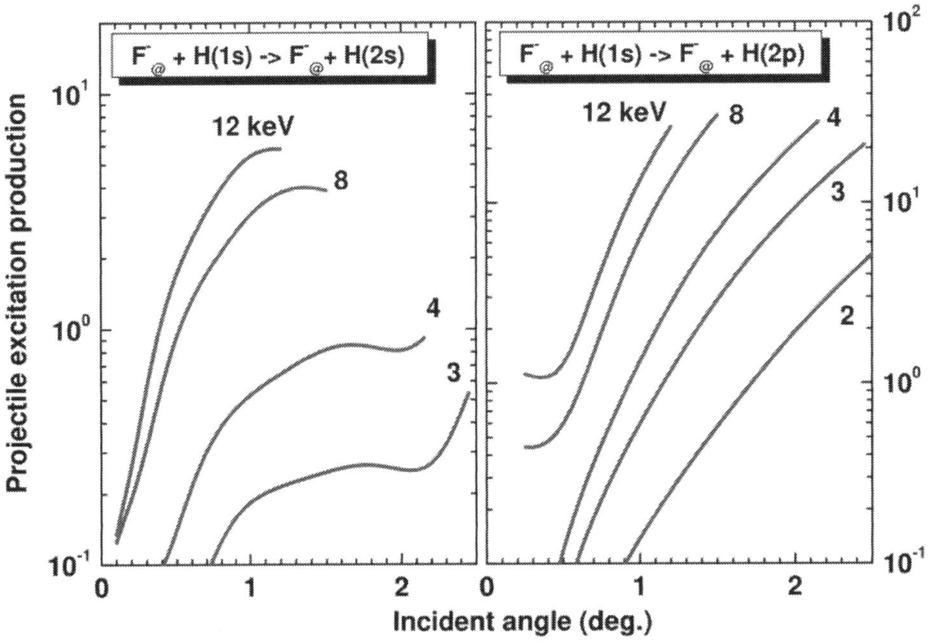

Figure 2. (Color online) Calculated number of projectile excitations produced by the impact of F@ on H(1s), as a function of the incident angle θ_i for different impact energies as indicated.

where v_i (M_P) is the velocity (mass) of the projectile, θ_i is the incident angle with respect to the surface and δ_s is the superficial density of targets. In Eq.(6) $Z_{\min}$ is the minimal distance to the surface and $V_i(Z)$ is the surface-atom potential under planar condition: this is at random azimuthal orientation. Calculated number of N_{2s} and N_{2p} projectile excitations by impact of $F^-_@$ and $Li^+_@$ are shown in the Figs. 2 and 3.

Let us now compare with the experiments by considering the impact of 3 keV H(1s) at 1.2 degree of incidence. From the figures 2 and 3, we estimate the total number of H($n = 2$) to be 3.5 which is not very far from the 4 electrons obtained in the experiments (see Fig.6 of Ref [4]). For higher energies our model improves. For 4 keV H(1s) impact we obtain up to 9 possible electrons while the experiments accounts around 6 (see Fig.8 of Ref [4]). The rest of electron excitations may go to exciton production via reaction (4)

Finally we should mention that to determine the electron emission statistics, Eq.(6) forces to use the Poisson distribution [8] instead of the binomial one used in Ref. [5]. Then the probability of producing n transitions is given by

$$p_n = \frac{N^n}{n!} e^{-N}, \tag{7}$$

which permits us a direct comparison with the statistical data.

In summary, instead of visualizing the negative ion formation as the precursor, we have proposed the formation of excited H(n=2) as a catalyzer to produce excitons and electrons.

Journal of Physics: Conference Series **583** (2015) 012031 doi:10.1088/1742-6596/583/1/012031

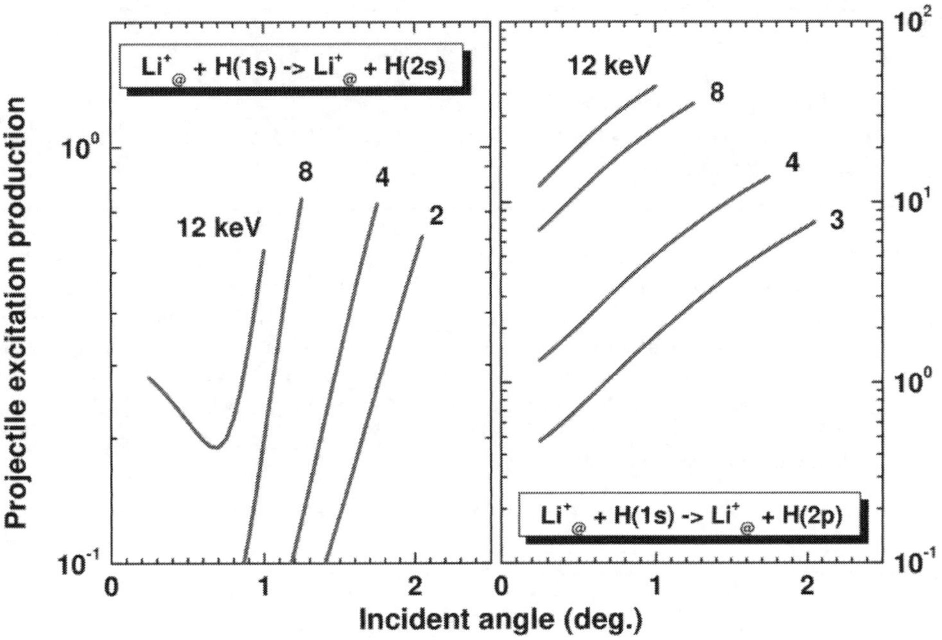

Figure 3. (Color online) Calculated number of projectile excitations produced by the impact of Li@ on H($1s$), as a function of the incident angle θ_i for different impact energies as indicated.

Acknowledgments
I like to thank Raul Baragiola for encouraging this work. This work was partially supported by the following Argentinean institutions: Consejo Nacional de Investigaciones Científicas y Técnicas, Agencia Nacional de Promoción Científica y Tecnológica, and Universidad de Buenos Aires.

3. References

[1] C Auth, A G Borisov, and H Winter, *Phys. Rev. Lett.* **75,** 2292 (1995).
[2] H Winter, C Auth, and A G Borisov, *Nucl. Inst. and Methods* B **115,** 45 (2003).
[3] P Roncin, J Villete, J P Atanas, and H Khemliche, *Phys. Rev. Lett.* **83,** 133 (1996).
[4] H Winter, S Lederer, K Maass, A Mertens, F Aumayr, and HP Winter, *J. Phys. B: At. Mol. Opt. Phys.* **35,** 3315 (2002).
[5] H Winter, A Mertens, S Ledered, C Auth, F Aumayr, and HP Winter, *Nucl. Inst. and Methods* B **212,** 45 (2003).
[6] J E Miraglia and M S Gravielle, *Phys. Rev.* A **78**, 052705 (2008).
[7] J E Miraglia and M S Gravielle, to be published.
[8] S Lencinas, J Burgdörfer, J Kemmler, O Heil, K Kroneberger, N Keller, H Rothard, and K O Groeneveld, *Phys. Rev.* A **41**, 1435 (1990).

Imaging the photons during transmission of ions through glass capillaries

N Akram and R Schuch[1]
Physics Department, Stockholm University, S-106 91 Stockholm, Sweden
E-mail: schuch@fysik.su.se

Abstract. Charged particles, in particular slow highly charged ions, can be guided through nanocapillaries in various insulating materials by self-organized charge patches. We present a technique for observing directly the formation of these patches in glass capillaries. It is based on imaging the emitted visible photons. We report here on tests with 4.5 keV Ar^{+}-ions transmitted through straight and tapered borosilicate capillaries, using a highly sensitive digital camera. Simultaneously, the ions transmitted through the capillaries were detected. The number of emitted photons decreases by increasing the tilt angle of the capillary. The time evolution of emitted photons revealed a change of location of the charge patches during charging-up the capillary walls.

1. Introduction

Highly charged ions (HCI) can be transmitted through nanocapillaries in insulating materials even when the injection angles are larger than the geometrical opening angle of the capillaries without changing charge and energy. A single glass capillary also exhibits a similar feature of transmitting incident HCI (for larger injection angles). Such a phenomenon has been termed as "guiding" and is credited to the self-organized formation of charge patches by the incident ions on the capillary inner walls which guide the subsequent ions along the capillary axis.

Since the first observation of guiding phenomenon in polyethylene teraphthalate (PET) nanocapillaries [1], several other groups have extensively studied the guiding of HCI in different insulating materials like Al_2O_3, SiO_2, polycarbonate and mica [2-7]. Also for a single glass capillary the guiding and transmission process was characterized[8, 9]. In particular, as single tapered glass capillary have a potential application to selectively damage a single biological cell [10] and for controlled surface modifications [11].

Guiding involves charging-up of the inner capillary walls which is a time dependent process and the steady state of transmission is achieved when equilibrium between deposition of charges and drainage of charges is attained. The time evolution of the transmitted ion intensity and direction gives an indirect insight to the charge patches formed on the capillary walls [5,6,12,13].

In this paper, we report first results from directly visualizing the formation of charge patches inside a glass capillary. This was done by imaging the emitted visible photons when 4.5 keV Ar^{+}-ions interact with the inner walls of a glass capillary, by means of a highly sensitive digital camera. When forming charge patches in the capillary to guide subsequent ions, the initially incident ions hit its wall and by that the incident ions and atoms in the wall are excited which results in emission of photons. By imaging these emitted photons we can infer information about the location of charge patches. In the first part, we present the images of emitted photons for different orientations of the capillary with respect to the incident beam direction. Later, the temporal evolution of the emitted visible photons is presented as images for 0° tilt angle with respect to the incident beam direction and capillary axis.

[1] To whom any correspondence should be addressed.

2. Experimental setup

The experiments were performed at S-EBIT laboratory of Stockholm University, Stockholm. Intense beams of Ar^{+}-ions were extracted at the energy of 4.5 keV from a Penning Ion Gauge (PIG) source and collimated on the way to the experimental chamber. The overall pressure was maintained at ~ 10^{-7} mbar. The glass capillary samples were mounted on a goniometer with three spatial and two-angular adjustments, to align it with the incident beam. To avoid charging-up effects at the entrance of the capillary, a grounded Al aperture of 0.4 mm diameter was placed there. A schematic of the experimental setup is shown in Fig.1. The beam current incident on the entrance aperture was around 50 nA. The transmitted ions were detected using a Faraday cup or a channeltron detector which was placed at 50 cm from the capillary exit.

We used two borosilicate glass capillary samples of length 54 mm with an outer diameter of 1.0 mm and inner diameter of 0.5 mm. But one capillary was tapered to have an exit size of 50 μm. The straight one had a geometrical opening angle of 0.53° or an aspect ratio of 110. The tapered one had a geometrical opening angle of 0.06°. It was prepared by heating and pulling a straight glass tube using a Sutter laser puller. Both glass capillary samples have same composition of SiO_2 72 %, B_2O_3 12%, Al_2O_3 7%, Na_2O 6%, K_2O 3%, CaO 0.5% and MgO 0.2%.

A Nikon D3s camera at an ISO setting of 12.8k was used to image the photons. It should have a quantum efficiency of ~ 53 % [14]. In order to avoid background photons, it was completely shielded from room light, as well as from photons from ions hitting apertures. In the selection of a lens, focal length and aperture, an optimum in transmitted light intensity and image size for the given geometry was found essentially from: relative transmitted light intensity $I = (a-f)^2/(aF)^2$. That optimum was for a =220 mm (distance to object) at a focal length of f=55mm and aperture number F=1.2. That gives an image ratio (magnification) of 0.4. The CCD chip has a size of 24 x 35 mm^2 with 12 x 10^6 pixels, thus, each pixel images an area of 21 x 21 μm^2 of the capillary. The total relative solid angle is (assuming the photon-emission is isotropic) 2 10^{-3}, and with the quantum efficiency of the D3s, the total detection efficiency per pixel is roughly 1×10^{-10}.

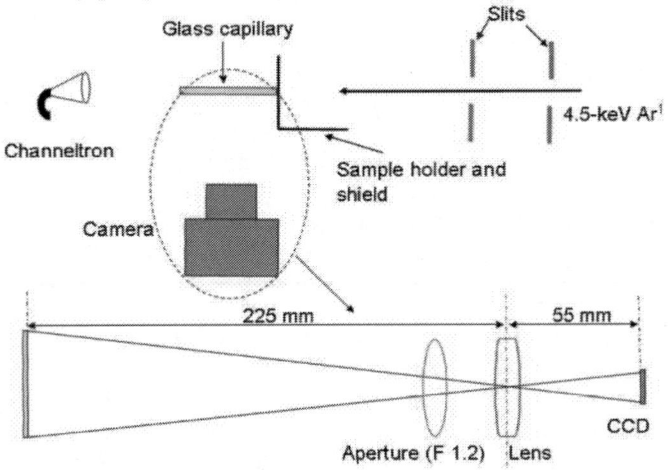

Figure 1. Schematic of the experimental setup. Tilt angle is the angle between capillary axis and the incident beam direction.

3. Results and discussions

3.1. Photon imaging at different tilt angles

First, we investigated the emitted visible photons with the tapered glass capillary for different tilt angles (see Fig. 1) in steps of 0.6°. Fig. 2 shows the images taken for an exposure time of 30 sec for each angle. As seen, with increasing tilt angle the radiant portion of the capillary is shortened towards

Journal of Physics: Conference Series **583** (2015) 012032 doi:10.1088/1742-6596/583/1/012032

the entrance side of the capillary. Also the photons from ions hitting the tapered exit of the capillary disappears already at 0.6°. This corresponds to the exposed length of the capillary to the incident beam (the geometrical acceptance angle is 0.53°).

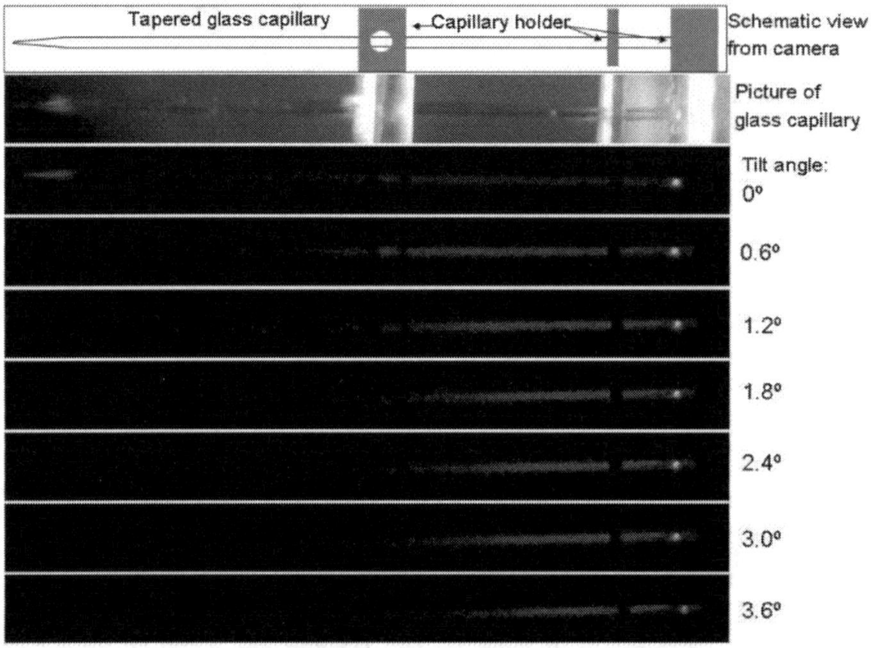

Figure 2. Images for an exposure time of 30 sec for different tilt angles with a tapered borosilicate glass capillary. First and second images show the mounted capillary (in colour on-line).

For quantitative analysis: we use that all ions, entering the capillary (3.2×10^9 ions/sec), interact with the capillary wall, because transmission through the capillary is less than 2 % in steady state. If we assume that every ion hitting the capillary wall contributes with the emission of a photon, then a rate of 3×10^6 photons/sec should be detected. If we assume additionally that the ions are lost uniformly along the inner surface of the capillary, in that case 5×10^4 photons/sec should be projected on a single pixel on the CCD. Counting the illuminated pixels in the images, we find a much lower rate. That could mean a much lower photon emission efficiency per ion and also that much more then one or two photons hit a pixel.

We also find that the normalized number of illuminated pixels decreases with increasing tilt angle of the capillary (nearly exactly inverse proportional). However, with increasing the tilt angle, the density of ions increases that hits the inner wall of the capillary which should result in an increase of the number of emitted photons from this area. Hence, the number of detected photons per pixel should increase there which can not be registered with such a camera and thus for longer exposure times the photon number hitting a pixel can be much larger then unity.

3.2. Time evolution of charge patches from imaged photons

The photon imaging method should allow "to see" the temporary behaviour of charging-up and formation of charge patches on the capillary walls. For this purpose we started with an initially uncharged glass capillary. The capillary was oriented at a tilt angle, 0°, and then left to discharge for 20h. The time evolution of emitted photon patterns for tapered glass capillary is shown in Fig. 4. It is clearly seen that the emitted photons come from different locations on the capillary with time when the wall is charged up.

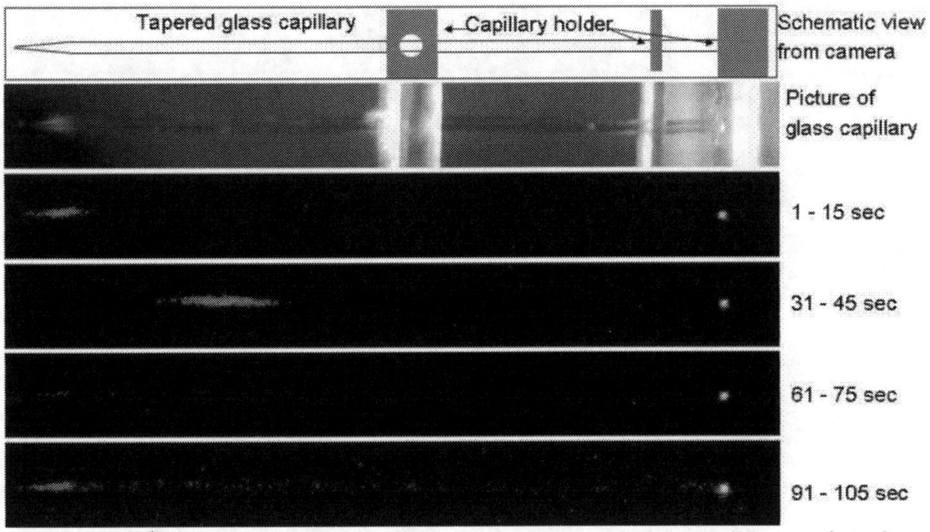

Figure 3. Images for 0° tilt angle, with initially uncharged tapered glass capillary as function of time (in colour on-line). The recording time intervals are stated to the right of each image.

We also performed measurements of the time evolution of emitted photons for straight borosilicate glass capillary. Fig. 4 shows the time evolution of emitted photons from the inner wall of the straight glass capillary when 4.5 keV Ar^{1+}-ions are transmitted through it for a tilt angle of 0°. As seen, the location of emitted photons varies slightly during the course of time.

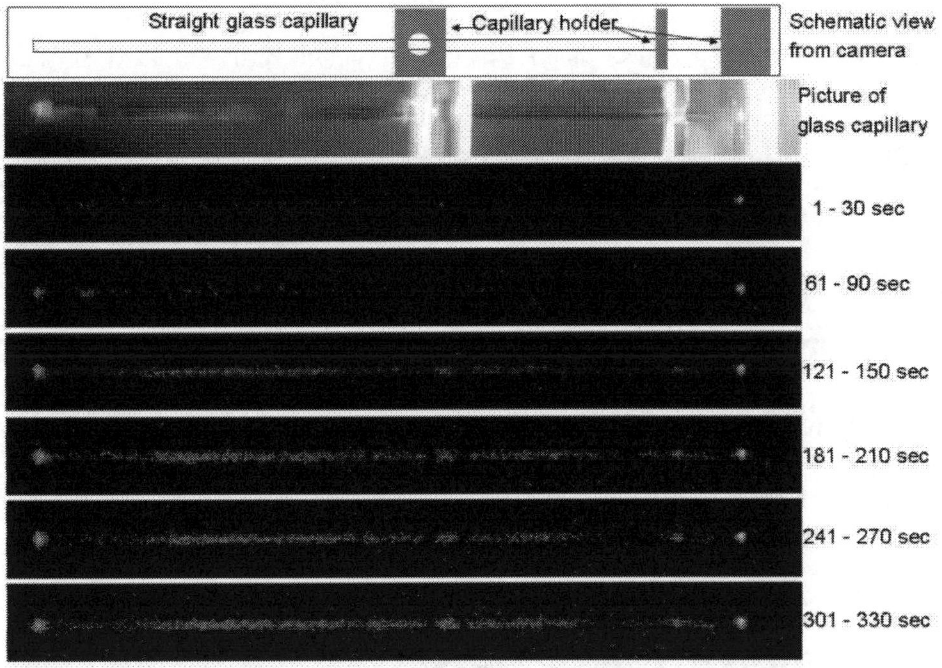

Figure 4. Images for 0° tilt angle and initially uncharged straight glass capillary as function of time.

It is seen in Fig. 4, the number of emitted photons increases with time, as also shown in Fig. 5a, whereas the transmitted current decreases with the deposition of charge inside the capillary as shown in Fig. 5b. We observe a short burst in transmitted current in the start and a decrease to a constant value relative to the incident current. A similar behaviour was also seen by other groups at 0° tilt angle[15]. As the increase of emitted photons is in accord with the decrease of transmitted current, we conclude that the ions are blocked inside the capillary and interact with the wall leading to higher photon emission.

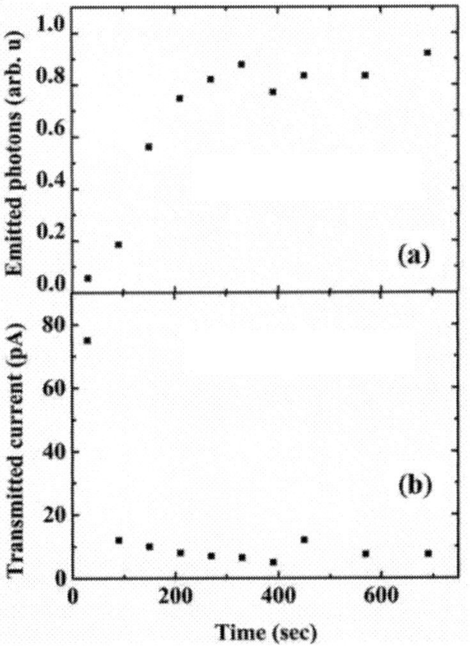

Figure 6. (a) Normalized counts of emitted photons as a function of time and **(b)** The average transmitted current as a function of time for initially uncharged straight glass capillary.

4. Summary

We report imaging of visible photons which are emitted during interaction of 4.5 keV Ar^{1+}-ions with inner walls of borosilicate glass capillary. This new method provides a direct visualization of charge patches formed inside the glass capillary during ion-transmission. The time evolution of the imaged visible photons reveals that the charge patches change their location in the process of reaching the steady state of transmission. It is also found that the intensity of emitted photons is reasonably consistent with that of the transmitted ions.

We acknowledge support by the K&A Wallenberg foundation and Swedish Research Council (VR).

References

[1] Stolterfoht N et al. 2002 *Phys. Rev. Lett.* **88**, 133201
[2] Sahana M B, Skog P, Víkor G, Rajendra Kumar R and Schuch R 2006 *Phys. Rev.* A **73**, 040901
[3] Mátéfi-Tempfli S et al. 2006 *Nanotechnology* **17**, 3915
[4] Skog P, Zhang HQ, and Schuch R 2008 *Phys. Rev. Lett.* **101**, 223202
[5] Kanai Y et al. 2009 *Phys. Rev.* A **79**, 012711
[6] Zhang HQ, Akram N, Skog P, Soroka I L and Schuch R 2012 *Phys. Rev. Lett.* **108**, 193202
[7] Lemell C, Burgdörfer J, Aumayr F 2013 *Progr. in Surface Sci.* **88,** 237
[8] Ikeda T et al. 2006 *Appl. Phys. Lett.* 89 (16) 163502
[9] Bereczky R J, Kowarik G, Aumayr F, Tőkési K 2009 *Nuclear Instr. Meth.* B **267** 317
[10] Iwai Y et al. 2008 *Appl. Phys. Lett.*, **92**(2):023509
[11] Lemell C et al. 2007 *Solid State Electr,* **51**, 1398
[12] Stolterfoht N et al. 2009 *Phys. Rev.* A **79**, 022901
[13] Cassimi A et al. 2009 *Nuclear Instr. Meth.* B **267**, 674
[14] http://forums.dpreview.com/forums/read.asp?forum=1019&message=34581894
[15] Kreller M, Zschornack G, Kentsch U 2011 *Nuclear Instr. Meth.* **269**, 1032

Journal of Physics: Conference Series **583** (2015) 012033 doi:10.1088/1742-6596/583/1/012033

Photoionization of diatomic molecules: effect of coulombic interactions

D I R Boll[1], **O A Fojón**[1,2]

[1]Instituto de Física Rosario, CONICET-UNR, Pellegrini 250, 2000 Rosario, Argentina
[2]Escuela de Ciencias Exactas y Naturales, Universidad Nacional de Rosario, Argentina.

E-mail: fojon@ifir-conicet.gov.ar

Abstract. We analyze the angular distribution of electrons in the laser-assisted photoionization of HeH^{2+} molecular targets. By means of a simple model, we obtain the observables for the reaction and we compare them with previous results obtained with more elaborated ones. The two-center interference effect and the asymmetric emission in the monochromatic ionization persist in the presence of a near infrared laser bath leading to a characteristic spectrum.

1. Introduction

The development of extreme ultraviolet (XUV) pulses with durations of a few hundreds of attoseconds (ATP) has opened the way to study the photoionization dynamics with a time resolution close to the electronic time scale. In the pump-probe experiments ([1] and references therein) where the reaction is assisted with a near infrared laser (NIR), it is possible to extract information about ATP [2] as well as monitoring and/or steering electron dynamics in molecules [3, 4]. In this perspective, both single ATP (SATP) and attosecond pulse trains (ATPT) turn out to be promising tools to control the electronic properties in molecular systems [4, 5].

The numerical resolution of the time dependent Schrödinger equation (TDSE) for pump-probe reactions with molecular targets represent a computational challenge [6, 7]. In this context, the use of a simplified model leading to reasonable results could be very helpful to understand the underlying physical processes. Among them, we could mention the Strong Field Approximation (SFA) in which the electron is supposed to evolve without interacting with the Coulomb field created by the target [8]. An improvement on this approximation is given by the Separable Coulomb-Volkov (SCV) model to treat atomic or molecular targets [9, 10], in which the coulombic interaction between the photoelectron and the residual target is included.

In this work, we show that the SCV model gives a qualitative agreement with ab initio calculations of photoionization cross sections [7] of HeH^{2+} molecules and we analyze the case where the reaction is assisted by a NIR. Employing approximate initial and final wavefunctions, analytical expressions for the observables of interest, namely, photoionization spectrum and photoelectron angular distributions (PADS), are obtained.

Atomic units are used otherwise explicitly stated.

2. Theory

In the following, we summarize the basic ingredients of the SCV model [10]. The coordinate system describing the fixed-in-space molecule is one where $\mathbf{R}$ is the internuclear vector pointing

Journal of Physics: Conference Series **583** (2015) 012033 doi:10.1088/1742-6596/583/1/012033

from nucleus 1 to 2 and $\mathbf{r}_i$ denotes the electron position vector with respect to the i-th nuclei. The electron coordinate with respect to the molecular geometric center is given by $\mathbf{r} = (\mathbf{r}_1 + \mathbf{r}_2)/2$.

The laser assisted photoionization amplitude within the dipole approximation in the velocity gauge reads [8],

$$M_{scv}(\mathbf{p}) = -i \int_{-\infty}^{\infty} dt \langle \Psi_f(\mathbf{r}, t) | \hat{\mathbf{p}} \cdot \mathbf{A}(t) | \Psi_i(\mathbf{r}, t) \rangle, \tag{1}$$

where $\hat{\mathbf{p}}$ is the momentum operator, and $\Psi_{i,f}(\mathbf{r}, t)$ are the initial and final wavefunctions, respectively. The vector potential $\mathbf{A}(t)$ representing the XUV pulse is given by,

$$\mathbf{A}(t) = (\mathbf{e}_z \cos\phi + i\mathbf{e}_z \sin\phi) A_0 e^{-i\Omega t} \exp\left[-\frac{(t - t_0)^2}{2\tau^2}\right], \tag{2}$$

where Ω, ϕ, A_0, t_0 and τ are the central frequency, a phase defining the polarization, the amplitude, the delay and the duration of the attosecond pulse, respectively. For a large Keldysh parameter ($\gamma \sim 9$ in our case), the multiphoton ionization of the target by the NIR is negligible as compared to the one-photon XUV case. Also, for the considered NIR intensities, the bound state polarization is expected to be small [11]. Therefore, we consider approximate *laser-free* wavefunctions given by,

$$\Psi_i(\mathbf{r}, t) = \psi_i^0(\mathbf{r}) \exp(iI_p t), \tag{3}$$

where I_p is the ionization potential of the molecule. We describe the continuum states of the electron by the Coulomb-Volkov *ansatz* [9] in which the interaction of the ejected electron with both the residual ionic target and the laser bath is taken into account, *i.e.*,

$$\Psi_f(\mathbf{r}, t) = \psi_f^0(\mathbf{r}) \exp\left\{-\frac{i}{2}\int^t [\mathbf{p} + \mathbf{A}_L(t)]^2 \, dt\right\}, \tag{4}$$

with $\mathbf{p}$ the asymptotic value of the photoelectron momentum and $\mathbf{A}_L(t)$ the potential vector describing the NIR. The wavefunctions $\psi_{i,f}^0(\mathbf{r})$ describe the bound and continuum molecular states in the absence of both the XUV and the NIR fields. As exact wavefunctions are in general available only in numerical form, we approximate them by analytical ones that simplify the calculation. For the bound states, we employ two-center expansions with Slater type orbitals (STOs) located on each molecular center whereas for the continuum states we used coulombic wavefunctions satisfying the correct asymptotic coulombic conditions,

$$\psi_i^0(\mathbf{r}) = \textstyle\sum_i c_i^{(1)} \phi_i(\mathbf{r}_1) + \sum_j c_j^{(2)} \phi_j(\mathbf{r}_2) \tag{5}$$

$$\psi_f^0(\mathbf{r}) = (2\pi)^{-3/2} e^{i\mathbf{p}\cdot\mathbf{r}} N_p^2 \, G(\mathbf{r}_1) G(\mathbf{r}_2) \tag{6}$$

where $\phi_i(\mathbf{r})$ are STOs, $N_p = \Gamma(1 + i\nu) \exp(\pi\nu/2)$ and $G(\mathbf{r}_i) = {}_1F_1(-i\nu_i; 1; -1(pr_i + \mathbf{p} \cdot \mathbf{r}_i))$ is the confluent hypergeometric function. The Sommerfeld parameter is $\nu_i = Z_i/p$ with Z_i being the residual effective charge of the i-esime atom.

The vector potential corresponding to the NIR field may be written as

$$\mathbf{A}_L(t) = -\frac{\mathbf{E}_1}{\omega_1} \sin(\omega_1 t), \tag{7}$$

where $\mathbf{E}_1$ and ω_1 are the NIR electric field and frequency, respectively.

Replacing Eqs. (2) and (7) into Eq. (1), and recalling that in the velocity gauge spatial and temporal integrals are separable [10], it can be shown that the matrix transition amplitude given by Eq. (1) is proportional to the monochromatic transition matrix elements $M_{ph}(\mathbf{p})$ corresponding to the absorption of one XUV photon of frequency Ω, which are given by

$$M_{ph}(\mathbf{p}) = \chi_1 M_1 + \chi_2 M_2, \tag{8}$$

where we have employed the Coulomb Continuum approximation [9, 10] in which the molecular interference factors denoted by $\chi_{1,2}$ are given by,

$$\chi_{1,2} = N_p^* e^{\pm i\mathbf{p}\cdot\mathbf{R}/2} \, {}_1F_1(i\nu_{2,1}; 1; i(pR \mp \mathbf{p}\cdot\mathbf{R})), \tag{9}$$

and the $M_{1,2}$ factors are the atomic photoionization amplitudes from the molecular centers 1, 2, respectively.

Therefore, the photoelectron spectrum is given by,

$$\frac{d^3 P_{scv}}{dp \, d\Omega_e} = p^2 |M_{scv}(\mathbf{p})|^2. \tag{10}$$

where $d\Omega_e = \sin\theta_e \, d\theta_e \, d\phi_e$ denotes the solid angle in the direction of $\mathbf{p}$.

3. Results

We compare our SCV results with more elaborated ones [7], where the TDSE of the HeH^{2+} ion exposed to an attosecond x-ray pulse is solved in two-center prolate spheroidal coordinates within the fixed nuclei approximation. At first, we test our monochromatic transition amplitudes by computing the PADs for the fixed-in-space HeH^{2+} molecular target. Setting the polarization vector $\mathbf{\Pi}(\phi) = \mathbf{e}_z \cos\phi + i\mathbf{e}_z \sin\phi$ with $\phi = 0$ $(\pi/2)$ we get the linear parallel (perpendicular) polarization case with respect to the internuclear vector $\mathbf{R}$. We also include the case of right circular polarization, which is obtained by setting $\phi = -\pi/4$.

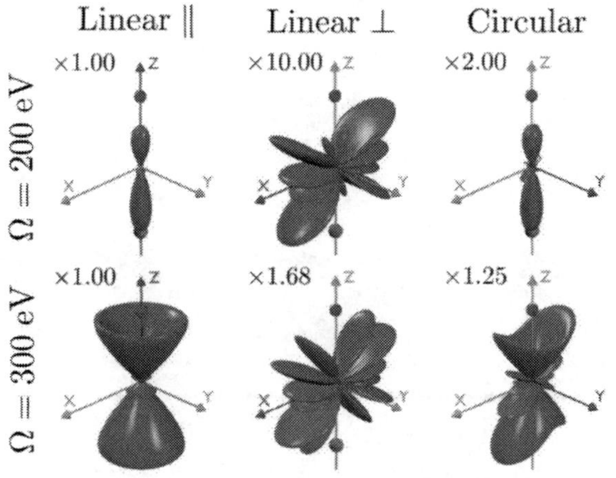

Figure 1. Monochromatic PADs for the excited bound $2p\sigma$ state of HeH^{2+} at R = 3.898 a.u., with XUV polarization parallel and perpendicular to $\mathbf{R}$ as well as circular polarization. The nuclei are indicated by an orange sphere for the α particle and a smaller dark gray sphere for the proton. In the linear case, the polarization vector is indicated by a different color axis (violet). For circular polarization, the violet axis indicates the incidence direction. Angular distributions (green-blue plots) are computed for photon energies of $\Omega = 200$ eV and 300 eV.

In Fig. 1, our computed PADs for photon frequencies of $\Omega = 200$ eV and 300 eV are shown. Both linear polarizations results are in qualitative agreement with *ab initio* ones [7]. As expected, in the parallel case asymmetric PADs are obtained due to the different charge of each nucleus in the target. It is worthy to mention that the SFA will fail to describe these asymmetries [7, 9].

For perpendicular polarization, the PADs are also asymmetric. The asymmetry is more clear in the completely parallel geometry ($\mathbf{p} \parallel \mathbf{\Pi}(\phi) \parallel \mathbf{R}$), however in this case asymmetries are concentrated for emission around the polarization direction.

In the case of circular polarization, our PADs exhibit a mixture of both linear parallel and perpendicular cases with an additional *torsion* [6] of the PADs, arising from the cross product between ionization transition amplitudes of the two independent components of the XUV. This is in agreement with the homonuclear results reported previously [6, 13]. For 200 eV, the perpendicular polarization PADs are too small compared to the parallel ones. Therefore this, leads to circular polarization PADs dominated by the later ones. On the contrary, for $\Omega = 300$

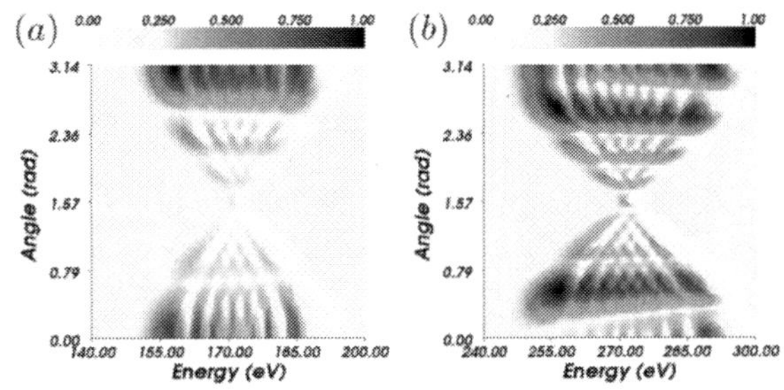

Figure 2. (a) Photoelectron spectrum in the xz plane for HeH^{2+} (R = 3.898 a.u.) ionized by an attopulse of Ω = 200 eV and τ = 500 attoseconds, linearly polarized parallel to the molecular orientation, in the presence of an equally linearly polarized NIR with a wavelength of 800 nm. The NIR intensity is 3.5×10^{12} W/cm^2. (b) Idem (a) but with Ω = 300 eV.

eV, the parallel and perpendicular PADs are of comparable size giving place to a more important torsion, particularly in the perpendicular direction.

In Fig. 2(a), we present the laser assisted photoelectron spectrum centered at frequency Ω= 200 eV for linear parallel polarization as a function of the photoelectron energy and the polar ejection angle θ_e of the photoelectrons. We observe typical sidebands located around the central frequency. These sidebands are produced by quantum interferences occurring during the interaction of the photoelectron with the laser NIR bath [1]. Sidebands are separated roughly by twice the NIR frequency. As the duration of the XUV pulses increases reaching the period of the NIR, the separation between sidebands is given by the NIR frequency [12]. As can be seen, the preferred ejection direction, *i.e.*, 0 and π radians coincides with the molecular orientation. The observed up-down asymmetry is a consequence of the monochromatic effects explained before.

In Fig. 2(b), the photoelectron spectrum show a marked decrease for the emission in the direction of the proton, while emission in the α particle direction remains important. Moreover, the emergence of new structures is clearly visible. In this way, the asymmetry effects observed in the case of monochromatic pulses leave their trace in the photoelectron spectrum when the NIR is added.

4. Conclusions

We have studied PADs for the monochromatic photoionization of fixed-in-space HeH^{2+} molecular targets. A good qualitative agreement with *ab-initio* results is found. For circularly polarized radiation, the PADs exhibit a mixture of the linear parallel and the linear perpendicular cases with an additional torsion. Moreover, the laser-assisted reaction spectra present sidebands with asymmetries that may be explained through the monochromatic contribution that include the asymptotic coulombic conditions. We note that a SFA model will fail to describe these asymmetries. The present study may be extended without great effort to more complex diatomic molecules. Work in this direction is in progress.

References

[1] Krausz F and Ivanov M 2009 *Rev. Mod. Phys.* **81** 163–234
[2] Laurent G, Cao W, Li H, Wang Z, Ben-Itzhak I and Cocke C L 2012 *Phys. Rev. Lett.* **109**(8) 083001
[3] Kling M F, *et al* 2006 *Science* **312** 246–248
[4] Kelkensberg F, *et al* 2011 *Phys. Rev. Lett.* **107**(4) 043002
[5] Singh K P, *et al* 2010 *Phys. Rev. Lett.* **104**(2) 023001
[6] Guan X, DuToit R C and Bartschat K 2013 *Phys. Rev.* A **87**(5) 053410
[7] Guan X, Secor E B, DuToit R C and Bartschat K 2012 *Phys. Rev.* A **86**(5) 053425
[8] Reiss H 1992 *Progress in Quantum Electronics* **16** 1 – 71
[9] Yudin G L, Chelkowski S and Bandrauk A D 2006 *J. Phys. B: At. Mol. Opt. Phys.* **39** 1537
[10] Yudin G L, Patchkovskii S, Corkum P B and Bandrauk A D 2007 *J. Phys. B: At. Mol. Opt. Phys.* **40** F93
[11] Kazansky A K and Kabachnik N M 2007 *J. Phys. B: At. Mol. Opt. Phys.* **40** 2163
[12] Kazansky A K, Sazhina I P and Kabachnik N M 2010 *Phys. Rev.* A **82**(3) 033420
[13] Boll D I R and Fojón O A 2014 *Phys. Rev.* A *in press*

Journal of Physics: Conference Series **583** (2015) 012034 doi:10.1088/1742-6596/583/1/012034

Exact photoionization and radiative recombination cross section in a Coulomb potential

C F Clauser and R O Barrachina

Centro Atómico Bariloche and Instituto Balseiro, Comisión Nacional de Energía Atómica and Universidad Nacional de Cuyo, Av. Bustillo 9500, 8400 Bariloche, Argentina.

E-mail: `cesar.clauser@ib.edu.ar`

Abstract. In this work, we analytically evaluate the exact matrix element for a non-relativistic Continuum–Bound radiative transition in a Coulomb potential. We apply this result to the photoionization (and radiative recombination) process, and calculate the total cross sections giving an analytical expression for the transition from the ground state. To our best knowledge the exact solution to this problem has never been addressed even though the analytical calculation is quite simple, and the final result is not more complicated than any of its approximations. We compare our result with the well-known plane wave and Coulombian dipolar approximations, in a wide range of energies.

1. Introduction

Atomic processes that involve photons are essential for a wide range of applications. For instance, they are customarily used for the study and diagnostic of astrophysical as well as laboratory plasmas [1, 2]. At present, most studies employ the dipolar approximation [3, 4], which is customarily assumed to be a fair approximation in the non-relativistic case [5], and whose evaluation has been even numerically extended into the relativistic regime [6].

In this work, we analytically evaluate the non-relativistic total cross section including all its multipoles contributions for the ground state of a Hydrogenic system. We show that it is not more complicated than the dipolar or other usual approximations. Finally, we demonstrate that it might differ significantly from any of these standard approximations at energies within the non-relativistic regime.

2. Photoionization and Radiative Recombination

Let us consider the photoionization of a bound state ϕ of a Hydrogenic atom or ion of effective charge Z by the absorption of a photon of wave vector $\mathbf{k}$ (angular frequency $\omega = kc$) and polarization $\hat{e}$ ($\perp \mathbf{k}$), with emission of an electron of momentum $\mathbf{p}$. By energy conservation, $\omega = Z^2/2n^2 + p^2/2$, with n the principal quantum number of the bound state (Atomic units are used throughout this article). In the velocity gauge, the non-relativistic transition matrix element reads

$$T = \langle \psi_{\mathbf{p}} \mid e^{i\mathbf{k}\cdot\mathbf{r}}\, \hat{e} \cdot \nabla_{\mathbf{r}} \mid \phi \rangle \, , \tag{1}$$

where

$$\psi_{\mathbf{p}}(\mathbf{r}) = \frac{1}{(2\pi)^{3/2}} \Gamma(1+i\mu)\, e^{-\pi\mu/2}\, {}_1F_1\left(-i\mu, 1; i(pr + \mathbf{p}\cdot\mathbf{r})\right)\, e^{i\mathbf{p}\cdot\mathbf{r}}$$

Journal of Physics: Conference Series **583** (2015) 012034 doi:10.1088/1742-6596/583/1/012034

represents the Coulomb continuum state of the electron emitted with momentum $\mathbf{p}$. Here we have defined the Sommerfeld parameter $\mu = Z/p$. The same transition matrix element describes the Radiative Recombination process [7], where an atom captures an electron with emission of a photon. The differential cross sections for the photoionization (PI) and radiative recombination (RR) processes read

$$\frac{d\sigma^{\mathrm{PI}}}{d\Omega_{\mathbf{p}}} = \frac{4\pi^2}{c^2}\frac{p}{k}|T|^2 \qquad \text{and} \qquad \frac{d\sigma^{\mathrm{RR}}}{d\Omega_{\mathbf{k}}} = \frac{4\pi^2}{c^2}\frac{k}{p}|T|^2 \, ,$$

By averaging on the polarization and integrating over the solid angle, the total cross sections σ^{PI} and σ^{RR} are obtained, which verify $k^2 \times \sigma^{\mathrm{PI}} = p^2 \times \sigma^{\mathrm{RR}}$.

3. Dipolar Approximation

The transition matrix element T is commonly evaluated in a dipolar approximation,

$$T_{\mathrm{D}} = \langle \psi_{\mathbf{p}} \mid \hat{e} \cdot \nabla_{\mathbf{r}} \mid \phi \rangle \, ,$$

which is valid for sufficiently low energies, namely $\omega \ll Zc/n$. For the ground state the total cross section reads [8],

$$\sigma_{\mathrm{D}} = N_{\mathrm{D}}(\mu)\, \sigma_o \qquad (2)$$

where we have defined

$$N_{\mathrm{D}}(\mu) = 2\pi\mu \, \frac{1+\mu^2}{1-e^{-2\pi\mu}} \, e^{-4\mu \cot^{-1}\mu}$$

and where σ_o should be replaced by

$$\sigma_o^{\mathrm{PI}} = \frac{p}{k}\frac{8\pi}{3c^2}\frac{p^2 Z^5}{\omega^4} \qquad \text{and} \qquad \sigma_o^{\mathrm{RR}} = \frac{k}{p}\frac{8\pi}{3c^2}\frac{p^2 Z^5}{\omega^4} \, , \qquad (3)$$

for the photoionization and radiative recombination processes, respectively [9]. Note that this cross section σ_o is obtained from Eq. (2) in the $\mu \to 0$ limit. The extension to arbitrary nlm states was analytically evaluated by Aaron *et al* [3].

4. Plane Wave Approximation

Another commonly employed approximation replaces the exact Coulomb continuum state in (1) by a plane wave, which is valid for photon (or electron) energies much larger than that of the ground state. The transition matrix element reads

$$T_{\mathrm{PW}} = \tilde{\phi}(\mathbf{q})\, \hat{e} \cdot \mathbf{p}$$

where $\tilde{\phi}(\mathbf{q})$ is the eigenstate of the Hydrogenic atom in the momentum representation and $\mathbf{q} = \mathbf{k} - \mathbf{p}$ is the momentum transfer. These functions were evaluated by Podolsky and Pauling in 1929 [10]

$$\tilde{\phi}_{nlm}(\mathbf{q}) = 2^{2l+3} Z^{l+3}\, l! \left(\frac{(n-l-1)!}{2\pi Z\,(n+l)!} \right)^{1/2} \frac{1}{q^2} \left(\frac{nq}{(nq)^2 + Z^2} \right)^{l+2} C_{n-l-1}^{l+1} \left(\frac{(nq)^2 - Z^2}{(nq)^2 + Z^2} \right) Y_{lm}(\hat{\mathbf{q}})$$

where $C_\nu^\mu(x)$ are the Gegenbauer polynomials. The corresponding total cross section for the ground state reads

$$\sigma_{\mathrm{PW}} = N_{\mathrm{PW}}(\mu, \nu)\, \sigma_o \, , \qquad (4)$$

where

$$N_{\mathrm{PW}}(\mu, \nu) = \left[\frac{(1+\mu^2)^2}{(1+\mu^2+\mu^2\nu^2)^2 - 2\mu^2\nu^2} \right]^2 ,$$

with $\nu = k/Z$. Note that σ_o is recovered from Eq. (4) in the $\nu \to 0$ limit.

Journal of Physics: Conference Series **583** (2015) 012034 doi:10.1088/1742-6596/583/1/012034

5. Exact non-relativistic Solution

For the ground state ϕ_{1s}, and after some simple algebra, we obtain

$$T_{\text{Exact}} = \hat{e} \cdot \nabla_{\mathbf{k}} \langle \psi_{\mathbf{p}} \mid \frac{1}{r} e^{i\mathbf{k}\cdot\mathbf{r}} \mid \phi_{1s} \rangle.$$

The matrix element can be analytically evaluated by means of Nordsieck's method [11, 12] in order to obtain

$$\langle \psi_{\mathbf{p}} \mid \frac{1}{r} e^{i\mathbf{k}\cdot\mathbf{r}} \mid \phi_{1s} \rangle = \frac{(2Z)^{1/2}}{\pi} Z^2 \, \Gamma(1 - i\mu) \, e^{-\pi\mu/2} \, \frac{\left(Z^2 + |\mathbf{k} - \mathbf{p}|^2\right)^{i\mu - 1}}{\left((Z - ip)^2 + k^2\right)^{i\mu}} \, ,$$

so that

$$T_{\text{Exact}} = \frac{(2Z)^{5/2}}{2\pi} \, \frac{\Gamma(2 - i\mu) \, e^{-\pi\mu/2}}{\left(Z^2 + |\mathbf{k} - \mathbf{p}|^2\right)^2} \left(\frac{Z^2 + |\mathbf{k} - \mathbf{p}|^2}{(Z - ip)^2 + k^2} \right)^{i\mu} \hat{e} \cdot \mathbf{p} \, .$$

Finally, the corresponding total cross section reads,

$$\sigma_{\text{Exact}} = e^{4\mu\Theta} \, \frac{\sigma_D \times \sigma_{PW}}{\sigma_o} \, , \tag{5}$$

where

$$\Theta = \frac{1}{2} \cot^{-1} \left(\frac{4c^2 + Z^2 - p^2}{2pZ} \right) \, .$$

Note that in the non-relativistic regime, $\Theta \approx 0$, and therefore,

$$\sigma_{\text{Exact}} \approx \frac{\sigma_D \times \sigma_{PW}}{\sigma_o} \, . \tag{6}$$

The generalization of these results to nlm states in terms of Nordsieck's integrals is straightforward [13, 14].

In figure 1 we compare this exact solution with the plane wave σ_{PW}, dipolar σ_{D} and plane wave + dipolar σ_o approximations. We see that the plane wave approximation overestimates the total cross section on the whole energy range and, as expected, improves for increasing energies. On the contrary, the dipolar approximation fairly agrees with the exact expression but starts to fail at high but still non-relativistic energies.

6. Conclusions

We presented an analytical formula for the photoionization and radiative recombination cross sections in the non-relativistic regime which, for the ground state of a Hydrogenic atom or ion, is not more complicated than the standard Dipole or Plane Wave approximations. In fact, we show that it is given by the product of these two approximations much commonly employed in the literature.

Acknowledgments

This work was supported by the Comisión Nacional de Energía Atómica, CNEA (Controlled Nuclear Fusion Program) and Universidad Nacional de Cuyo (Grant 06/C416). ROB is also member of the Consejo Nacional de Investigaciones Científicas y Técnicas (CONICET), Argentina. CFC acknowledges the support of CONICET.

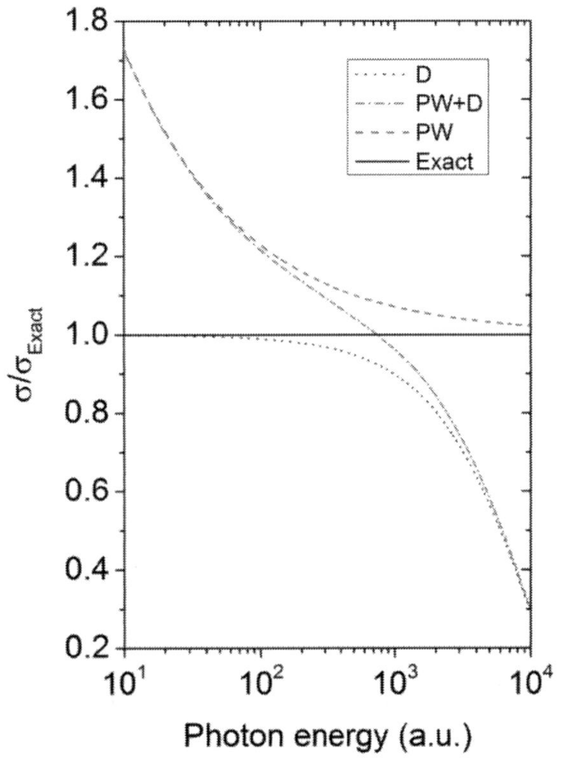

Figure 1. Non-relativistic cross section for the photoionization or radiative recombination processes (with $Z = 1$) as a function of the photon energy. The Dipolar (D), Plane Wave + Dipolar (PW+D) and Plane Wave (PW) approximations are compared with the Exact solution (Exact), as given by Eqs. (2), (3), (4) and (5), respectively.

References

[1] National Research Council 2007 *Controlling the Quantum World: The Science of Atoms, Molecules, and Photons* (Washington DC: The National Academic Press).
[2] Kunze H J 2009 *Introduction to Plasma Spectroscopy* (Berlin, Heidelberg: Springer Series on Atomic, Molecular and Plasma Physics)
[3] Aaron F D, Costescu A and Dinu C. 1993 *J. Phys. II France* **3** 1227
[4] Hahn Y 1997 *Rep. Prog. Phys* **60** 691
[5] Hall H 1936 *Rev. Mod. Phys.* **8** 358
[6] Eichler J and Stöhlker T 2007 *Phys. Rep.* **439** 1
[7] Shevelko V P and Vainshtein L A 1993 *Atomic Physics for Hot Plasmas* (Bristol: Institute of Physics Pub.)
[8] Gottfried K 1966 *Quantum Mechanics, Volume I* (NY: W A Benjamin Inc)
[9] Nagy L, Borbély S and Póra K 2004 *Phys. Lett.* A **327** 481
[10] Podolsky B and Pauling L 1929 *Phys. Rev.* **34** 109
[11] Nordsieck A 1954 *Phys. Rev.* **93** 785
[12] McDowell M R C and Coleman J P 1970 *Introduction to the theory of ion-atom collisions* (London: North Holland Pub. Co.)
[13] Datta S 1985 *J. Phys.* B **18** 853
[14] Colavecchia F D, Gasaneo G and Garibotti C R 1997 *J. Math. Phys* **38** 6603

Journal of Physics: Conference Series **583** (2015) 012035 doi:10.1088/1742-6596/583/1/012035

W^{63+} and W^{64+} ionization by protons and photons

S D López [1,3], **J M Randazzo**[2,3], **R Della Picca**[2,3] **and F D Colavecchia**[2,3]

[1] Instituto de Astronomía y Física del Espacio, Casilla de Correo 67 Suc. 28, C1428ZAA Ciudad Autnoma de Buenos Aires, Argentina.

[2] División Física Atómica, Molecular y Óptica, Centro Atómico Bariloche, 8400 S. C. de Bariloche, Río Negro, Argentina.

[3] Consejo Nacional de Investigaciones Científicas y Técnicas (CONICET), Argentina.

E-mail: sebastlop@iafe.uba.ar

Abstract. In this work we investigate the ionization by proton impact, and photo ionization, involving highly charged ions of W, in particular the Ne-like W^{64+} and Na-like W^{63+} which are more likely to appear in ITER plasma. Total cross sections are calculated in the continuum-distorted-wave-eikonal-initial-state (CDW-EIS) approximation for ion-impact, while two completely different methods are used for photoionization: a perturbative dipolar aproach and the recently introduced Sturmian model.

1. Introduction

Tungsten (W) is the material of choice for use as a key component in the divertor of the International Thermonuclear Experimental Reactor, ITER [1]. For this reason, the properties of tungsten, as a wall material and as an impurity within the plasma are of great interest. Also, data for radiative and collisional atomic processes involving tungsten ions interacting with the plasma are scarce, and need to be computed and measured. Among the variety of reactions within the plasma, we focus our attention in the ionization of highly charged tungsten ions by proton and photon impact. We investigate these collisions with the Ne-like W^{64+} and Na-like W^{63+}, which are the most common ionic states of Tungsten in ITER plasma [2, 3]. Singly differential and total cross sections are calculated in the continuum-distorted-wave-eikonal-initial-state (CDW-EIS) approximation for ion-impact [4, 5]. The initial bounded and final electronic wave functions are obtained with the Optimized Potential Model (OPM) presented by Talman [4].

2. Ionization by proton impact

Many perturbative theories have been developed to describe the single ionization produced by charged ion impact, being the distorted waves methods the most reliable and accurate at differential levels for intermediate to high impact energies. Particularly, the CDW-EIS model has proven to be very accurate for the ion-atom ionization collisions since early 80's [4]. This approximation is a two center approach, which includes the interaction of an active electron with both ions, the projectile and the target core, via Coulombic Distorted Waves. Moreover, the theory can accomodate modified Coulomb waves, which are solutions of a model potential that includes the screening effect of passive electrons [5]. This screening is included in the problem as

Journal of Physics: Conference Series **583** (2015) 012035 doi:10.1088/1742-6596/583/1/012035

a short range potential whereas keeping the long range interatiction for large distances between the electron and the residual dressed ionic core. The initial state considered is:

$$\chi_i(\mathbf{r}_T, \mathbf{r}_P) = \phi(\mathbf{r}_T) E_{-\mathbf{v}}(\mathbf{r}_P) e^{i\mathbf{K}_i \cdot \mathbf{R}}, \tag{1}$$

where $\mathbf{r}_T$, $\mathbf{r}_P$ and $\mathbf{R}$ are the electron- target core, electron- projectile and projectile- target core coordinates respectively. Here, E represents an eikonal wave, with $\mathbf{v}$ the projectile velocity. We define $\mathbf{K}_i$ ($\mathbf{K}_f$) as the initial (final) projectile momentum relative to the target core. The wave function $\phi(\mathbf{r}_T)$ is the initial bound state of the electron:

$$\phi(\mathbf{r}_T) = R_{NL}(r)\, Y_{L,M}(\Omega). \tag{2}$$

The wave function proposal for the final state is:

$$\chi_f(\mathbf{r}_T, \mathbf{r}_P, \mathbf{R}) = \psi_{\mathbf{k}}(\mathbf{r}_T) D_{\mathbf{k}-\mathbf{v}}(\mathbf{r}_P) e^{i\mathbf{K}_f \cdot \mathbf{R}}, \tag{3}$$

where k is the final electron momentum relative to the target core, D represents the usual Coulomb continuum distortion and $\psi_{\mathbf{k}}(\mathbf{r}_T)$ is the continuum wave function of the electron in the target core field

$$\psi_{\mathbf{k}}(\mathbf{r}_T) = \sum_{m,\ell} f_\ell(k,r)\, Y^*_{\ell,m}(\Omega_k)\, Y_{\ell,m}(\Omega). \tag{4}$$

As can be seen, the main problem to deal with, is to solve the eigen-states of the electron in the target core field. We use the Optimized Model Potential of Talman [6] to obtain effective central potentials for the active electron. With these potentials, we calculate radial wave functions $R_{NL}(r)$ ($f_\ell(k,r)$) for the bound (continuum) states, respectively, by solving a one active electron Schrödinger equation. The use of these radial functions leads to a CDW-EIS theory free of Post-Prior discrepancy [7].

In this frame, the total cross section (TCS) reads:

$$\sigma = (2\pi)^4 \mu^2 \frac{K_f}{K_i} \int |T_{fi}|^2 d\Omega_{K_f} d\mathbf{k}, \tag{5}$$

where T_{fi} is the transition matrix. Here μ is the projectile-target reduced mass and ε_i is the ionization energy of the target. Here we present results of the total cross section of the W^{+63} and W^{+64} ions as a function of the energy of the projectile, from the 3s, $2p^0$ and $2p^1$ levels of the W^{+63} ion and 2s, $2p^0$ and $2p^1$ of the W^{+64} ion, see Fig. 1. Cross sections were computed for impact energies between 150 keV and 3 MeV. For energies smaller than 150 keV, the CDW-EIS approximation is not correct, and for energies above 3 MeV, a simpler First Born Approximation could be used [4]. Overall all the cross sections present a similar behavior within the projectile energy range computed, not reaching any maximum in this region. To the best of our knowledge, these cross sections were never reported in the literature before.

3. Ionization by photon impact

We proceed now to evaluate photo ionization of the W^{+63} and W^{+64} ions considering the same initial levels that we use for ionization by proton impact. The cross section of photo ionization of q electrons from the shell NL due to the impact of photons with frequency ω is (see for example Eq. (3) of [8]):

$$\sigma = \frac{q(2\pi)^2}{2L+1} \alpha\omega k \sum_M \int d\Omega_k\, |\langle \psi_{\mathbf{k}} | \hat{\epsilon} \cdot \mathbf{r} | \phi \rangle|^2 = \frac{q(2\pi)^2}{3(2L+1)} \alpha\omega k \left[I_{L-1} L + I_{L+1}(L+1) \right] \tag{6}$$

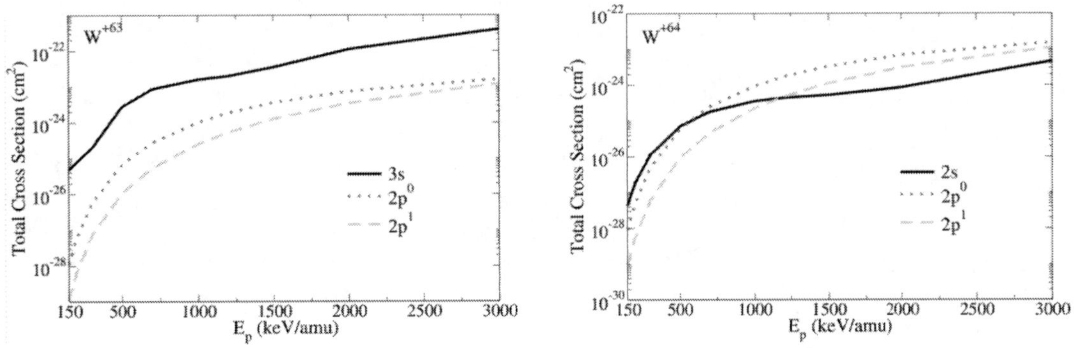

Figure 1. Total cross sections for the ionization from the $2p^0$, $2p^1$ and $3s$ level of the W^{+63} (left panel) and from the $2p^0$, $2p^1$ and $2s$ level of W^{+64} ions (right panel) as a function of the proton impact energy.

where the final electron momentum $\mathbf{k}$ is determined by the conservation of the total energy: $k^2/2 = \hbar\omega + E_{NL}$. The wave function ϕ is the initial state, Eq. (2), while $\psi_{\mathbf{k}}$ represents the one electron final state, Eq. (4). The radial integrals are defined by:

$$I_{L\pm1} = \int_0^\infty dr\, f^*_{L\pm1}(k,r)\, r^3 R_{NL}(r) \tag{7}$$

where f_L and R_{NL} are the radial part of final and initial electronic wave functions defined above.

We also make use of an alternative approach to evaluate the scattering wave function for a single photon absorption solution of the following non homogeneous equation:

$$\left[-\frac{1}{2}\nabla^2_{\mathbf{r}} + V(r) - E\right]\Psi^+_{V,sc}(\mathbf{r}) = \frac{1}{2}(\hat{\varepsilon}.\nabla_{\mathbf{r}})\Phi_0(\mathbf{r}) \tag{8}$$

in the velocity gauge, while in the length gauge $\nabla_{\mathbf{r}}$ must be replaced by $\omega\mathbf{r}$. This calculation can be performed using the Generalized Sturmian Functions model presented recently [9]. In this method, the wave functions $\Psi^+_{V,sc}(\mathbf{r})$ and $\Phi_0(\mathbf{r})$ are expanded in terms of a Sturmian radial basis set, and the equation (8) is solved for the coefficients of that expansion. The key role of the basis is that all its elements include the correct long-range boundary condition of the problem under scrutiny. The asymptotic behavior of the scattering wave functions must have the same behavior in both velocity and length gauges and has the form:

$$\Psi^+_{sc}(\mathbf{r}) \simeq A(\hat{\mathbf{r}})\frac{e^{i(kr-\frac{Z}{k}\log 2kr)}}{r}. \tag{9}$$

The cross section is a balance between the electronic flux $2i\,j_e = \Psi^{+*}_{sc}\nabla\Psi^+_{sc} - \Psi^+_{sc}\nabla\Psi^{+*}_{sc}$, and the photon flux j_s given by the Poynting vector associated to the electromagnetic wave:

$$\frac{d\sigma}{d\Omega} = \frac{j_e}{j_s} = \frac{2\mu_0 ck}{\omega}|A(\hat{\mathbf{r}})|^2. \tag{10}$$

We present results of the total ionization cross section of the W^{+63} and W^{+64} ions as a function of the energy of the ejected electron (which depends on the photon energy), from the 3s, $2p^0$ and $2p^1$ levels of the W^{+63} ion and 2s, $2p^0$ and $2p^1$ of the W^{+64} ion, see Fig. 2. Contrary to the proton impact case, there are non-dipolar full relativistic calculations of photoionization cross sections, recently reported by Trzhaskovskaya and collaborators [10, 11, 12]. We note that both methods presented here agree with each other. Furthermore, it is clear from the figure that our results deviates only a few percent from the relativistic ones in the high energy regime (beyond 30 keV of photon impact).

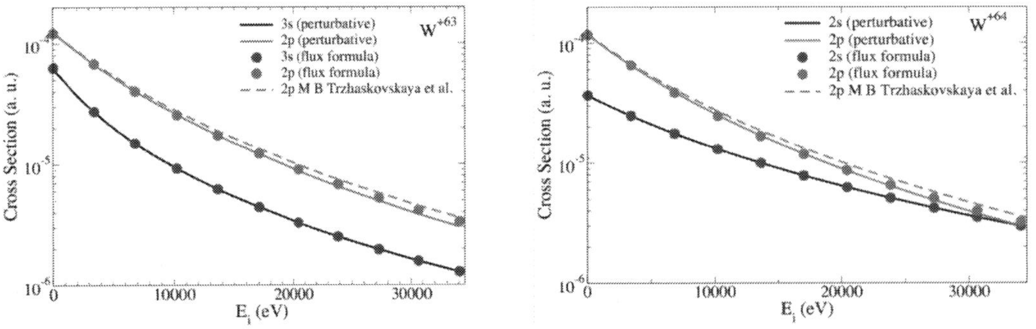

Figure 2. Total cross sections for the photo ionization from the 3s and 2p levels of the W^{+63} ion (left) and from the 2s and 2p levels of the W^{+64} ion (right) as a function of the energy of the emitted electron. Non-dipolar relativistic calculations (dashed lines) of Ref. [11, 12].

4. Summary and outlook

We have computed the total cross section of ionization of W^{+63} and W^{+64} ions by proton, which have not been considered before; and photon impact, that were presented this year in the literature. The total ionization cross sections for proton impact increase monotonically for all the sub-shells up to 3 MeV. The cross sections for photoionization agree well with their relativistic counterparts, suggesting that it is possible to consider a simpler dipolar approximation for these highly charged targets in the energy regime up to a few keVs. Further theoretical developments will be pursued along two directions: study other highly charged tungsten ions, and focus on multiple ionization processes.

Acknowledgments

We would like to thank the National Atomic Energy Comission of Argentina (CNEA), through the Integral Program of Controlled Nuclear Fusion. We also thanks the financial support of the National University of Cuyo, through the program 06/C424.

References

[1] www.iter.org
[2] Pütterich T *et al.* 2008 *Plasma Phys. Control. Fusion* **50** 085016
[3] Hirai T *et al.* 2013 *Fusion Eng. Design* **88** 1798
[4] Crothers D S F and McCann J F 1983 *J. Phys.* B **16**, 3229
[5] Gulyas L, Fainstein P D and Salin A 1995 *J. Phys.* B **28** 245
[6] Talman J D 1989 *Comp. Phys. Comm.* **54** 85
[7] López S D, Fiori M and Garibotti, C R 2011 *Phys. Rev.* A **83** 032716
[8] Della Picca R, Fainstein P D and Dubois A, 2011 *Phys Rev* A **84** 033405
[9] Gasaneo G *et al Adv. Quant. Chem.* **67** 153
[10] Trzhaskovskaya M B, Nikulin V K and Clark R E H 2008 *At Data Nucl Data Tables* **94** 71
[11] Trzhaskovskaya M B and Nikulin V K 2013 *At Data Nucl Data Tables* **99** 249
[12] Trzhaskovskaya M B and Nikulin V K 2014 *At Data Nucl Data Tables* **100** 986

17th International Conference on the Physics of Highly Charged Ions IOP Publishing
Journal of Physics: Conference Series **583** (2015) 012031 doi:10.1088/1742-6596/583/1/012031

Modeling of the M X-ray line structures for tungsten and L X-ray line structures for molybdenum

K Słabkowska[1,2], M Polasik[1,2], Ł Syrocki[1,3], E Szymańska[1,2], J Rzadkiewicz[1,4], N R Pereira[5]

[1] Institute of Plasma Physics and Laser Microfusion, 01-497 Warsaw, Poland
[2]Faculty of Chemistry, Nicolaus Copernicus University, 87-100 Toruń, Poland
[3]Faculty of Physics, Astronomy and Informatics, Nicolaus Copernicus University, 87-100 Toruń, Poland
[4]National Centre for Nuclear Research, 05-400 Otwock, Poland
[5]Ecopulse Inc., 7844 Vervain Court, Springfield , VA 22152, USA

katarzyna.slabkowska@umk.pl

Abstract. Modeled within the Collisional-Radiative approach, for parameters relevant to plasmas in the centre of JET, the M X-ray line structures for tungsten and L X-ray line structures for molybdenum both occur in the wave length range 5.0-5.35 Å; and, their strengths are comparable. Therefore, the spectra obtained with the upgraded high-resolution X-ray spectrometer KX1 on JET should include both tungsten and molybdenum in their interpretation. The same will be true for the high-resolution X-ray diagnostic on tokamaks such as WEST and ITER, where tungsten plasma-facing components will be implemented.

1. Introduction

The theoretical background for the future high-resolution X-ray diagnostics of tokamak plasmas is an active topic of research [1-5]. The urgent need for such studies comes primarily from the choice of tungsten as a plasma facing component in the future ITER fusion reactor. Highly charged tungsten ions appearing in the central plasma due to the plasma-wall interaction [6] are therefore a natural source of X-rays for high-resolution X-ray diagnostics that monitor ion temperature, rotational velocity and W impurity level in tokamak fusion reactors. The structures of tungsten X-ray spectra emitted over a wide wave length range are extremely complex, and require exhaustive theoretical and experimental studies that provide guidance for X-ray diagnostics of the plasma planned for next-generation tokamaks.

In this work we present the results of modeling of the M X-ray line structures for tungsten in the 5.0-5.35 Å wavelength range, and the L X-ray line structures for molybdenum in the same wave length range, expected from high-temperature plasmas with parameters relevant to the centre of the plasma in JET, and the outside plasma expected on ITER (r/a~0.8-0.9). The simulations are performed using the Flexible Atomic Code (FAC) package, within the framework of Collisional-Radiative Model (CRM) approach [4, 7, 8]. FAC calculates the atomic structure with a fully relativistic Dirac-Fock-Slater method and the configuration interaction method. The modeling performed assumes a plasma electron temperature of 4.5 keV and electron density of 6×10^{13} cm^{-3}. For tungsten this predicts ions will be ionized from W^{+44} to W^{+47} and their X-rays are primarily from the M-shell, in the 5.0-5.35 Å wavelength range. For molybdenum the ions will be mostly from Mo^{+30} to Mo^{+35} and these primarily emit L-shell radiation, which happens to be in the same wave length range.

2. Theory - Flexible Atomic Code

The simulations have been performed using the FAC package within the framework of CRM approach [4, 7, 8]. In the package the full-relativistic Dirac-Fock-Slater iteration method and the configuration interaction method are used to calculate of the atomic structures, and the distorted-wave approximation method is used for electron collision cross sections. Briefly [7], the energy levels of an atomic ion with N electrons are obtained by diagonalizing the relativistic effective Hamiltonian H. In atomic units this reads

$$H = \sum_{i=1}^{N} h_D(i) + \sum_{j>i=1}^{N} \frac{1}{r_{ij}},$$

where $h_D\ (i)$ is the single-electron Dirac Hamiltonian for the potential due to the nuclear charge. The basis states Φ_v, which are usually referred to as configuration state functions (CSF), are antisymmetric sums of products of N one-electron Dirac spinors $\varphi_{n\kappa m}$

$$\varphi_{n\kappa m} = \frac{1}{r}\begin{pmatrix} P_{n\kappa}(r)\chi_{\kappa m}(\theta,\phi,\sigma) \\ iQ_{n\kappa}(r)\chi_{-\kappa m}(\theta,\phi,\sigma) \end{pmatrix},$$

where $\chi_{\kappa m}$ is the usual spin-angular function. n is the principal quantum number, κ is the relativistic angular quantum number, which is related to the orbital and total angular momentum.

The FAC code not only computes atomic data, such as energy levels and transition probabilities for radiative transitions and auto-ionization, but also cross sections for excitation and ionization by electron impact and the inverse processes (e.g., radiative recombination and dielectronic capture). With these data, and within the approximations of the CRM (e.g., [9]), FAC can also compute fractional ionizations and the resulting X-ray spectrum, with rates computed for specific electron energy distributions.

3. Results

The preliminary results in this paper are done for an initial interpretation of the X-rays emitted by the plasma in the JET tokamak operated in the ITER-like wall configuration, as registered by the upgraded high-resolution crystal spectrometer (KX1) [10, 11]. Below we present some results for the X-ray spectra for tungsten and molybdenum, with electron temperature and density relevant to central JET; these parameters are similar to those expected for high-radius ITER plasmas. Therefore, based on the shape of X-ray spectrum registered KX1 spectrometer on JET [12] we established a line width in Figure 1 and Figure 2 equal to 2 eV. It should be emphasized that in our simulation we applied original distribution of tungsten and molybdenum ions predicted by the most advanced version of CR model applied in FAC package.

3.1 M X-ray spectra for tungsten

Figure 1 presents the M X-ray line structures for tungsten in plasmas with electron temperature of 4.5 keV and electron density of 6×10^{13} cm^{-3}. The modeling includes the weak 4d→3p and other low intensity transitions in tungsten ions from W^{+44} to W^{+47} that occur in the 5.0-5.35 Å wavelength range. Their intensity is about 20-30 times lower than for the most intense lines Mα_1 and Mβ_1 (i.e. for 4f→3d transitions). However, for higher electron temperature, e.g. 7.0 keV, the M X-ray line structures will contain significant contributions of the much stronger M X-ray lines Mα_1 and Mβ_1 that originate from the more highly ionized tungsten ions W^{+48} to W^{+54} [13].

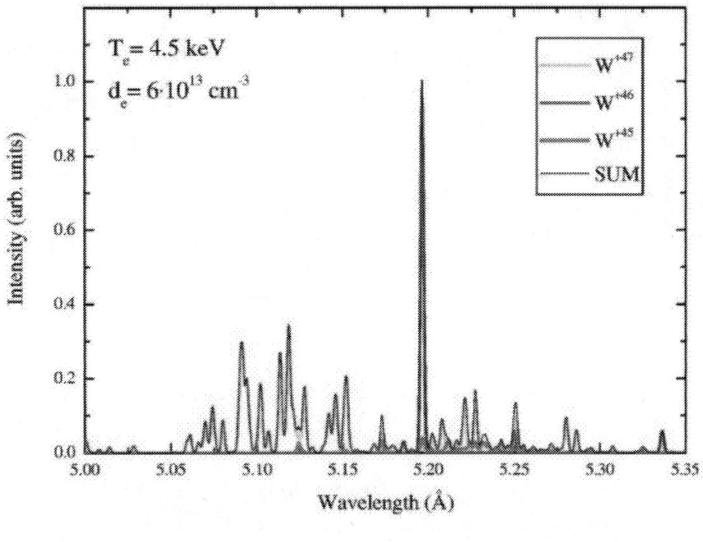

Figure 1. M X-ray line structures of tungsten ions (from W^{+44} to W^{+47}) for electron temperature of 4.5 keV and electron density of 6×10^{13} cm^{-3}.

3.2. L X-ray spectra for molybdenum

Figure 2 shows the corresponding results for the L X-ray line structures for molybdenum ions from Mo^{+30} to Mo^{+35} expected in this same plasma (4.5 keV electron temperature, 6×10^{13} cm^{-3} electron density). Over the entire wavelength range the spectrum may be less complicated than for W, but is still sufficiently complex to be potentially confusing.

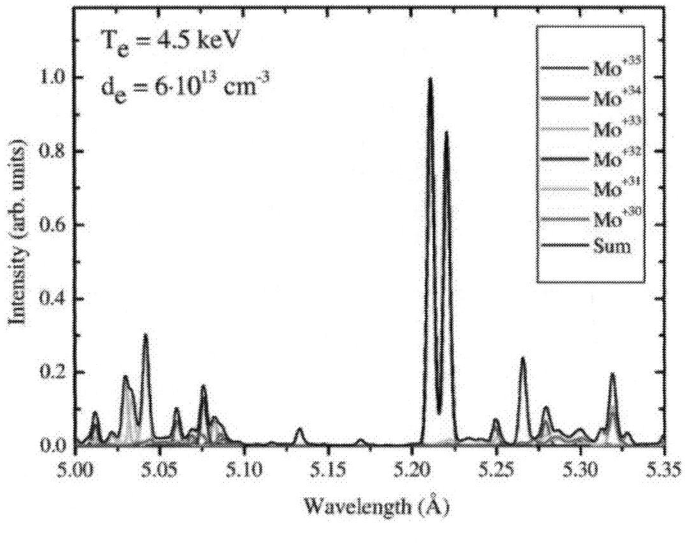

Figure 2. L X-ray line structures of molybdenum ions (from Mo^{+30} to Mo^{+35}) for electron temperature of 4.5 keV and electron density of 6×10^{13} cm^{-3}.

The upgraded KX1 spectrometer [10, 11] almost always finds that the JET plasma emits these molybdenum spectra. One of the possible molybdenum sources is the marker made from molybdenum tile installed as an inner vertical divertor target [12]. Therefore, both elements (tungsten and molybdenum) should be taken into account in the analysis of high-resolution X-ray spectra in this wave length regime. Moreover, presented in Figure 1 and Figure 2 results should reliably reproduce the measured KX1 spectrometer spectra.

4. Summary and conclusions

In this work we present the results of modeling of the M X-ray line structures for tungsten in the 5.0-5.35 Å wavelength range, and the overlapping L X-ray line structures for molybdenum, for an electron temperature and density that is relevant to central JET and high-radius ITER plasmas. The L X-ray spectra for molybdenum have a significant spectral contribution in the same wavelength region as the M X-ray spectra for tungsten, hence both elements should be included in the interpretation of the high-resolution X-ray spectra registered on JET. The obtained results are equally useful for the high-resolution X-ray diagnostics on the ITER and WEST tokamaks, where tungsten X-ray spectra will overlap the radiation emitted by other plasma facing components that will most likely include molybdenum as well.

Acknowledgments

This work has been carried out within the framework of the EUROfusion Consortium and has received funding from the Euratom research and training programme 2014-2018 under grant agreement No 633053. The views and opinions expressed herein do not necessarily reflect those of the European Commission. This scientific work was partly supported by Polish Ministry of Science and Higher Education within the framework of the scientific financial resources in the year 2014 allocated for the realization of the international co-financed project. This work was also done under Grant No. 2011/01/D/ST2/01286 of the Polish National Science Centre.

References

[1] Neu R, Fournier K B, Bolshukhin D and Dux R 2001 *Phys. Scr.* **T92** 307
[2] O'Mullane M G, Summers H P, Whiteford A D et al. 2006 *Rev. Sci. Instrum.* **77** 10F520
[3] Ralchenko Y, Tan J N, Gillaspy J D and Pomeroy J M 2006 *Phys. Rev. A* **74** 042514
[4] Clementson J, Beiersdorfer P, Brown G V and Gu M F 2010 *Phys. Scr.* **81** 015301
[5] Słabkowska K, Polasik M, Szymańska E, Starosta J, Syrocki Ł, Rzadkiewicz J and Pereira N R 2014 *Phys. Scr.* **T161** 014015.
[6] Beiersdorfer P et al. 2010 *J Phys.***B43** 144008
[7] Chambers D M, Pinto P A, Hawreliak J et al. 2002 *Phys. Rev. E***66** 026410
[8] Gu M F 2008 *Can. J. Phys.* **86** 675
[9] Kano K, Suzuki M and Akatsuka H 2000 *Plasma Sources Sci. Technol.* **9** 314
[10] Rzadkiewicz J et al. 2013 *Nucl. Instrum. Methods in Phys. Research A* **36** 720
[11] Schumack A et al. 2014 *Rev. Sci. Instrum.* **85** 11E425
[12] Nakano T et al. 2014 41st EPS Conference on Plasma Physics P1.019
[13] Słabkowska K, Syrocki Ł, Szymańska E, Polasik M, Rzadkiewicz J to be published

Journal of Physics: Conference Series **583** (2015) 012037 doi:10.1088/1742-6596/583/1/012037

Zero energy resonances in atomic processes within a plasma

C F Clauser and R O Barrachina

Centro Atómico Bariloche and Instituto Balseiro, Comisión Nacional de Energía Atómica and Universidad Nacional de Cuyo, Av. Bustillo 9500, 8400 Bariloche, Argentina.

E-mail: cesar.clauser@ib.edu.ar

Abstract. We investigate different atomic processes that might occur in the interior of plasmas. In particular we apply the Final-State Interaction theory in order to study the behavior of the corresponding cross sections when the relative energy of a pair of charged particles in the initial or final states vanishes. Through this analysis we uncover the presence of zero - energy resonances for particular configurations of density and temperature. For fusion plasmas, these effects might have an important impact on the working conditions and performance of the reactor.

1. Introduction

Let us consider an atomic transition taking place within a plasma [1]. Naturally, the corresponding transition matrix element $\mathcal{T}$ depends on the parameters that characterize the atomic partners before or after the transition. Furthermore, it also depends on the given conditions of density and temperature of the plasma, since they can affect the corresponding interactions. A well-known example is given by the screening length λ of the interaction between charged particles.

In many cases of interest, the situation where the relative momentum k of a pair of charged partners in the initial or final configuration vanishes can be relevant for the transition under study. When $k\lambda \gg 1$, the screened potential amplitudes converge to those for a pure Coulomb interaction, but this condition should fail for small enough values of k.

In this article we discuss on very general grounds the kind of consequence that such a screening might have on the atomic transitions taking place within the plasma, and whether they might be observable or lead to sizeable effects.

2. Final State Interaction Theory

According to the Final-State Interaction (FSI) theory [2, 3], the $k \to 0$ limit of $\mathcal{T}$ is dominated by the inverse of the s-wave Jost function $f_0(k,\lambda)$ [4] (Atomic Units are used throughout this article, except where otherwise stated)

$$f_0(k,\lambda) = 1 + \frac{2m}{k}\int_0^\infty e^{ik\lambda} V_\lambda(r)\, \phi_{0\,k}(r)\, \mathrm{d}r\ . \tag{1}$$

Here $V_\lambda(r)$ is the screened Coulomb potential between the active particles of reduced mass m and relative momentum k and distance r, while $\phi_{0\,k}(r)$ represents the $\ell = 0$ regular solution of

Journal of Physics: Conference Series **583** (2015) 012037 doi:10.1088/1742-6596/583/1/012037

the radial equation. For small values of $k\lambda$, this wave function becomes linear in k, and therefore

$$f_0(k,\lambda) \propto \frac{1-ika}{a}\,, \tag{2}$$

where a is the scattering length for the potential $V_\lambda(r)$.

For a pure Coulomb interaction $V(r) = -Z/r$, *i.e.* for $\lambda \to \infty$, we obtain

$$f_0(k,\infty) = \frac{e^{-\pi\mu/2}}{\Gamma(1-i\mu)}\,, \tag{3}$$

with $\mu = mZ/k$ the so-called Gamow factor. Here we are considering only attractive (*i.e.* $Z > 0$) potentials, since they are the ones that -as it is shown in this article- are prone to produce the larger distortions of the transition $\mathcal{T}$-matrix elements through zero-energy resonances. The case of repulsive potentials ($Z < 0$) can be readily obtained *mutatis mutandis* from the present results.

When factored out of the $\mathcal{T}$-matrix element for small values of k, $f_0(k,\lambda)$ would produce an enhancement factor $\mathcal{F}(k,\lambda) = 1/|f_0(k,\lambda)|^2$, with very different behaviors for screened or unscreened potentials. For finite values of λ, we obtain,

$$\mathcal{F}(k,\lambda) \propto \frac{a^2}{1+k^2a^2} \qquad \text{for} \qquad k \ll 1/\lambda\,, \tag{4}$$

while for a pure Coulomb potential $\lambda = \infty$,

$$\mathcal{F}(k,\infty) \approx \frac{2\pi m Z}{k} \qquad \text{for} \qquad k \ll mZ\,. \tag{5}$$

We clearly see that the two limits $k \to 0$ and $\lambda \to \infty$ do not commute [5, 6]. This behavior is similar to the one observed on the half-energy shells for the Coulomb off-energy-shell transition matrix element [7, 8, 9, 10].

One of the most striking effects produced by this diverging enhancement factor [11] is the Electron Capture to the Continuum (ECC) peak that can be observed in the momentum distribution of electrons emitted in atomic collisions when the velocity of the electron matches that of the projectile, *i.e.*, when the relative electron-projectile momentum k vanishes. This effect was observed for outgoing ions [12, 13] and neutral atoms [14, 15, 16]. More recently, a similar effect was observed in positron-atom ionization collisions [17, 18].

3. Hulthèn potential

As a working example of the general theory sketched in the previous section, let us consider the Hulthèn potential

$$V_\lambda^H = -\frac{Z}{\lambda}\frac{e^{-r/\lambda}}{1-e^{-r/\lambda}}. \tag{6}$$

The s-wave Jost function can be analytically evaluated for this exponentially screened Coulomb potential [5],

$$f_0(k,\lambda) = \frac{\Gamma(1-2ik\lambda)}{\Gamma(1+i\beta)\,\Gamma(1-i\beta-2ik\lambda)} \tag{7}$$

with

$$\beta = k\lambda\left[\left(1-\frac{2\mu}{k\lambda}\right)^{1/2} - 1\right]. \tag{8}$$

Except for a logarithmic distortion of the phase, we recover the pure Coulomb case for $k\lambda \gg 2\mu$ [5],

$$f_0(k,\lambda) = f_0(k,\infty)\, e^{-i\mu \ln(2k\lambda)} . \tag{9}$$

On the other hand, for $k\lambda \ll 2\mu$,

$$\mathcal{F}(k,\lambda) = \frac{2mZ}{\lambda} \frac{a^2}{1+k^2a^2} , \tag{10}$$

with

$$a = \frac{\pi\lambda}{\sin(\pi\sqrt{2mZ\lambda})} . \tag{11}$$

We see that this enhancement factor would produce a Lorentzian-shape peak with a width or the order of $1/a$. However, the scattering length becomes infinite whenever [6],

$$\lambda = \frac{n^2}{2mZ} \qquad \text{with } n = 1, 2, 3, \cdots \tag{12}$$

At these zero energy resonances, the enhancement factor diverges like k^{-2}, and is responsible for the observation of the ECC effect even by neutral outgoing projectiles [14, 15, 16]. Any of these behaviors are extremely different from the k^{-1} divergence produced by a pure Coulomb potential. These results are not unique to the Hulthèn potential, but are generally valid for any exponentially screened or cut-off Coulomb potential [6]. Hence, we variationally adjust the charge Z in order to correctly match the spectral properties of the Jost function for the Huthèn potential to those corresponding to a Yukawa potential. The analysis of the energy spectrum of the bound $s-$states [19] suggests to change $Z \to \alpha Z$ with $\alpha \approx 0.58$.

4. Radiative Recombination Processes

In order to illustrate the kind of effects that a zero-energy resonance might produce on an atomic process within a plasma, let us apply the previous results to a simple radiative recombination process, where an electron is captured by an ion into a $1s$ bound state with emission of a photon. For any given value of λ, we variationally adapt the typical radius of the final hydrogenic $1s$ state to reproduce a Yukawa $s-$state and, according to the FSI theory, incorporate on the cross section σ the effects of the screening on the initial continuum state by means of the enhancement factor $\mathcal{F}(k,\lambda) = 1/|f_0(k,\lambda)|^2$. We observe that for the particular case of a Hulthèn potential, the values of λ at which the zero energy resonances occur, as given in Eq. (12), are in agreement with those obtained in recent numerical calculations of photoionization of hydrogen and hydrogen-like helium [20, 21] that employ a Debye screening potential within the framework of the complex-coordinate rotation method. For instance, the resonances observed in Figs. 1 and 2 of Ref. [20] and in Figs. 2 and 4 of Ref. [21] match those of Eq. 12 for $n = 2$. Even the successive resonances evaluated in Ref. [21] for H and He^+ with a exponential-cosine-screened Coulomb potential verify the Z^{-1} and n^2 scaling rules predicted by Eq. 12. These results substantiate the validity of our simple and general approach. Furthermore, they seem to indicate that the observed resonances might correspond to zero-energy effects and not to shape-resonances resulting from quasi-bound np states, as stated in Ref. [21].

As we have already mentioned, the screening length λ depends on the plasma temperature T and the particle density ρ of the plasma electrons. For a static limit of the ion velocity, we can relate them in a very simple way by means of the standard interpolation [22, 23], which in atomic units reads,

$$\lambda^2 \approx \frac{1}{4\pi\rho}\left(T + \frac{(3\pi^2\rho)^{2/3}}{3}\right) . \tag{13}$$

For simplicity, we consider a static limit, where the ion velocity is assumed to be much lower than the electron thermal velocity. Then, the reaction frequency can be calculated from

$$\nu = \rho \langle \sigma v \rangle = \rho \int d^3k f(k)\sigma(k)k \,, \tag{14}$$

where $f(k)$ is the distribution function of the plasma electrons. We use a Maxwellian distribution function which is valid for high temperature plasmas [24], and σ is the total cross section for the radiative recombination process [25].

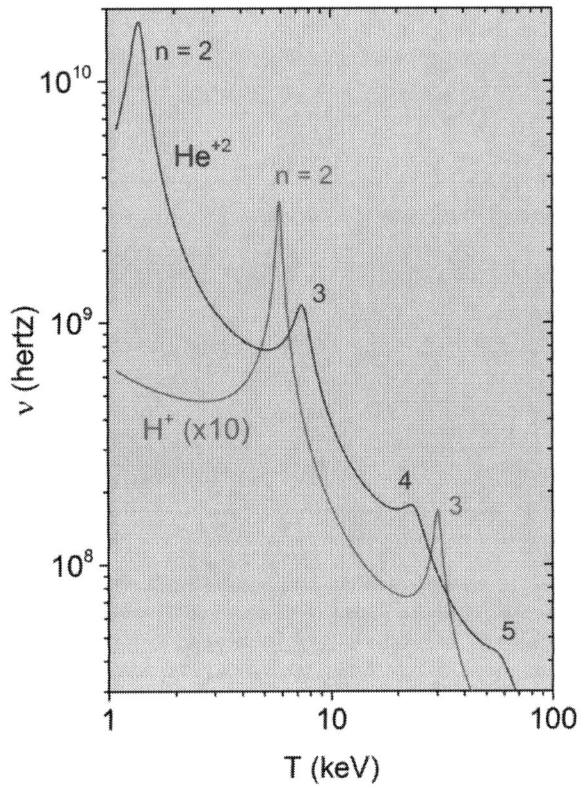

Figure 1. Reaction frequency for the radiative recombination of H^+ and He^{2+} in a dense plasma as those used in inertial confinement fusion (with $\rho = 10^{25}\text{cm}^{-3}$) as a function of the temperature T. A Hulthèn potential was employed to model the screened electron-nucleus interaction. The index n characterizes the zero energy resonances, as described in the text.

In Fig. 1 we show the reaction frequency for the radiative recombination of H^+ and He^{2+} in the case of inertial confinement (with $\rho = 10^{25}\text{cm}^{-3}$) as a function of the temperature T of the plasma. As described in the previous sections, a Hulthèn potential was employed to model the screened electron-nucleus interaction. We clearly observe that the averaging on the momentum k of the plasma electrons is not enough to wash out the zero-energy resonances. Actually, the reaction frequency increases by as much as an order of magnitude or even more whenever the temperature T matches the zero energy resonance condition given by Eq. (12).

5. Conclusions

In this article we have uncovered the presence of zero energy resonances in plasmas for particular configurations of density and temperature. We applied our model to the particular case of a radiative recombination process, but the same effects will be present in any atomic process, as for instance photoionization or electron-impact ionization, whenever the relative momentum of a pair of charged partners in the initial or final configuration vanishes. The relevance that these resonances might have on the working conditions and performance of a fusion reactor for inertial or magnetic confinement is to be analyzed and pondered.

Acknowledgments

This work was supported by the Comisión Nacional de Energía Atómica, CNEA (Controlled Nuclear Fusion Program) and Universidad Nacional de Cuyo (Grant 06/C416). ROB is also member of the Consejo Nacional de Investigaciones Científicas y Técnicas (CONICET), Argentina. CFC acknowledges the support of CONICET.

References

[1] Salzmann D 1998 *Atomic Physics in Hot Plasmas* (New York: Oxford University Press)
[2] Gillespie J 1964 *Final State Interactions* (San Francisco: Holden-Day).
[3] Barrachina R O 1997 *Nucl. Instr. Meth.* B **124** 198
[4] Taylor J R 1972 *Scattering Theory: The Quantum Theory on Nonrelativistic Collisions* (New York: John Wiley & Sons)
[5] Barrachina R O and Garibotti C R 1983 *Lett. Nuov. Cim.* **36** 583
[6] Garibotti C R and Barrachina R O 1983 *Phys. Rev.* A **28** 2792
[7] Ford W F 1964 *Phys. Rev.* **133** B1616
[8] Ford W F 1966 *J. Math. Phys.* **7** 626
[9] Chen J C Y and Chen A C 1972 *Advances un Atomic and Molecular Physics* **8** (New York)
[10] Kolsrud M 1978 *J. Phys.* A **11** 1271
[11] Fiol J, Barrachina R O and Rodríguez V D 2002 *J. Phys.* B **35** 149
[12] Crooks G B and Rudd M E 1970 *Phys. Rev. Lett.* **25** 1599
[13] Macek J 1970 *Phys. Rev.* A **1** 235
[14] Barrachina R O 1990 *J. Phys.* B **23** 2321
[15] Báder A, Sarkadi L, Víkor L, Kuzel M, Závodszky P A, Jalowy T, Groeneveld K O, Macri P A and Barrachina R O 1997 *Phys. Rev.* A **55** R14
[16] Macri P and Barrachina R O 1998 *J. Phys.* B **31** 1303
[17] Kövér Á and Laricchia G 1998 *Phys. Rev. Lett.* **80** 5309
[18] Fiol J, Rodríguez V D and Barrachina R O 2001 *J. Phys.* B **34** 933
[19] Stubbins C 1993 *Phys. Rev.* A **48** 220
[20] Zhao L B and Ho Y K 2004 *Phys. Plasmas* **11** 1695
[21] Lin C Y and Ho Y K 2010 *Eur. Phys. J.* D **57** 21
[22] Neufeld J and Ritchie R H 1955 *Phys. Rev.* **98** 1632
[23] Clauser C F and Arista N R 2013 *Phys. Rev.* E **88** 053102
[24] S. Ichimaru 1973 *Basic Principles of Plasma Physics* (Reading: Benjamin)
[25] Clauser C F and Barrachina R O 2014 Exact photoionization and radiative recombination cross section in a Coulomb potential *J. Phys.: Conf. Series* this volume

17th International Conference on the Physics of Highly Charged Ions IOP Publishing
Journal of Physics: Conference Series **583** (2015) 012038 doi:10.1088/1742-6596/583/1/012038

The cooling of naphthalene cations studied within an electrostatic storage ring

C Ortéga[1], R Brédy[1], L Chen[1], J Bernard[1], M Ji[1], G Montagne[1], A R Allouche[1], A Cassimi[2], C Joblin[3], and S Martin[1]

[1]Institut Lumière Matière, UMR5306 Université Lyon 1-CNRS,Villeurbanne, France
[2]CIMAP, GANIL Université de CAEN, Bd. H. Becquerel, Caen, France
[3]Université de Toulouse, UPS-OMP, IRAP, Toulouse, France

E-mail: celine.ortega@univ-lyon1.fr

Abstract. Fast population decay of naphthalene cations ($C_{10}H_8{}^+$) has been observed in a compact electrostatic storage ring, the Mini-Ring, up to 5 ms. Laser induced dissociation due to single-photon absorption was used to probe the internal energy distribution (IED) of the stored molecular ions as a function of the storage time. To determine the energy distribution of naphthalene cations, the experimental neutral decay curve was analysed with a model including the competition between dissociation and radiative cooling. Fast cooling rates from about 70 s^{-1} at the internal energy 5.6 eV to 140 s^{-1} at 6.2 eV were measured and compared with the data in the literature. This fast cooling mechanism is attributed to the fluorescence from thermally excited electrons and may have important implications in astrophysics for the lifetime and the critical size of Polycyclic Aromatic Hydrocarbons (PAH) in the interstellar medium.

1. Introduction

The abundant presence of PAHs in the interstellar medium was proposed almost thirty years ago [1, 2] in order to explain the unidentified infrared (UIR) emission band. Since then, numerous laboratory studies have been initiated to understand their origin and stability in the astrophysical environments [3, 4]. The photophysics of isolated PAHs irradiated by UV photons has been studied taking into account the dissociation and IR cooling of the molecule [5]. However, an efficient and faster cooling process than IR emission was also proposed by Leger and Boissel with the name of Poincaré fluorescence [6]. This fast process involves inverse internal conversion (IIC) and relies on the fast electronic fluorescence from low lying electronic states. Recently, evidence of this fast electronic cooling mechanism was observed using electrostatic storage ring (ESR) in the case of anthracene cation ($C_{14}H_{10}{}^+$) [7] and hot small molecular anions C_6^- [8]. In earlier works, many other molecules and clusters have been studied within ESRs, like ELISA in Aarhus, which appeared to be very efficient to study the relaxation dynamics of complex systems in a long time range from microsecond to second [9]. In this paper, we study the cooling of naphthalene cations stored in our small electrostatic storage ring, the Mini-Ring. A numerical method is used to analyze the experimental data in order to determine the population decay rates of the stored ions for several excitation energies.

2. Experimental set-up

The experiment consists in storing naphthalene cations ($C_{10}H_8^+$) in an electrostatic storage ring, the Mini-Ring, whose characteristics were described in details in papers [10, 11]. Naphthalene molecules were ionized and heated in an Electron Cyclotron Resonance (ECR) ion source with a broad initial internal energy distribution, and accelerated to 12 keV. The singly charged naphthalene molecular ions were selected using a 90° magnet and the ion beam was pulsed into bunches of 1 μs width every 5 ms. These ion bunches were injected into the Mini-Ring and stored with a 5.5 μs revolution period. The Mini-Ring vacuum chamber was pumped down to a pressure of 2.10^{-9} mbar to limit the probability of collision with the residual gas molecules. The stored naphthalene ions were expected to dissociate mainly via the emission of C_2H_2 and H. These neutral fragments due to unimolecular dissociation were detected at each turn by a position sensitive detector (PSD) facing one of the straight lines of the stored ion trajectory. After dissociation, the daughter ions could not remain on a stable trajectory because their kinetic energy did not fulfill the stable storage conditions and so, they were quickly ejected out of the ring. Nanosecond laser pulses with a photon energy of 2.33 eV (second harmonic of a 1 kHz Nd-YAG OPO laser) were sent at a different precisely controlled storage time $t_{laser} = 0.5$, 1.5, 2.5, 3.5, and 4.5 ms to excite a fraction of the stored ions in order to probe the internal energy distribution (IED) of the ion bunch. The intensity of the laser pulse was about 300 μJ per pulse, corresponding to a low enough photon flux to ensure the single-photon absorption condition.

3. Results

The number of emitted neutral particles (with laser pulses) is plotted in the Figure 1 as a function of the storage time, integrated over 3.10^6 injected bunches. Without laser excitation, after a storage time of 0.5 ms, the dissociation probability of the excited ions from the source was negligible and the detected neutrals were due to collisions with residual gas molecules. Therefore, for storage time longer than 0.5 ms, further evolution of the IED is assumed to be due to radiative cooling only.

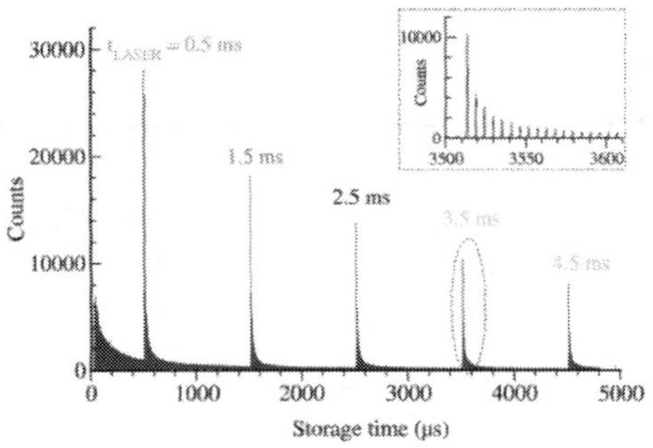

Figure 1. Evolution of the number of neutral particles emitted as a function of the storage time. Laser pulses were sent to excite the ions from 0.5 ms to 4.5 ms every 1 ms.

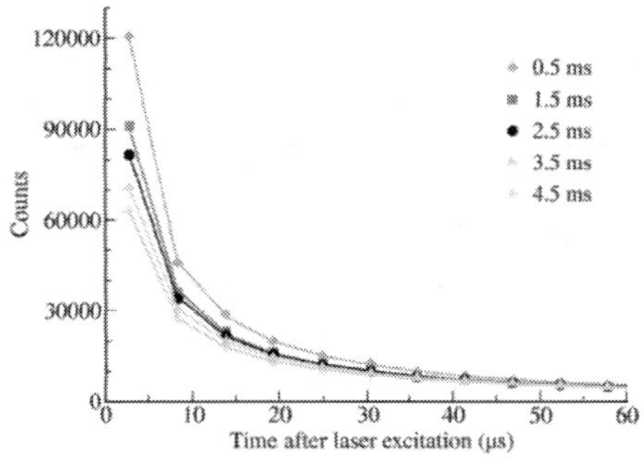

Figure 2. (colored online) Symbols : neutral yield integrated over each revolution period as a function of time after laser excitation, taking t_{laser} as the time reference. Full lines: simulated decay curves adjusted to the experimental data.

Journal of Physics: Conference Series **583** (2015) 012038 doi:10.1088/1742-6596/583/1/012038

The strong increase of neutral numbers at t_{laser} is characteristic of the laser shots. After each laser pulse, about 1% of the ions in the bunch absorbed a photon. With their increased internal energies, these molecules had a higher probability to dissociate. An enlargement inserted in the Figure 1 shows the emission period of neutral particles, which corresponds to the naphthalene cation revolution period. The integral of the neutral yield over each revolution period is plotted in the Figure 2 as a function of the storage time, taking t_{laser} as the time reference.

For each t_{laser}, a neutral decay curve was simulated using a code written in C language. The simulations started with an initial IED of the stored ions at t_{laser} after the photon absorption. The number of emitted neutrals was calculated as a function of time considering the energy dependency of the dissociation rate [12] as well as the probability for the electronic fluorescence emission, which was considered to be negligible for the energies larger than 7.5 eV [13]. Each experimental decay curve was then fitted independently with the modeled decay curve by varying the parameters of the initial IED. The best fit provided the IED after laser excitation. The IED of the stored ions at different t_{laser} prior to laser absorption was then obtained by subtracting the energy of the photon $h\nu = 2.33$ eV (Figure 3). It is noteworthy that due to the sensitive energy window of the experiment (see [11] for more details) the choice of the modeled initial IED is not unique for the low energy part. However the decay curve is very sensitive to the position of the high energy edge of the IED, and not to the low energy shape. For this work a Gaussian shape has been chosen to model the high energy part of the IED, but other shapes with similar high energy edges [7] would provide similar results.

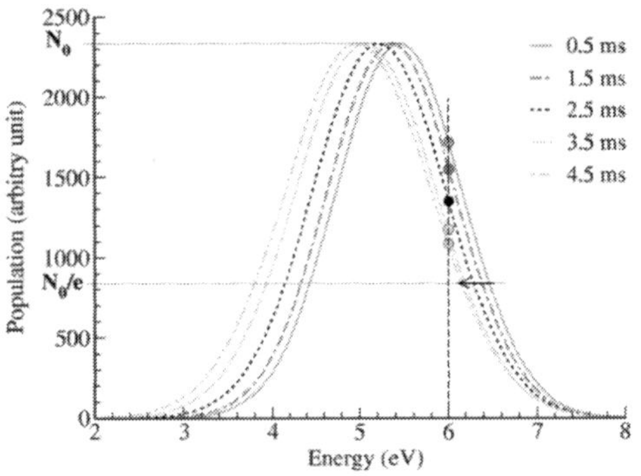

Figure 3. (colored on line) IED of the stored ions before laser excitation. The vertical line shows the population decay from 0.5 ms to 4.5 ms at a given energy.

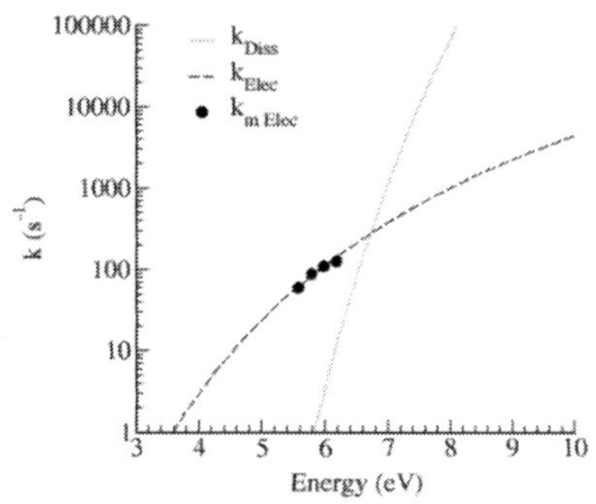

Figure 4. The measured electronic rates k_m compared to the evolution of the theoretical dissociation rate k_{Diss} [12] and fluorescence emission rate k_{Elec} [13].

The time evolution of the energy distribution is distinctly visible in Figure 3. An average shift $(\frac{\Delta E}{\Delta t})_m = 100\ \mathrm{eV \cdot s^{-1}}$ of the high energy edge at N_0/e to lower energy values was measured at about 6.5 eV. This shift is faster than the value $(\frac{\Delta E}{\Delta t})_{IR} = 13\ \mathrm{eV \cdot s^{-1}}$ for infrared emission at about 6.5 eV deduced from Figure 2 of reference [13]. From Figure 3 a measured population decay rate k_m can be determined by fitting with an exponential law the evolution of the population as a function of time for a given energy (not shown). A k_m of 120 s^{-1} was deduced from the fit of the five crossing points at E = 6.0 eV (vertical line in Figure 3). The same method was used to measure the population decay rate at 5.6, 5.8

and 6.2 eV. To extract the measured electronic transition rate k_{mElec} from k_m, it is needed to take account of the contribution from IR emissions. As k_m is proportional to $(\frac{\Delta E}{\Delta t})_m$, k_{mElec} was deduced by subtracting 10% from k_m. The deduced k_{mElec} values are plotted in Figure 4 in comparison with the evolution of the theoretical dissociation rate k_{Diss} and fluorescence emission rate k_{Elec}, as a function of the naphthalene cation internal energy.

Our measured electronic transition rates are in good agreement with the theoretical values calculated with an electronic transition energy of 2.1 eV [14] and an oscillator strength of 0.052 [15]. In the sensitive energy window, k_{Diss} is negligible comparing to k_{Elec}. This demonstrates that the cooling of the hot naphthalene cations stored in the Mini-Ring in the time range from 0.5 to 4.5 ms is mainly due to the electronic radiative transition, as already observed with the anthracene cations [7].

4. Conclusion

Fast population decay rates of naphthalene cations were found to be ranging from about 70 s^{-1} at 5.6 eV to 140 s^{-1} at 6.2 eV. The measurements were performed using a small electrostatic storage ring and laser induced dissociation to probe the evolution of the internal energy distribution (IED) of the stored molecular ions from 0.5 to 4.5 ms storage time. The IED were obtained by fitting the measured decay curves of the emitted neutrals due to unimolecular dissociation. The measured population decay rates were obtained at different internal energies from the evolution of these IED and the measured electronic transition rates were deduced by subtracting the contribution of the IR emission. The fluorescence emission from low lying electronic states is found to be the most efficient radiative cooling channel. This process is strongly dependent on the energy gap from the first electronic state to the electronic ground state of the cation. Further studies with different PAH are planned in order to study the evolution of the fast electronic emission with the molecule properties, which may have an important role in the lifetime and critical size of PAH in interstellar clouds.

References

[1] Allamandola L J, Tielens A G G M and Barker J R 1985 *Astrophys. J.* **290** L25
[2] Tielens A G G M 2013 *Rev. Mod. Phys.* **85** 1021–1081
[3] Leger A and Puget J L 1984 *Astron. Astrophys.* **137** L5–L8
[4] Allain T, Leach S and Sedlmayr E 1996 *Astron. Astrophys.* **305** 602
[5] Leger A, D'Hendecourt L, Boissel P and Desert F X 1989 *Astron. Astrophys.* **213** 351–359
[6] Leger A, Boissel P and d'Hendecourt L 1988 *Phys. Rev. Lett.* **60** 921
[7] Martin S, Bernard J, Brédy R, Concina B, Joblin C, Ji M, Ortega C and Chen L 2013 *Phys. Rev. Lett.* **110** 063003
[8] Ito G, Furukawa T, Tanuma H, Matsumoto J, Shiromaru H, Majima T, Goto M, Azuma T and Hansen K 2014 *Phys. Rev. Lett.* **112** 183001
[9] Andersen J U, Gottrup C, Hansen K, Hvelplund P and Larsson M O 2001 *Eur. Phys. J.* D **17** 189–204
[10] Bernard J, Montagne G, Brédy R, Terpend-Ordacière B, Bourgey A, Kerleroux M, Chen L, Schmidt H T, Cederquist H and Martin S 2008 *Rev. Sci. Instrum.* **79** 075109–8
[11] Chen L, Ji M, Bernard J, Brédy R, Concina B, Joblin C, Ortega C, Montagne G and Martin S 2014 *J. Phys.: Conf. Series* **488** 012039
[12] Ho Y P, Dunbar R C and Lifshitz C 1995 *J. Am. Chem. Soc.* **117** 6504–6508
[13] Boissel P, de Parseval P, Marty P and Lefèvre G 1997 *J. Chem. Phys.* **106** 4973
[14] Sánchez-Carrera R S, Coropceanu V, da Silva Filho D A, Friedlein R, Osikowicz W, Murdey R, Suess C, Salaneck W R and Brédas J L 2006 *J. Phy. Chem.* B **110** 18904–18911
[15] Tan X, Bernstein L, Cami J and Salama F 2011 *Astrophys. J.* **728** 62

Journal of Physics: Conference Series **583** (2015) 012039 doi:10.1088/1742-6596/583/1/012039

Hyperfine splitting of the $2s_{1/2}$ and $2p_{1/2}$ levels in lithium-like Pr^{56+}

E Träbert[1,2], P Beiersdorfer[1], G V Brown[1], J Clementson[1], D B Thorn[1], M H Chen[1], K T Cheng[1] and J Sapirstein[3]

1 Physics Division, Lawrence Livermore National Laboratory, Livermore, CA 94550-9234, U.S.A.
2 Astronomisches Institut, Ruhr-Universität Bochum, D-44780 Bochum, Germany
3 University of Notre Dame, Notre Dame, IN 46556, U.S.A.

E-mail: traebert@astro.rub.de

Abstract. Measurements of hyperfine splittings in highly charged ions are sensitive to details of the nuclear structure and the nuclear magnetic field distribution, but the proper interpretation of the measurements requires that the atomic structure is understood in sufficient detail. We discuss the reasoning behind various recent experiments and what advantage is offered by the study of the Li-like ion of a mid-Z element such as praseodymium.

1. Introduction

The hyperfine splitting of spectral lines has been recognized in the 1930s as a window for the study of nuclear properties, and a large number of nuclear spins have been determined mostly from highly resolved optical spectra. Systematic studies of isotopes and isobars were helped by laser spectroscopy since the 1970s and have yielded hyperfine constants of many nuclei which were tried to interpret by models of nuclear shape and magnetization. A key problem in the quest for determining the nuclear properties, however, was the fact that the actual observations were performed on multi-electron atoms. This problem is aggravated by the Z^3 dependence of hyperfine splittings on the atomic number (nuclear charge) Z, which calls for high-Z elements in the most sensitive measurements of hyperfine splittings. High-Z elements imply relativistic effects in the electronic shells and notably large quantum electrodynamic (QED) contributions from the strong Coulomb field the electrons experience near the nucleus, where they sense the nuclear magnetic field distribution, in addition to the mounting problems of describing such a multi-electron system *per se*.

Theory and atomic structure computations have made great strides to cope with such a complex problem, but it seems advantageous to develop experimental techniques that significantly reduce the complexity by massively reducing the number of electrons involved. The ultimate system here is a hydrogen-like heavy ion, that is a high-Z nucleus with a single electron. The hyperfine splitting in the ground state (n=1) of such atomic systems has been studied by optical spectroscopy at the Livermore SuperEBIT [1, 2, 3, 4, 5, 6, 7] and by laser spectroscopy at the Darmstadt heavy-ion storage ring ESR [8, 9, 10]. In various aspects the agreement with theory has not been fully convincing [11]. Of course, theory can be adjusted to match experimental findings, but then theories that seem to do well for n=1 levels may be tested

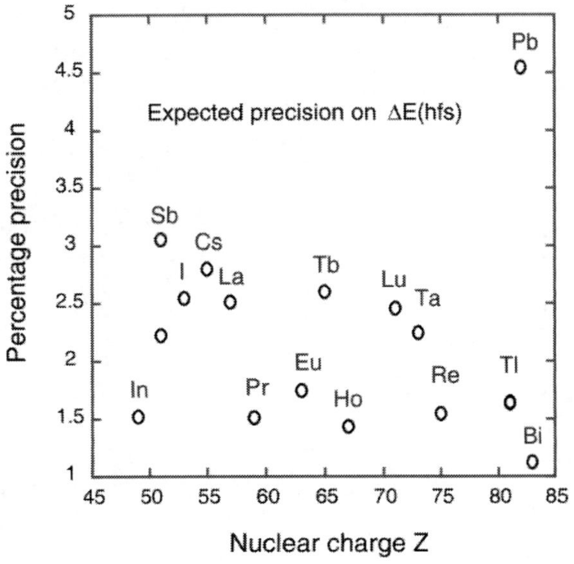

Figure 1. Expected precision of determining the hyperfine splittings in EUV observations of few-electron ions with a flat-field grating EUV spectrometer [12] at Livermore. Similar distributions apply to other instruments, because the relation of the hyperfine splitting to the overall transition energy matters. Incidentally, isotopes of Bi, Pb, Tl, Re, Ho, and now Pr, have been measured by various techniques, and In and Eu are likely candidates for future work.

by measurements of the n=2 hyperfine structure. High-resolution observations of transitions in the n=2 shell of H-like ions suffer from the short level lifetimes involved which leads to lifetime broadening of the spectral lines. This effect can be largely avoided by blocking any decays to the n=1 shell; in Li-like ions the ground state is a n=2 level, at the cost of theory having to deal with a three-electron system (which is much simpler though than, say, a 79-electron system in the case of Pb), with the added benefit of a much higher production yield of the Li-like atomic species (compared to H-like ions) in the experiment.

2. Choice of Li-like ions

S electrons are the best probes of the nuclear structure, because their wave function does not vanish at r=0. In Li-like ions, the transitions with the largest contribution from hyperfine structure are the 2s-2p transitions. The fine structure intervals (such as $2p_{1/2}$ - $2p_{3/2}$) increase with Z^4; in high-Z elements, the 2s-$2p_{1/2}$ transitions lie in the extreme ultraviolet (EUV), while the 2s-$2p_{3/2}$ transitions fall into the X-ray range. The sensitivity of the transition energy measurement to hyperfine effects thus is much larger for the former. The sensitivity is much higher again for observations of the transition between the hyperfine levels of the $2s_{1/2}$ level, but these transitions are in the infrared where the search for such a resonance in a laser experiment suffers directly from the uncertainty of theoretical predictions. It is for this reason that the laser experiment at GSI Darmstadt [10] has taken many years and several attempts to find a resonance in Bi^{80+}, even after X-ray and EUV observations at Livermore [5, 6] had narrowed the parameter space.

In general, one can estimate the sensitivity of any such experiments by the ratio of the hyperfine splitting to the observable transition energy of, for example, observations of the 2s-$2p_{1/2}$ transitions in Li-like ions. The hyperfine splitting depends on the nuclear spin and thus on

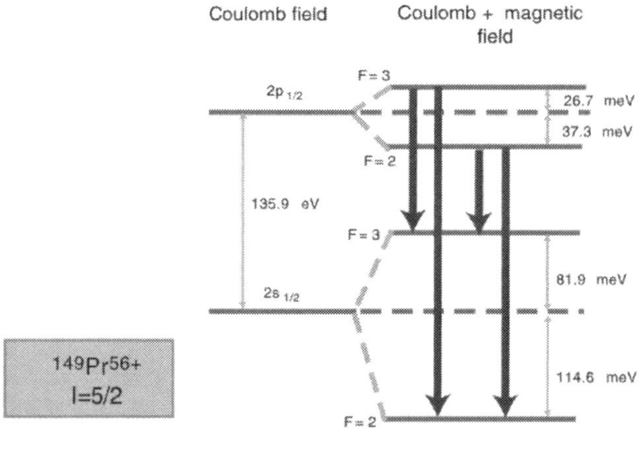

Figure 2. Calculated $2s_{1/2}$ and $2p_{1/2}$ level structure and 2s-2p transitions of $^{141}_{59}$Pr with the fine structure transition on the left and with the hyperfine intervals indicated on the right.

the nuclear properties of each isotope (and calculated data are available for a number of these), while the 2s-$2p_{1/2}$ transition energies usually are calculated without taking nuclear properties into account. For a number of isotopes and 2s-$2p_{1/2}$ transitions in Li-like ions such estimates are shown in fig. 1. Similar estimates pertain to hyperfine effects on transitions in the n=1 shell, the hyperfine splitting of the electronic levels falling off roughly as n^{-3}.

QED contributions scale approximately with Z^4, quite similarly to the fine structure splitting and not very much differently from hyperfine effects. QED effects depend on nuclear size and shape. The doubly magic nucleus $^{208}_{82}$Pb is very close to spherical and assumed to be well known in radius, whereas until recently the shape of the $^{238}_{92}$U nucleus was considered to cause a notable uncertainty in the QED calculations. (Such models and the confidence on their validity are, of course, time dependent.) Hence QED effects are larger in uranium, but possibly under better control in lead. Similar considerations about the nuclear magnetization distribution have lead Shabaev, Volotka and coworkers [13, 14] to suggest that one should measure hyperfine splittings in H- and Li-like ions (and possibly B-like ions as well); some of the less certain terms in the nuclear structure calculations are expected to (largely) cancel in the comparison. Another consideration is that of the relative magnitude of QED effects and hyperfine structure. In the middle of the periodic table (mid-Z elements) the QED contributions are assumed to be small enough and well enough known, so that their uncertainty should be smaller than that of the hyperfine structure.

3. Measurement of Li-like Pr^{56+}

Following the above discussion, we have chosen the isotope $^{141}_{59}$Pr (nuclear spin I=5/2) for studies of the hyperfine effects in Li- and Be-like ions [7]. This isotope has closed neutron shells but an open proton shell that challenges nuclear theory way beyond the 'single particle outside of a closed shell' model, and which results in an above-average hyperfine splitting conducive of reaching a high measurement sensitivity. The calculated 2s-$2p_{1/2}$ level structure and transition array is depicted in fig. 2. A sample of our SuperEBIT measurements in the EUV, using our high-resolution flat-field EUV spectrograph and a cryogenically cooled CCD camera, is shown in fig. 3, representing the sum of more individual exposures than displayed in [7], where the technical details of our experiment have been described.

The results of our measurements on the isotope $^{141}_{59}$Pr, such as the 196.5 ± 1.2 meV $2s_{1/2}$ splitting, agree with theory within the mutual error bars. This indicates a high level of

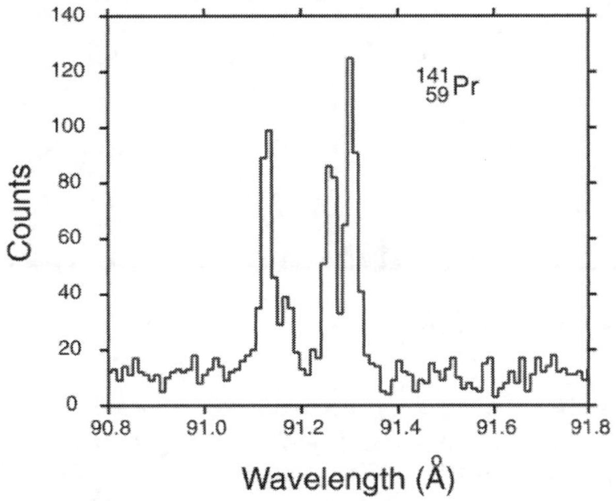

Figure 3. Spectrum of the $2s_{1/2}$ - $2p_{1/2}$ transition array observed with the isotope $^{141}_{59}$Pr in the Livermore SuperEBIT. Several individual spectra have been individually calibrated and then summed.

understanding reached for the atomic structure, which should carry over to higher-Z systems in which relativistic, QED, and nuclear magnetization contributions play a much larger role. An example are the aforementioned experiments on highly charged bismuth ions in which several approaches are pursued in trying to reduce the experimental uncertainties to values smaller than the scatter of the theoretical estimates. Now that the dominant atomic structure part of such calculations has been tested, the sensitivity of those experiments to the QED and nuclear contributions to theory has been enhanced.

Acknowledgements
This work was performed under the auspices of the U.S. Department of Energy by Lawrence Livermore National Laboratory under Contract DE-AC52-07NA27344. E.T. acknowledges support by the German Research Association (DFG) (grants Tr171/18 and Tr171/19).

References

[1] Beiersdorfer P, Behar E, Boyce K R, *et al* 2003 *Nucl. Instrum. Meth.* B **205** 173
[2] Crespo López-Urrutia J R, Beiersdorfer P, Savin D W and Widmann K 1996 *Phys. Rev. Lett.* **77** 826
[3] Crespo López-Urrutia J R, Beiersdorfer P, Widmann K, *et al* 1998 *Phys. Rev.* A **57** 879
[4] Beiersdorfer P, Utter S B, Wong K L, *et al Phys. Rev.* A **64** 032506
[5] Beiersdorfer P, Osterheld A L, Scofield J H, Crespo López-Urrutia J R and Widmann K 1998 *Phys. Rev. Lett.* **80** 3022
[6] Beiersdorfer P, Träbert E, Brown, G V, *et al* 2007 *Abstracts XXVth Int. Conf. on Photonic, Electronic and Atomic Collisions (ICPEAC), Freiburg (Germany) 2007, Abstract Mo179.*
[7] Beiersdorfer P, Träbert E, Brown G V, *et al* 2014 *Phys. Rev. Lett.* **112** 233003
[8] Klaft I, Borneis S, Engel T, *et al* 1994 *Phys. Rev. Lett.* **73** 2425
[9] Seelig P, Borneis S, Dax A, *et al* 1998 *Phys. Rev. Lett.* **81** 4824
[10] Nörtershäuser W, Lochmann M, Jöhren R, *et al* 2013 *Phys. Scr.* **T156** 014016
[11] Beiersdorfer P, Crespo López-Urrutia J R, Utter S B, *et al* 2003 *Nucl. Instrum. Meth.* B **205** 62
[12] Beiersdorfer P, Magee E W, Träbert E, *et al* 2004 *Rev. Sci. Instrum.* **75** 3723
[13] Shabaev V M, Artemyev A N, Yerokhin V A, Zherebtsov O M and Soff G 2001 *Phys. Rev. Lett.* **86** 3959
[14] Volotka A V, Glazov D A, Andreev O V, *et al* 2012 *Phys. Rev. Lett.* **108** 073001

Journal of Physics: Conference Series **583** (2015) 012040 doi:10.1088/1742-6596/583/1/012040

Above threshold ionization of Argon atoms by multicolor XUV radiation

R Della Picca[1] **and E Lindroth**[2]

[1]CONICET and Centro Atómico Bariloche, Av. Bustillo 9500, 8400 Bariloche, Argentina
[2]Department of Physics, Stockholm University, Alba Nova University Center, SE-106 91 Stockholm, Sweden

E-mail: renata@cab.cnea.gov.ar

Abstract. We analyse theoretically the Argon photoelectron spectra produced by strong and extreme ultraviolet radiation of six colors: from the 11th to the 16th harmonics of ω_0 (800nm). In particular we concentrate in the range of the spectra where absorption of two photons occurs. The combination of photons of different frequencies results in eleven peaks that are separated by ω_0. We point out that their relative intensities are very sensitive to the laser pulse parameters and target description. We also compare the theoretical description with experimental results finding good qualitative agreement.

1. Introduction

High Harmonic Generation (HHG) is one of the most interesting phenomena discovered for intense lasers interacting with matter. Since the first observation of the surprising intensity plateau for a wide range of harmonic orders, great progress has been made concerning the understanding and control of the process. Nowadays, the HHG radiation is commonly used as a light-source due to its attractive properties of coherence, very short duration (femto or attoseconds), high intensities and frequencies (see for example [1, 2, 3]). These characteristics of the radiation are necessary for initiating and probing ultrafast electron motion in matter.

In a recent experiment the energy-resolved photoelectron (PE) spectra produced via extreme-ultraviolet-(XUV)-photon above-threshold ionization (ATI) of argon were observed [2]. The XUV radiation consisted of higher-order harmonics (11th to 16th) of the original laser ω_0 ($\lambda = 800$ nm) generated by the process of the relativistic oscillating mirror (so-called ROM) in high-peak-power laser-pulse interaction with solid targets. In Ref. [2] it was possible to measure the ATI peaks in the PE spectra corresponding to different combinations of two harmonic photons leading to the same excess energy.

The theoretical description of this type of process with few-cycles pulses is commonly obtained neglecting the multielectronic nature of the target. For example, the numerical solution of the Time Dependent Schröedinger Equation (TDSE) is mostly performed in the single active electron approximation, also the Strong Field Approximation (SFA) neglects the interaction of the emitted electron with the residual multielectronic target [4]. Alternatively, in the Coulomb-Volkov approach (CV) [6] it is possible to include this interaction in the final state. Additionally, when multiphotonic regime is reached the PE spectra can be factored into two contributions separating the roles of the laser pulse and the parent ion. This approximation, which is based on

Journal of Physics: Conference Series **583** (2015) 012040 doi:10.1088/1742-6596/583/1/012040

the CV approach is here called DipA [5]. The yield factorization allows for the study of complex atoms or molecules since the target contribution, that is proportional to the monochromatic one-photon cross section, can be computed accurately independently of the laser-matter process involved.

In this work we study the above-threshold ionization (ATI) of argon atoms in the context of the DipA approximation and use approximately the same light-pulse parameters as the experiment [2]. Atomic units are employed, except where otherwise stated.

2. Theory

In this section we summarize the CV and DipA approximations [5, 6]. The transition matrix in length gauge reads:

$$T_{fi} = -\mathrm{i} \int_0^\tau \langle \Psi_f^-(t) | \boldsymbol{F}(t) \cdot \boldsymbol{r} | \phi_i(t) \rangle \, \mathrm{d}t \tag{1}$$

where $\Psi_f^-(t) = \phi_f^-(r) \exp\left[-\mathrm{i}(S(\boldsymbol{k},t) - \boldsymbol{A}(t)\cdot \boldsymbol{r} + E_f t)\right]$ is the CV wave function for the final electronic state with ingoing boundary conditions, $\phi_i(t)$ is the asymptotic target wave function with energy E_i in absence of external fields and S is part of the Volkov phase: $S(\boldsymbol{k},t) = \boldsymbol{k}\int^t \boldsymbol{A}(t')dt' + \frac{1}{2}\int^t A^2(t')dt'$. The ejected electron has momentum $\boldsymbol{k} \equiv (k, \Omega)$ corresponding to an energy of $E_f = k^2/2$.

The vector potential is obtained from the electric field of the ionizing radiation pulse as $\boldsymbol{A}(t) = -\int^t \boldsymbol{F}(t\prime)dt'$. In the present work we consider the sum of the 11th to the 16th harmonics of ω_0 ($\lambda = 800$ nm), each with an amplitude F_n, a delay t_n and a duration τ_n:

$$\boldsymbol{F}(t) = \hat{\varepsilon} \sum_{n=11}^{16} F_n \cos(n\omega_0(t - t_n - \tau_n/2)) \sin^2(\pi(t - t_n)\tau_n). \tag{2}$$

Each contribution vanishes outside the time interval $t_n < t < \tau_n$ and has its maximum amplitude at the middle of the symmetric pulse. The linear polarisation is given by the vector $\hat{\varepsilon}$. The duration τ of all the pulse is defined as the mayor value of $\tau_n + t_n$. For a coherent pulse we fix $t_n = 0$, $\tau_n = \tau$ and $F_n = F_0$ for all n.

The differential ionisation probability in energy is obtained from the transition matrix magnitude as

$$\frac{dP^{CV}}{dE} = k \int d\Omega \, |T_{fi}|^2 \tag{3}$$

In the high frequency regime $A \sim F_0/\omega$ is small, therefore the term $\boldsymbol{A}(t)\cdot\boldsymbol{r}$ in the Volkov phase can be neglected given rise to the DipA approximation. In this case temporal and radial integrals are decoupled leading to $T \simeq \boldsymbol{L}(\boldsymbol{k}) \cdot \boldsymbol{M}(\boldsymbol{k})$. Here $\boldsymbol{L} = \langle \varphi_f^- | \boldsymbol{r} | \varphi_i \rangle$ depends only on the target structure. Then, the approximate spectra can be obtained as:

$$\frac{dP^{DipA}}{dE} = k \int d\Omega \, |\boldsymbol{L}(\boldsymbol{k})|^2 \, |\boldsymbol{M}(\boldsymbol{k})|^2 \tag{4}$$

both terms inside the integral depend on the electronic momentum, however an additional approximation can be made to achieve total decoupling (DipA2); that is to average over the angular dependence of the factor $\boldsymbol{M}$:

$$\frac{dP^{DipA2}}{dE} = k \int d\Omega \, |\boldsymbol{L}(\boldsymbol{k})|^2 \, \frac{1}{4\pi} \int d\Omega \, |\boldsymbol{M}(\boldsymbol{k})|^2 \tag{5}$$

Finally, the contribution that accounts for the effect of the electromagnetic field on the free-electron final state is:

$$\boldsymbol{M}(\boldsymbol{k}) = \int_0^\tau \mathrm{d}t \, \boldsymbol{F}(t) \, \mathrm{e}^{\mathrm{i}S(\boldsymbol{k},t)} \, \mathrm{e}^{\mathrm{i}\omega_{fi} t} \tag{6}$$

Journal of Physics: Conference Series **583** (2015) 012040 doi:10.1088/1742-6596/583/1/012040

The square of the dipole matrix element, $|\boldsymbol{L}(\boldsymbol{k})|^2$ in Eq. (5), which depends on the atomic structure is calculated within the Random-Phase approximation, as implemented in Ref. [7]. This approximation, which includes certain classes of many-body effects such as target polarization and ground state correlation, is known to reproduce the main feature of the one-photon cross section for rare gas atoms.

3. Results

3.1. Checking CV, DipA and DipA2 approximations

In a previous work the accuracy of the CV description was analyzed by comparison with TDSE results for ionization of H(1s) [5]. In fact the DipA-approximation reproduces exactly the ionization spectra in the region where the one-photon process is dominant (first ATI peak), and therefore also describes accurately the total rate. However, it underestimates the high-order ATI peaks.

For the present study $A \sim F_0/\omega \ll 1$ and DipA should be applicable. To check this, we present in Fig. 1 the PE spectra for ionization of H from the $3p_0$ orbital with a effective charge Z_{eff} (producing the correct Argon ionization potential). We compare the CV, DipA and DipA2 results. DipA and DipA2 reproduce exactly the first CV-ATI peaks and underestimates with one and two order of magnitude the 2nd and 3rd ATI peaks respectively. Despite this, the shape of the spectra remains unchanged: when we add an arbitrary factor in each region to fit the spectra, the CV results are exactly reproduced. Thus, we expect that both DipA and DipA2 should yield reasonable spectra, although lacking the correct relative normalization for the different multi-photon processes.

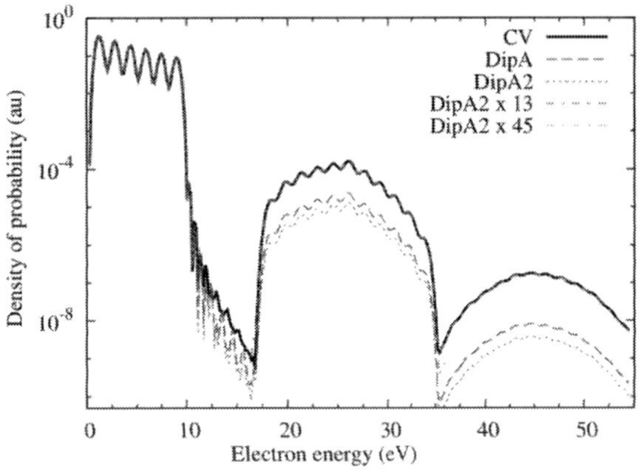

Figure 1. Comparison of CV, DipA and DipA2 spectra for H($3p_0$) ionization due to a coherent laser pulse eq. (2) with ω_0 corresponding to 800nm, $F_n = 0.01$ and τ =8fs. The frequencies varies from $11\omega_0 = 0.63$ to $16\omega_0 = 0.91$ a.u. An effective charge Z_{eff} =3.23 a.u. was used. Three regions can be distinguished: from 0 to 10eV, 15 to 35eV and 35 to 55eV, corresponding to process of absorption of only one, two and three photons respectively.

3.2. Argon spectra

In Fig. 2 we present the Argon spectra obtained with same conditions as above. We show in colors the ionization spectra for each harmonic. The first six peaks are positioned at $E = n\omega_0 - 15.76$eV corresponding to the absorption of only one photon of frequency $n\omega_0$. The first ionization potential of Argon is 15.76eV. As explained in [2], the eleven peaks with energies ranging from 22 to 32 ω_0 correspond to absorption of two photons, $E_{nn'} = n\omega_0 + n'\omega_0 - 15.76$eV where different combinations of n and n' lead to the same excess energy, thus contributing to the same ATI peak. In the right figure we have plotted this part of the spectra in linear scale. The results are presented in the DipA approach for H-like (as in previous section) and Argon. The two spectra show clearly different relative heights of the peaks, showing the sensitivity to the target description. On the other hand, DipA2 and DipA differ only with a global factor. This indicates that the laser field couples uniformly with electrons ejected in different angles.

The Argon spectrum, (red line in the right panel of Fig 2) qualitatively agrees whit the experimental result of Ref. [2].

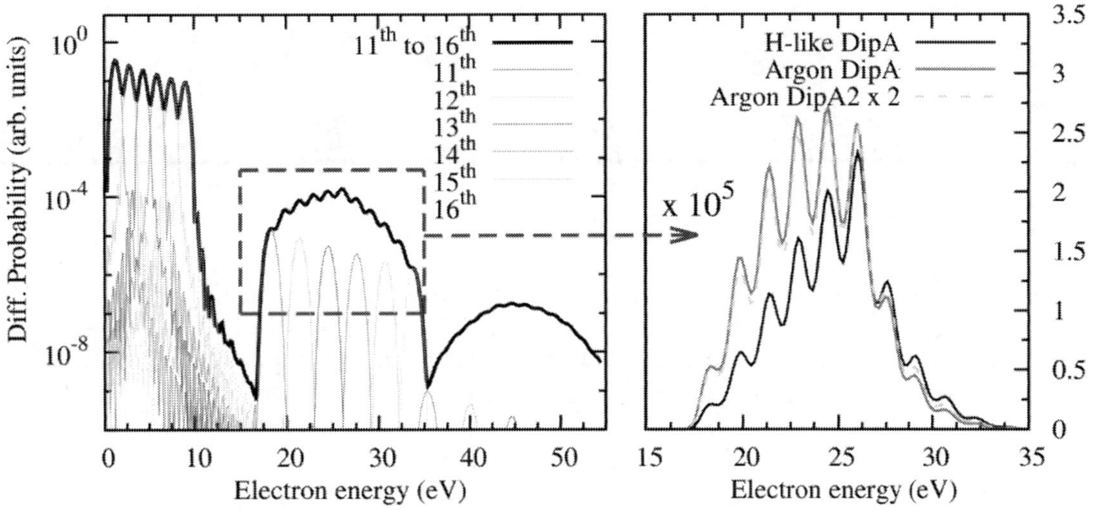

Figure 2. Argon $3p_{m=0}$ ionization spectra as obtained with an 8fs multicolor laser pulse. Black: result from the sum of the harmonics. Colors: contribution from each monochromatic component. Right: zoom of the two-photon absorption region (linear scale).

4. Conclusions

In the present work we analyze theoretically the argon PE spectra, using experimental light-pulse parameters. Due to the high frequency of the multicolor XUV radiation, the DipA approximation in the context of Coulomb Volkov approach should be adequate. We have shown that the applicability of DipA and DipA2 is well justified. These approximations allow to describe the spectra as the contribution of two factors: one accounts for the target information and the other for the electron-laser interaction. In the context of DipA approximation we have calculated the Argon spectra, in particular in the range where absorption of two photons occurs. The combination of photons of different frequencies results in eleven peaks that are separated by ω_0. We observe a variation in the relative intensity of these peaks depending on the description of the target. The theoretical description shows good qualitative agreement with recent experimental results for two-extreme-ultraviolet-photon above-threshold ionization of Argon atoms.

Acknowledgments

This work was performed with the support of the FP7-PEOPLE-2010-IRSES program 269243 DWBQS.

References

[1] Krausz F and Ivanov M 2009 *Rev. Mod. Phys.* **81** 163
[2] Heissler P, Tzallas P, Mikhailova J M, Khrennikov K, Waldecker L, Krausz F, Karsch S, Charalambidis D and Tsakiris G D 2012 *New. J. Phys.* **14** 043025
[3] Billaud P, Géléoc M, Picard Y J, Veyrinas K *et al* 2012 *J. Phys.* B **45** 194013
[4] Milosevic D B, Paulus G G, Bauer D and Becker W 2006 *J. Phys.* B **39** R203
[5] Della Picca R, Fiol J and Fainstein P D 2013 *J. Phys.* B **46** 175603
[6] Duchateau G, Cormier E and Gayet R 2002 *Phys. Rev.* A **66** 23412
[7] Dahlström J M and Lindroth E 2014 *J. Phys.* B **47** 124012

Journal of Physics: Conference Series **583** (2015) 012041 doi:10.1088/1742-6596/583/1/012041

Compton polarimetry using double-sided segmented x-ray detectors

G Weber[1,2], K-H Blumenhagen[1,3], H Bräuning[2], H Ding[1,3], S Fritzsche[1,4], S Hess[2], R Märtin[1,2], U Spillmann[2], A Surzhykov[1,4], S Trotsenko[1,2], D F A Winters[2], V A Yerokhin[5], and Th Stöhlker[1,2,3]

[1] Helmholtz-Institut Jena, 07743 Jena, Germany
[2] GSI Helmholtzzentrum für Schwerionenforschung, 64291 Darmstadt, Germany
[3] Institut für Optik und Quantenelektronik, Friedrich-Schiller-Universität, 07743 Jena, Germany
[4] Theoretisch-Physikalisches Institut, Friedrich-Schiller-Universität, 07743 Jena, Germany
[5] Saint-Petersburg State Polytechnical University, St. Petersburg 195251, Russia

E-mail: g.weber@gsi.de

Abstract. Hard x-ray polarimetry of radiation emitted in collisions of heavy ions, electrons or photons with matter provides detailed information on the collision dynamics as well as of the atomic structure in the presence of extreme field strengths. Moreover, it also opens a route for polarization diagnosis of spin-polarized ion and electron beams which, for example, might be useful in future parity non-conservation studies. Owing to recent progress in the development of highly segmented solid-state detectors, a novel type of polarimeter for the hard x-ray regime has become available. Applied as Compton polarimeters, two-dimensional position-sensitive x-ray detectors now allow for precise and efficient measurements of x-ray linear polarization properties. In this report recent polarimetry studies using such detector systems are reviewed.

1. Introduction

The study of particle and photon polarization phenomena occurring in the interaction of fast ion and electron beams with matter is of particular relevance for the understanding of cosmic and laboratory plasmas where high temperatures, high atomic charge-states and high field strengths prevail. In addition, polarization-sensitive studies of radiative processes in highly-charged, heavy ions may provide detailed insights in both relativistic particle dynamics as well as QED effects and other atomic structure properties at extreme electromagnetic field strengths [1]. Moreover, x-ray polarimetry was proposed as a tool for diagnosis of spin-polarized ion beams [2]. Owing to the recent progress in x-ray detector technology, accurate measurements of the linear polarization for hard x-ray photons as well as the determination of the polarization orientation have become possible. In this report we briefly present the technique of Compton polarimetry and we review a series of recent polarimetry studies that were mainly performed by the atomic physics division of GSI, Darmstadt.

2. Double-sided segmented x-ray detectors applied as Compton polarimeters

While in the soft x-ray regime up to roughly 10 keV linear polarization studies, addressing for example atomic transitions (see, e.g., [3, 4] and references therein) or betatron radiation [5], are routinely performed using crystal optics, hard x-ray polarization measurements were often hampered by the lack of adequate polarization analyzers. Compton polarimetry is the standard technique for hard x-rays, which is based on the fact that the scattered photon is preferably emitted perpendicularly to the incident photon electric field vector, whereas emission in the parallel direction is less probable. Thus, the degree of linear polarization as well as the orientation of the polarization axis of the incident photons can be obtained from the azimuthal emission pattern of the Compton scattered photons. A detailed discussion of this technique can be found in [6].

Though several hard x-ray polarimeter schemes based on the Compton effect were developed during the last decades, in particular in the field of x-ray astronomy [6, 7], most of those instruments are designed for very specific applications and experimental setups, thus being not well-suited (or not affordable) as general-purpose polarimeters for typical x-ray sources in the laboratory. This situation changed with the recent development of highly-segmented planar x-ray detectors, which provide good detection efficiency, energy and time resolution together with millimeter to submillimeter two-dimensional position resolution and a large detection area [8]. When an incident photon undergoes Compton scattering inside such a detector, the scattered photon may be detected at a different position within the same detector crystal. Thus each part of the detector crystal serves both as a scatterer and as an absorber for the scattered x-rays, resulting in a high efficiency due to a big active area and a good coverage of the azimuthal scattering angle. Combining energy, position and time information of each interaction then allows a reconstruction of the Compton scattering events. As a consequence, such detectors are well-suited for the application as Compton polarimeters, enabling precise and efficient linear polarization studies [9, 10, 11]. Depending on the crystal material and the thickness, which can typically reach up to about 2 cm, polarimeters for x-ray energies ranging from a few dozen of keV up to the MeV regime can be constructed.

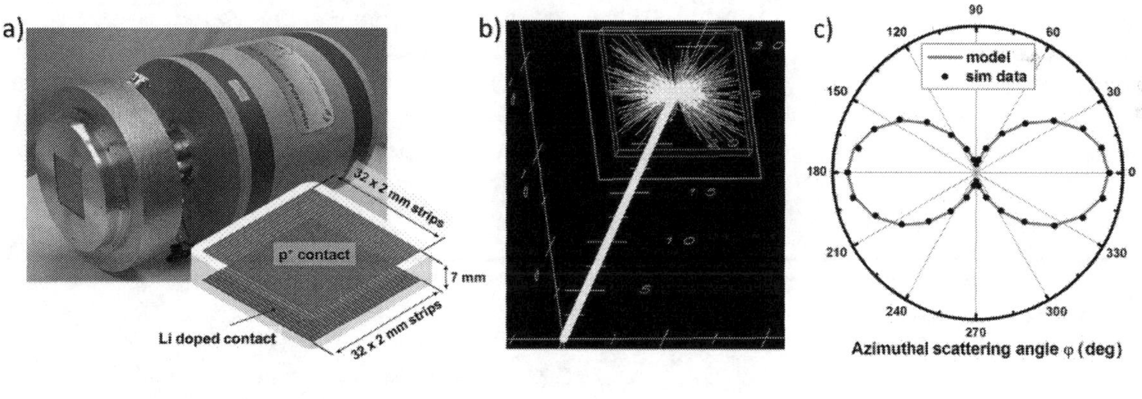

Figure 1. a) Photograph of the 2D Si(Li) polarimeter and a sketch of the crystal segmentation. b) Monte Carlo simulation of 100 % linearly polarized 100 keV photons undergoing Compton scattering inside the detector crystal. c) Azimuthal scattering distribution of the reconstructed Compton events inside the detector. The red line shows an adjustment of the theoretical scattering distribution with the degree of linear polarization as a free parameter.

In recent years several position-sensitive x-ray detectors were developed within a collaboration of the atomic physics division of GSI, the detector laboratory of FZ Jülich and the company Semikon, see [12] for details. The main working horse in terms of polarization measurements is a Si(Li) detector which consists of a 7 mm thick planar crystal with an active area of 64 mm × 64 mm whose electrodes are segmented into 32 horizontal strips on the front and 32 vertical strips on the back side, resulting in a structure of 1024 quadratic pseudo-pixels. Each segment of the detector crystal is connected to a charge sensitive preamplifier acting as an individual detector which provides time ($\Delta t = 50$ to 100 ns) and energy (FWHM = 2 keV at 60 keV) information for the local energy deposition [11]. This detector system was build as a dedicated Compton polarimeter for the SPARC collaboration [13] and is presented in figure 1a, while its measurement principle of x-ray linear polarization is illustrated in figure 1b and 1c. In figure 1b the Compton scattering of a 100 % linearly polarized beam of 100 keV photons is simulated inside the detector crystal using the EGS5 Monte Carlo package [14] while in figure 1c the azimuthal angular distribution of these scattered photons is shown together with an adjustment of the theoretical scattering distribution to the simulated data with the degree of incident photon linear polarization as a free parameter.

In the following we present three significant experimental results in terms of hard x-ray polarimetry in atomic physics that were recently obtained by the GSI atomic physics division.

3. Polarization studies of radiative electron capture

The capture of electrons into bound states of ions is of significant importance for both experiment and theory in the fields of atomic and plasma physics as well as for astrophysics. The capture process is called radiative if it is accompanied by the emission of a photon that carries away the initial electron's kinetic energy and the binding energy of the final state it is captured into. If the initial electron is considered to be free, the capture process is referred to as radiative recombination (RR) [15], being the time-reversal of the photoelectric effect [16], whereas the capture of a bound electron is called radiative electron capture (REC) [17, 18, 19].

The REC process is a prominent charge-changing process for fast, highly-charged ions interacting with dedicated target materials or with residual gas being present in the beamlines of accelerators and storage rings [20]. Moreover, when low- to medium-Z targets and heavy, highly-charged projectile ions are considered, the to-be-captured electrons can be treated as free particles having a momentum distribution equal to the one of the bound target states. This so-called impulse approximation reduces the REC description to the RR cross section convoluted with the incident electron momentum distribution. Consequently, both the REC and the RR process as well as the photoeffect can be treated within the same theoretical framework. Moreover, when compared to the photoeffect, the RR/REC process offers several experimental advantages, such as a more uniform emission pattern due to the partial cancelation of retardation and Lorentz transformation for a moving emitter system and the fact that x-rays, in contrast to electrons, can typically leave the target zone unaffected by secondary-collision effects. These facts motivated various REC measurements aiming for a deeper insight into the photoeffect while exploiting the advantageous experimental conditions present for the study of electron capture into fast, highly charged ions, see [19] and references therein.

A first study of the linear polarization of REC photons was published in 2006 [21] where a 4×4 pixel Ge(i) detector was used for Compton polarimetry of x-rays emitted in collisions of bare uranium ions with a N_2 target at the experimental storage ring (ESR) of GSI. The experimental findings are presented together with theory values in figure 2a. Having only 16 pixels, the relatively low granularity of the detector resulted in a poor angular resolution of the Compton scattering distribution, which limited the experimental accuracy to an uncertainty between ± 5 % and ± 10 % with respect to the degree of linear polarization. The much higher granularity of the newly developed Si(Li) polarimeter (see figure 2b) enables more precise studies and this instrument was already applied in a series of test measurements also addressing the REC radiation [22, 23]. Data analysis is still in progress and we expect an experimental uncertainty for these new polarization studies in the order of ± 1 % of the degree of linear polarization. With regard to future experiments at the new FAIR facility it is worth noting that while the existing ESR is limited to typical ion energies not higher than 400 MeV/u for beams of heavy ions,

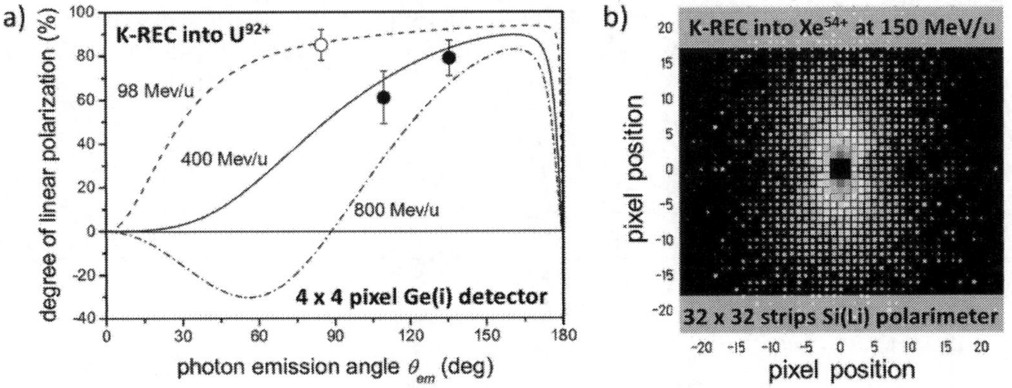

Figure 2. a) Degree of linear polarization of the radiative electron capture (REC) into the K-shell of bare uranium projectiles measured with a 4×4 pixel Ge(i) detector applied as a Compton polarimeter, taken from [21]. b) Position distribution of Compton scattered K-REC photons inside the Si(Li) polarimeter for the capture into bare xenon ions. Data analysis with respect to the degree of linear polarization is ongoing. The much higher granularity of the new detector is expected to enable significantly more precise measurements compared to the 16 pixel detector.

the planned high-energy storage ring (HESR) will reach up to approximately 5 GeV/u for bare uranium [24, 25]. With the extended energy range it will become possible to probe the cross-over effect in the degree of REC photon linear polarization which is predicted to occur at collision energies above 600 MeV/u and for forward emission angles, see the theory data for 800 MeV/u in figure 2a. In terms of the photoeffect this feature indicates that the initially bound electron is no longer preferentially ejected in the direction of the incident photon electric field vector, instead emission along the magnetic field vector is dominant.

4. Determination of atomic multipole-mixing using combined angular distribution and linear polarization measurements

Hydrogen-like ions are the simplest and most fundamental atomic systems whose study along the isoelectronic sequence provides detailed information about the effects of relativity and quantum-electrodynamics on the atomic structure. Since these effects are largest for the 1s ground state, experimental studies of $L \rightarrow K$ transitions are of paramount importance for such investigations. This is in particular true for the domain of high-Z ions, where precision measurements of the Lyman-α_1 (Ly-α_1: $2p_{3/2} \rightarrow 1s_{1/2}$) transition characteristics are a powerful technique to test the theory of strong-field QED [26] and atomic structure [27, 28], as well as to probe relativistic effects on the population processes of the excited state [29, 30]. The latter studies are based on the fact that both the linear polarization and the angular distribution are sensitive to a non-statistical population, i.e. alignment, of the excited states (see figure 3a), which provides subtle details on the dynamics of the respective population processes. However, due to the lack of efficient polarimeter systems for the hard x-ray regime, studies in the high-Z domain were restricted to measurements of the spectral and angular distribution of the emitted radiation.

By applying both the Si(Li) detector described above and a highly segmented two-dimensional position-sensitive Ge(i) detector as Compton polarimeters we recently performed a study of the linear polarization of the Ly-α_1 radiation following radiative electron capture into initially bare uranium ions in collisions with H_2 molecules at the gas target of the ESR. This was the first polarization measurement of a characteristic K-shell transition in a hydrogen-like high-Z system, namely U^{91+}, whose main results were reported in [31]. In accordance with theory [32], a significant depolarization due to the interference between the dominant E1 and the M2 decay branches of the Ly-α_1 radiation was observed, see figure 3c. It is interesting to note that while the linear polarization is decreased by this so-called multipole mixing, the angular distribution is becoming more anisotropic when compared to the pure E1 case.

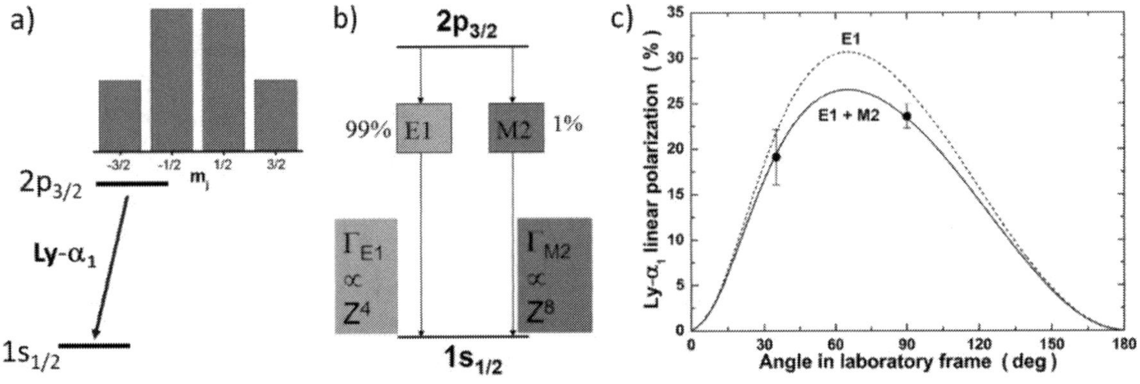

Figure 3. a) Illustration of the non-statistical population of the $2p_{3/2}$ state in a hydrogen-like system. b) Schematic representation of the interference between the dominant E1 and the M2 decay branch. Due to its strong scaling with the atomic number Z, for the heaviest nuclei the M2 branch contributes with about 1 % to the total decay rate. c) Confirmation of the predicted depolarization of the Ly-α_1 radiation as a result of E1-M2 interference in U^{91+}, taken from [31].

Besides the confirmation of the predicted depolarization, it was found that the combination of linear polarization data with a measurement of the Ly-α_1 angular distribution allows to decouple the effects of the collision-induced alignment of the excited $2p_{3/2}$ state and the E1-M2 interference on the transition properties. Thus, a model-independent and precise determination of both the alignment, which is sensitive to the dynamics of the population process, and the ratio of the E1 and M2 transition amplitudes, and consequently the transition rates, is feasible (see [31] for details). Here, in contrast to previous studies based only on the observation of the angular distribution [28], no theoretical assumption about the population mechanism for the excited state or the E1-M2 transition ratio is required. Moreover, with a Γ_{M2}/Γ_{E1} transition rate ratio of 0.00689 with a relative uncertainty of $\pm 2.8\,\%$ a remarkable precision in the determination of the M2 contribution was achieved. This is due to the fact that the experimental observables are sensitive to the ratio of amplitudes rather than transition rates as it is the case for lifetime studies of excited states. This achievement recently motivated a detailed study of the effects of QED on the Ly-α_1 radiation in high-Z systems where they are expected to be roughly of the same order as the experimental uncertainty in the present measurement [33].

5. Polarization transfer in bremsstrahlung emission by spin-polarized electrons

Electron-atom bremsstrahlung, sometimes also referred to as ordinary bremsstrahlung, is one of the basic photon-matter interaction processes. Recent theoretical and experimental studies were focussed on the influence of the incident electron spin-polarization on the bremsstrahlung polarization properties [34, 35, 36]. Due to the spin-orbit interaction a non-zero electron polarization breaks the rotational symmetry of the bremsstrahlung emission with respect to the collision axis, which is defined by the incident electron direction. In terms of bremsstrahlung linear polarization, this gives rise to a rotation of the photon polarization vector with respect to the reaction plane and to an enhanced degree of linear polarization in comparison to the

Journal of Physics: Conference Series **583** (2015) 012041 doi:10.1088/1742-6596/583/1/012041

unpolarized electron case. While a slight azimuthal anisotropy of the bremsstrahlung emission pattern as a result of incident electron spin-polarization was already observed about 20 years ago [37], the effect on the bremsstrahlung polarization, which was predicted to be more pronounced, could not be studied due the lack of adequate hard x-ray polarimeters.

Recently, two experiments were performed at the polarized electron source SPIN of the TU Darmstadt to study the linear polarization of bremsstrahlung arising from the collision of polarized 100 keV electrons with thin gold targets [38, 39]. The latter one of these measurements was performed by members of the atomic physics division of GSI using the Si(Li) polarimeter described above. The detector was located under 130° with respect to the electron beam axis and recorded the degree and the orientation of bremsstrahlung photon polarization between the short-wavelength limit, i.e. the incident electron kinetic energy, and roughly 70 keV, which is the lower threshold of this polarimeter. In figure 4 the influence of the transverse electron spin-polarization on the azimuthal scattering distribution of the Compton scattered photons is presented. In accordance with theory, a significant rotation of the bremsstrahlung polarization axis due the spin-polarization of the incident electron beam is observed. Also a slight increase of the degree of bremsstrahlung polarization was found.

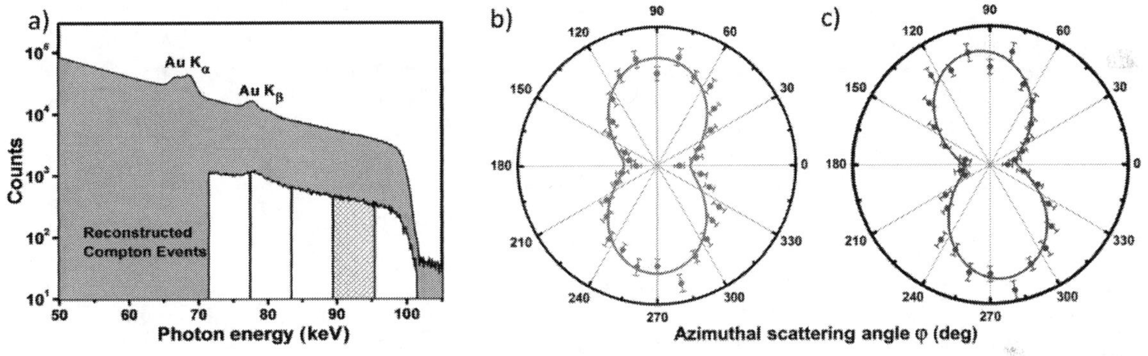

Figure 4. a) Bremsstrahlung spectral distribution arising from 100 keV electrons impinging on a thin gold foil. The data were recorded using a Si(Li) polarimeter located under an observation angle of 130°. For about 1 % of the detected photons a Compton scattering event inside the detector can be reconstructed. For an energy-dependent analysis of the linear polarization features, these events were aggregated into 6 keV broad bins. b) The azimuthal Compton scattering distribution of bremsstrahlung photons with energies of 92.5 ± 3 keV stemming from unpolarized electrons. c) The same distribution for 75 % transversally spin-polarized electrons. A rotation of the polarization axis is visible and the distribution is slightly more anisotropic, indicating a higher degree of linear polarization. All figures were taken from [39].

Such a simultaneous measurement of the degree and the direction of the linear photon polarization as a function of photon energy and electron polarization grants complete insight into the bremsstrahlung polarization correlations, and hence into details of relativistic electron-photon interactions. Beyond the purely scientific interest, there is also a practical application, since similar polarization transfer effects were predicted for the radiative recombination process and are currently discussed as a diagnostic method for spin-polarized heavy ion beams [40]. Using polarized-electron bremsstrahlung as a test case, the present measurements proved the applicability of this technique. Moreover, it was found that bremsstrahlung polarization is surprisingly sensitive to the target thickness, leading to a significant depolarization even for target foils with a thickness of just a few tens of nm [41, 42].

Journal of Physics: Conference Series **583** (2015) 012041 doi:10.1088/1742-6596/583/1/012041

6. Summary

Owing to recent progress in the development of highly segmented solid-state detectors, novel two-dimensional position-sensitive x-ray detectors with time and energy resolution have become available. When applied as Compton polarimeters, such detector systems enable efficient and precise linear polarization studies of photons in the hard x-ray regime. As a consequence, in recent years a variety of important radiative processes were revisited with respect to the linear polarization of the emitted x-rays, enabling deeper insight into basic photon-matter interactions and the atomic structure of high-Z systems. Moreover, it was demonstrated that x-ray polarimetry can serve as a valuable tool for the control of spin-polarized beams of charged particles.

References

[1] Stöhlker Th *et al.* 2009 *The European Physical Journal Special Topics* **169** 5
[2] Bondarevskaya A *et al.* 2011 *Physics Reports* **507** 1
[3] Shlyaptseva A S *et al.* 2001 *Review of Scientific Instruments* **72** 1241
[4] Robbins D L *et al.* 2006 *Phys. Rev. A* **74** 022713
[5] Schnell M *et al.* 2013 *Nature Communications* **4** 2421
[6] Lei F, Dean A and Hills G 1997 *Space Science Reviews* **82** 309
[7] McConnell M and Ryan J 2004 *New Astronomy Reviews* **48** 215
[8] Vetter K 2007 *Annual Review of Nuclear and Particle Science* **57** 363
[9] Kroeger R, Johnson W, Kurfess J and Phlips B 1999 *Nucl. Instr. Meth. Phys. Res. A* **436** 165
[10] Spillmann U *et al.* 2008 *Review of Scientific Instruments* **79** 083101
[11] Weber G *et al.* 2010 *Journal of Instrumentation* **5** C07010
[12] Weber G *et al.* 2012 *AIP Conference Proceedings* **1438** 73
[13] Stöhlker Th *et al.* 2011 *AIP Conference Proceedings* **1336** 132
[14] Weber G *et al.* 2011 *Physica Scripta* **T144** 014034
[15] Oppenheimer J R 1928 *Phys. Rev.* **31** 349
[16] Scofield J H 1989 *Phys. Rev. A* **40** 3054
[17] Schnopper H W *et al.* 1972 *Phys. Rev. Lett.* **29** 898
[18] Ichihara A *et al.* 1994 *Phys. Rev. A* **49** 1875
[19] Eichler J and Stöhlker Th 2007 *Physics Reports* **439** 1
[20] Stöhlker Th *et al.* 1998 *Phys. Rev. A* **58** 2043
[21] Tashenov S *et al.* 2006 *Phys. Rev. Lett.* **97** 223202
[22] Hess S *et al.* 2009 *Journal of Physics: Conference Series* **194** 012025
[23] Hess S *et al.* 2009 *Journal of Physics: Conference Series* **163** 012072
[24] Stöhlker Th *et al.* 2013 *Physica Scripta* **T156** 014085
[25] Stöhlker Th *et al.* 2014 *Hyperfine Interactions* **227** 45
[26] Gumberidze A *et al.* 2005 *Phys. Rev. Lett.* **94** 223001
[27] Stöhlker Th *et al.* 1997 *Phys. Rev. Lett.* **79** 3270
[28] Surzhykov A, Fritzsche S, Gumberidze A and Stöhlker Th 2002 *Phys. Rev. Lett.* **88** 153001
[29] Gumberidze A *et al.* 2013 *Phys. Rev. Lett.* **110** 213201
[30] Tashenov S *et al.* 2014 *Phys. Rev. Lett.* **113** 113001
[31] Weber G *et al.* 2010 *Phys. Rev. Lett.* **105** 243002
[32] Surzhykov A, Fritzsche S and Stöhlker Th 2003 *Hyperfine Interactions* **146-147** 35
[33] Surzhykov A *et al.* 2014 *manuscript in preparation*
[34] Shaffer C D, Tong X M and Pratt R H 1996 *Phys. Rev. A* **53** 4158
[35] Yerokhin V A and Surzhykov A 2010 *Phys. Rev. A* **82** 062702
[36] Jakubassa-Amundsen D and Surzhykov A 2011 *The European Physical Journal D* **62** 177
[37] Mergl E, Prinz H T, Schröter C D and Nakel W 1992 *Phys. Rev. Lett.* **69** 901
[38] Tashenov S *et al.* 2011 *Phys. Rev. Lett.* **107** 173201
[39] Märtin R *et al.* 2012 *Phys. Rev. Lett.* **108** 264801
[40] Surzhykov A, Fritzsche S, Stöhlker Th and Tashenov S 2005 *Phys. Rev. Lett.* **94** 203202
[41] Weber G *et al.* 2012 *Nucl. Instrum. Meth. B* **279** 155
[42] Märtin R *et al.* 2013 *Physica Scripta* **T156** 014070

Journal of Physics: Conference Series **583** (2015) 012042 doi:10.1088/1742-6596/583/1/012042

PAH radiative cooling and fragmentation kinematics studied within an electrostatic ring

R Brédy[1], C Ortéga[1], M Ji[1], L Chen[1], J Bernard[1], A R Allouche[1], C Joblin[2], A Cassimi[3] and S Martin[1]

[1]Institut Lumière Matière, UMR5306 Université Lyon 1-CNRS, Université de Lyon 69622 Villeurbanne cedex, France
[2]Université de Toulouse, UPS-OMP, IRAP, Toulouse, France
[3]CIMAP, GANIL Université de CAEN, Bd. H. Becquerel, Caen, France

E-mail: richard.bredy@univ-lyon1.fr

Abstract. Radiative cooling of polycyclic aromatic hydrocarbon (PAH) cations has been studied using a compact electrostatic ion storage ring, the Mini-Ring, in a time range up to 8 ms. The time evolution of the internal energy distribution of the ensemble of stored ions shows evidences of fast cooling which is attributed to the fluorescence from thermally excited electronic states. The internal energy distribution was probed by inducing unimolecular dissociation with single-photon absorption at given storage times. Information on the fragmentation kinematics and the dissociation channels were obtained by analyzing the image of the emitted neutrals detected with a time and position sensitive detector.

1. Introduction

Interests for polycyclic aromatic hydrocarbons (PAH) have been developed since almost thirty years as these molecules are considered as good candidates to explain the unidentified infrared emission bands observed in some interstellar regions [1]. However, this commonly accepted PAH hypothesis could not yet be confirmed unambiguously and complex related PAH species including grains [2], fullerenes [3] and organic nanoparticles [4, 5] are considered as well. Numerous experimental and theoretical studies have been conducted to explain formation and survival of PAH in the interstellar medium [6, 7]. For vibrationally excited molecules competition between the internal energy relaxation mechanisms such as fragmentation and radiative cooling is a key parameter to understand their survival in a highly isolated environment as in space. It is now well known that the main dissociation channels of PAH are the loss of hydrogen (H, H_2) and acetylene (C_2H_2) depending on the size and internal energy of the molecule [8-10]. Regarding the radiative cooling mechanisms of small molecules, infrared (IR) emission due to vibrational transitions in the electronic ground state is usually considered as a slow process. For PAH the average IR radiative cooling rate has been calculated to be of the order of 2 s^{-1} nearly independent on the size and internal energy of the PAH [11]. However, evidences of fast radiative cooling have been observed very recently for anion clusters [12, 13] and anthracene cations [14] using ion storage techniques. This fast cooling was attributed to fluorescence emission from low lying electronic excited states predicted by Leger et al. [7, 11]. This process is

Journal of Physics: Conference Series **583** (2015) 012042 doi:10.1088/1742-6596/583/1/012042

induced by inverse internal conversion (IIC) which converts vibrational excitation energy stored in the electronic ground state to electronically excited states. Due to the high spontaneous electronic transition rates, in the order of 10^6-10^7 s^{-1} for anthracene cation, fast electronic transition can occur and the energy release by the emitted fluorescence photon ($h\nu_{elec}$=1.7eV for anthracene) is much larger than for the emission of a single IR photon (typically 0.1-0.2 eV). A first indirect evidence of the contribution of the electronic transition to the cooling has been provided for anthracene cation by Boissel et al. [11]. They needed to include the electronic transition process in their numerical simulations in order to reproduce the measured photofragmentation efficiency curve of the cold trapped ions after irradiation with a continuous and filtered visible light emitted from a Xenon arc lamp. The fast cooling rate of anthracene cation could be measured only recently by following the time evolution of the internal energy distribution (IED) of an ensemble of ions stored inside a storage ring [14, 15].

In this paper we report on the measurement of the population decay of anthracene cations at a given internal energy by analyzing the shift of the IED as a function of time. This population decay is directly related to the radiative cooling mechanism and could be compared with the calculated fluorescence emission rate due to the electronic transition. The IED of the ensemble of the stored ions was probed using laser induced dissociation and preliminary results on the fragmentation kinematics of other PAH cations (naphthalene, pyrene) are also presented.

2. Radiative cooling of anthracene cations

The radiative cooling of PAH was studied using an electrostatic ion storage ring, the Mini-Ring [14-17]. Anthracene molecules ($C_{14}H_{10}$) were ionized in an Electron Cyclotron Resonance (ECR) ion source creating an ensemble of ions with a broad IED. After extraction and acceleration of the ion beam to 12keV kinetic energy, a bunch of anthracene cations (1μs width) was injected inside the ring and stored up to 8ms with a 6.5 μs revolution period. The pressure inside the Mini-Ring vacuum chamber was maintained under 2 10^{-9} mbars to limit probability of collision with the residual gas hence ensuring a storage lifetime up to 300 ms. The evolution of the IED of the stored ions is complex and includes the depletion of the high energy population due to the delayed dissociation as well as the shift of the population toward lower energy governed by the radiative cooling mechanisms. Neutral fragments emitted due to unimolecular dissociation were detected as a function of the storage time by a time and position sensitive detector (PSD) facing one of the straight line of the stored ion trajectory. After dissociation, the daughter ions were quickly ejected out of the ring due to the decrease of their kinetic energy which did not fulfil the stable storage conditions anymore. For anthracene cations previous results indicates that the dissociation process is quenched by the radiative cooling for time t>3ms [14, 15]. Therefore, if collisions with the residual gas is neglected, for time t>3ms further evolution of the high energy part of the IED can be attributed to the radiative cooling. In this time range, the IED was probed at a controlled storage time t_{laser} by merging a nanosecond laser UV pulse ($h\nu$=3.5eV, 3rd harmonic of a 1kHz Nd-YAG OPO Laser) with the stored ion bunch in order to induce dissociation. To ensure single photon absorption condition, the pulse energy was controlled to be about 300 μJ/pulse. As discussed in [15] the laser wavelength has to be chosen carefully in order to bring the high-energy edge of the IED inside the sensitive energy window of the experiment. Indeed, the decay of the enhanced neutral yield induced by the photon absorption and detected with the Mini-Ring setup was sensitive only to dissociation processes of anthracene cations with energy between 9 and 10.5 eV. The IED after laser excitation was determined by analyzing the decay profile. A simulated decay curve was calculated based on the statistical dissociation rate as a function of the internal energy of anthracene [18] and a modeled IED $g^*(E,t)$ after laser excitation. The experimental neutral decay curve was fitted by varying the parameters of the modeled $g^*(E,t)$. Assuming the energy dependence of the absorption efficiency to be rather weak in the considered energy range, the IED before laser excitation $g(E,t)$ was then obtained by subtracting the energy $h\nu$=3.5eV of the absorbed photon $g(E,t) = g^*(E,t) - h\nu$.

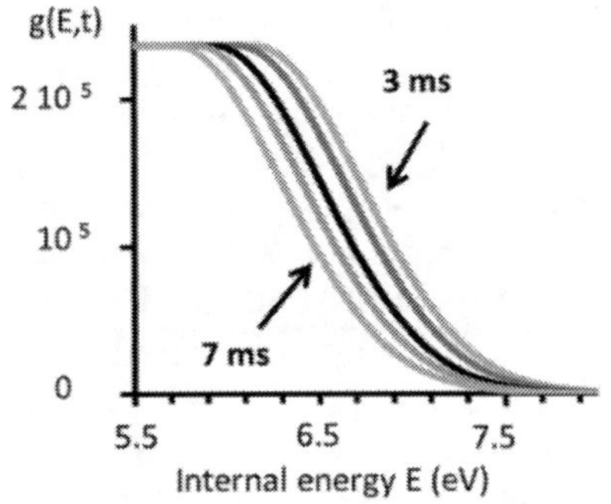

Figure 1. Internal energy distribution g(E,t) of the stored ions at 3, 4, 5, 6 and 7 ms of storage time.

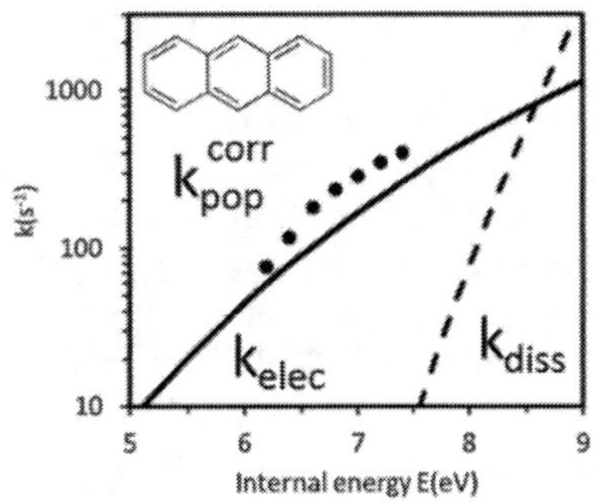

Figure 2. Symbols: population decay rates corrected from the IR contribution, full line: theoretical electronic transition rate for fluorescence emission, dash line: dissociation rate

In this work, the IED was modelled using a half Gaussian function for the high energy part and a constant function for the low energy part. The shape of this modelled IED is not unique and different functions were tested. However the fit of the experimental data is very sensitive to the shape of the high-energy edge of the IED and the half Gaussian shape led to the best fit. The evolution of the IED before photon absorption g(E,t) is presented in figure 1 for storage time between 3 and 7 ms. As expected, g(E,t) shifted toward lower energy with increasing storage time. In the considered time range a shift ΔE=-0.5eV was measured at an average internal energy $<E>$=6.8 eV leading to a mean cooling rate $k_{cool}=<E>^{-1} |\Delta E/\Delta t| = 18s^{-1}$. This cooling rate is much higher than the expected cooling rate for IR emission $k_{cool,IR} = 2s^{-1}$ calculated by Boissel et al. [11]. At 7eV, the cooling $[\Delta E/\Delta t]_{IR}$ = 14eV/s due to IR emission represents roughly 10% of the total measured value $[\Delta E/\Delta t]_{mes}$ = 125eV/s. Therefore the observed fast cooling cannot be attributed only to emission from vibrational transitions which appeared to be a minor process at the considered energies. The shift toward lower energy of the IED induced ion population decay at a given internal energy E. The population decays measured at E=6.2 up to 7.2 eV follow an exponential decay law and the population decay rates k_{pop} range from $85s^{-1}$ up to $390s^{-1}$, respectively. The contribution $k_{pop,IR}$ of the IR emission to this population decay rate depends on the energy but also on the shape of the IED. It was found that $k_{pop,IR}$ is proportional to k_{pop} with a coefficient corresponding to the ratio $[\Delta E/\Delta t]_{IR}/[\Delta E/\Delta t]_{mes} = 0.1$. The population decay rates corrected from the IR contribution $k_{pop}^{corr} = k_{pop} - k_{pop,IR}$ are plotted in figure 2. Considering radiative cooling due to the electronic transitions, the emission of a fluorescence photon of energy $h\nu_{elec}$ induces population decay at the energy E and an increase of the population at the energy E- $h\nu_{elec}$. As a first approach, the increase of the population at the energy E due to fluorescence from molecular ions at energy E+$h\nu_{elec}$ is considered negligible due to the high dissociation probability in the considered energy range (7+1.7=8.7eV). Considering only the first electronic excited state, the theoretical rate k_{elec} of emission of a fluorescence photon $h\nu_{elec}$=1.7eV was calculated by the product of the probability of presence in the excited state by the Einstein coefficient of the transition [11]. Results are potted in figure 2 together with the dissociation rate for comparison. From figure 2 it can be noticed that the fast measured population decay rate k_{pop}^{corr} is much higher than the dissociation rate k_{diss} and is in fairly good agreement with the theoretical electronic transition rate k_{elec}. Therefore the observed fast cooling is attributed to the predicted fluorescence emission from thermally excited electronic states. It is noteworthy that this cooling process is expected to depend strongly on the electronic structure of the molecular ion. Results obtained on naphthalene cation corroborated this interpretation. The importance

Journal of Physics: Conference Series **583** (2015) 012042 doi:10.1088/1742-6596/583/1/012042

of low lying electronic states in the radiative cooling process was also confirmed very recently by the observation of a drastic change in the cooling of small anion clusters C_6^- and C_6H^- [13].

3. Fragmentation kinematics

Determination of the IED through the decay of the neutral signal as a function of time after the laser excitation relies on a simple unimolecular statistical dissociation model. For PAHs, the main fragmentation scheme presents a strong competition between the loss of H and C_2H_2. The knowledge of the branching ratios of the competitive channels as well as the evolution of the dissociation rates with the internal energy is therefore crucial to perform the analysis. In the following, preliminary results on the fragmentation kinematics of PAH are presented using the image of the neutrals impacting the PSD. Figure 3(a) presents the PSD image for neutrals emitted by stored pyrene cations ($C_{16}H_{10}^+$). Figure 3(b) and figure 3(c) are the projections of the PSD images selected for neutrals emitted only during the first revolution period after laser excitation at t_{laser} = 0.5ms for anthracene ($C_{14}H_{10}^+$) and pyrene cations, respectively. The detection efficiency of a neutral fragment by the PSD strongly depends on its kinetic energy. Assuming the velocity of the neutral fragment equals to that of the parent ion, the kinetic energy of C_2H_2 (m=26 a.m.u.) is 26 time higher than for H (m=1 a.m.u.). For anthracene cation (m=178 a.m.u.) stored at 12keV kinetic energy, the detection efficiency of the lighter fragment H at lower kinetic energy (<100eV) is estimated to be less than 5% [19] while it is estimated to about 40% for C_2H_2 [20]. Moreover, the collection efficiency for the H loss channel is also reduced due to the kinetic energy release (KER). The average KER for this channel was measured from mass-analyzed ion kinetic-energy spectra (MIKES) to be 0.35eV for pyrene [21]. Based on this value, the diameter of the spot of the neutrals H hitting the PSD was calculated to be about 4cm, a value larger than the diameter (2.5 cm) of the detector. For anthracene cation, comparable dissociation rates are observed for both H and C_2H_2 channels [18]. Due to the very low detection and reduced collection efficiencies for H fragment, impacts recorded with the PSD were mainly attributed to the emitted C_2H_2 neutrals. Based on this assumption, a maximum KER of 0.5±0.2 eV was measured from the spot diameter on the PSD [22]. This value is in good agreement with previous MIKES measurement [18] and support the attribution of the detected fragments to C_2H_2 emission for anthracene cation (figure 3(b)). For pyrene cation the PSD image and its projection clearly exhibits two components. The main intense spot is attributed to C_2H_2 emission, similar as the anthracene case. The second component, superimposed to the main spot is attributed to H emission. Indeed branching ratios of 1.3%/98.5% was measured for the C_2H_2/H channels [23]. Despite the low detection and collection efficiencies of the emitted H fragment, this channel could be detected in the case of pyrene due to its high branching ratio. For two other PAH molecules, we have noticed that the PSD image

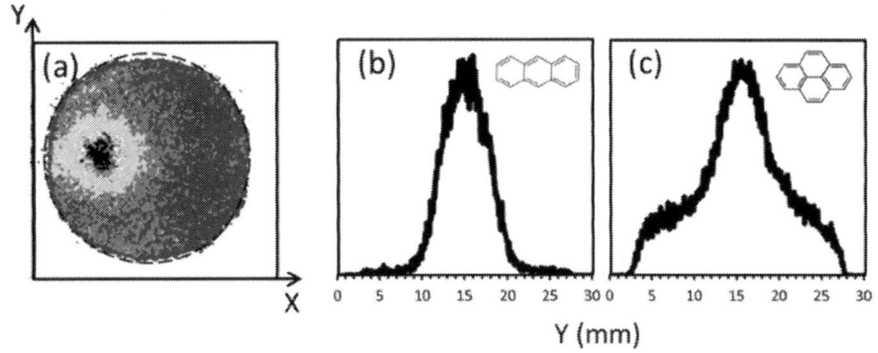

Figure 3. (a) PSD image for neutrals emitted by stored pyrene cations ($C_{16}H_{10}^+$) (b) projection of the PSD image for stored anthracene cations ($C_{14}H_{10}^+$) (c) projection of the PSD image for stored pyrene cations ($C_{16}H_{10}^+$).

obtained for naphthalene ($C_{10}H_8^+$) presents the same characteristic as for anthracene while for coronene ($C_{24}H_{12}^+$) the intensity of the signal attributed to C_2H_2 is drastically reduced. For the cata-condensed PAH, naphthalene and anthracene, the emission of C_2H_2 is indeed considered as an important channel while for the more compact structure of the peri-condensed pyrene and coronene molecules the loss of H is the dominant channel.

Successive loss of H from pyrene cation $C_{16}H_{10}^+$ has been investigated by storing successively $C_{16}H_{10-n}^+$ (n=0 - 4) within the Mini-Ring. The mass spectrum of ions extracted from the ECR source is presented in Figure 4(a). Figures 4(b)-(f) present the partial projections of the PSD images obtained by recording the emitted neutrals during 70μs after laser excitation of the stored ions at t_{laser} = 0.5ms. The projections have been normalized to the intensity of the main peak attributed to the detection of C_2H_2 fragments. The ratio between the intensity of the main peak and the intensity of the component attributed to the loss of H provides qualitative information on the C_2H_2/H branching ratios of the $C_{16}H_{10-n}^+$ parent ion. The intensity of the signal attributed to the H loss channel decreases with the dehydrogenation of the parent molecular ion except for $C_{16}H_7^+$. For $C_{16}H_7^+$ this intensity is close to the one observed for $C_{16}H_9^+$. This is interpreted as a higher dissociation probability via H loss for $C_{16}H_9^+$ and $C_{16}H_7^+$. This interpretation is corroborated by recent calculations on the dissociation energies for the successive loss of H from pyrene cation [23]. The dissociation energies exhibit an odd-even effect with lower dissociation energy for $C_{16}H_9^+$ and $C_{16}H_7^+$ (table 1) indicating a lower stability of these molecular ions. This feature is expected as the lowest energy structures with an even number of empty hydrogen sites always have them paired [23]. This is also observed in the alternation of the peak intensity of the mass spectrum (figure 4(a)). The decay curve of the neutral signal associated to H emission can be obtained using a gate on the PSD (not shown). It is noteworthy that the same odd-even effect is observed with a faster decay of the neutral yield for $C_{16}H_9^+$ and $C_{16}H_7^+$. This observation supports our interpretation.

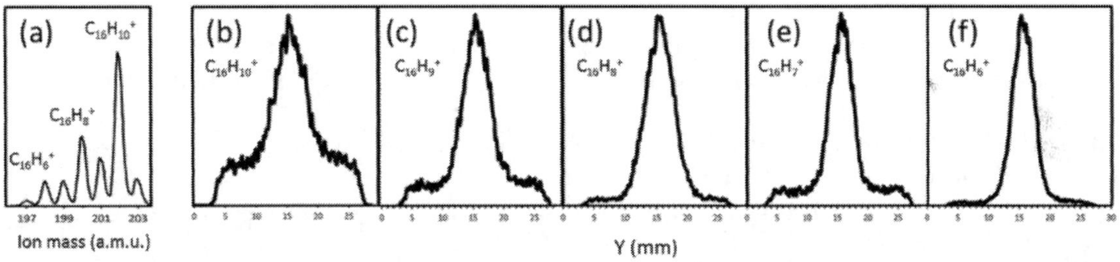

Figure 4.(a) mass spectrum for pyrene ions, (b)-(f) partial projections of the PSD image for neutrals emitted by stored $C_{16}H_{10-n}^+$ (n=0 - 4). The projections have been normalized to the intensity (arbitrary units) of the main peak.

Table 1. Calculated dissociation energies (eV) for H and C_2H_2 loss from $C_{16}H_{10-n}^+$ (n=0-4) parent ions (from [23]).

Parent ion	H loss	C_2H_2 loss
$C_{16}H_{10}^+$	5.10	6.02
$C_{16}H_9^+$	3.50	
$C_{16}H_8^+$	4.48	6.00
$C_{16}H_7^+$	4.11	
$C_{16}H_6^+$	4.45	

Journal of Physics: Conference Series **583** (2015) 012042 doi:10.1088/1742-6596/583/1/012042

4. Conclusion

PAH radiative cooling and fragmentation kinematics were studied using an electrostatic storage ring, the Mini-Ring. The time evolution of the internal energy distribution of the stored ions was probed using laser induced photo-dissociation. The observed fast radiative cooling is attributed to fluorescence from thermally excited electronic state populated via inverse internal conversion. By detecting the emitted neutral fragments with a position sensitive detector, information on the fragmentation channels and their qualitative branching ratios were obtained. Coincidence measurements between the neutral fragment and the charged daughter ion are planned in the future to unambiguously assign the observed decay channels.

Acknowledgments

This work has been supported by the ANR 2010-042601"ANNEAU".

References

[1] Tielens A G G M 2013 *Rev. Mod. Phys.* **85** 1021
[2] Draine B T and Li A 2001 *Astrophys. J.* **551** 807
[3] Cami J, Bernard-Salas J, Peeters E and Malek S E 2010 *Science* **329** 1180
[4] Kwok S and Zhang Y 2011 *Nature* **479** 80
[5] Li A and Draine B T 2012 *Astrophys. J. Lett.* **760** L35
[6] Allain T, Leach S and Sedlmayr E 1996 *Astron. Astrophys.* **305** 602
[7] Léger A, Boissel P and d'Hendecourt L 1988 *Phys. Rev. Lett.* **60** 921
[8] Jochims H W, Ruhl E, Baumgartel H, Tobita S and Leach S 1994 *Astrophys. J.* **420** 307
[9] Dyakov Y A, Ni C, Lin S H, Lee Y T and Mebel A M 2006 *Phys. Chem. Chem. Phys.* **8** 1404
[10] Holm A I S, Johansson H A B, Cederquist H and Zettergren H 2011 *J. Chem. Phys.* **134** 044301
[11] Boissel P, de Parseval P, Marty P and Lefèvre G 1997 *J. Chem. Phys.* **106** 4973
[12] Andersen J U, Gottrup C, Hansen K, Hvelplund P and Larsson M O 2001 *Eur. Phys. J. D* **17** 189
[13] Ito G, Furukawa T, Tanuma H, Matsumoto J, Shiromaru H, Majima T, Goto M, Azuma T and Hansen K 2014 *Phys. Rev. Lett.* **112** 183001
[14] Martin S, Bernard J, Brédy R, Concina B, Joblin C, Ji M, Ortega C and Chen L 2013 *Phys. Rev. Lett.* **110** 063003
[15] Chen L, Ji M, Bernard J, Brédy R, Concina B, Joblin C, Ortega C, Montagne G and Martin S 2014 *J. Phys.: Conf. Series* **488** 012039
[16] Bernard J, Montagne G, Brédy R, Terpend-Ordacière B, Bourgey A, Kerleroux M, Chen L, Schmidt H T, Cederquist H and Martin S 2008 *Rev. Sci. Instrum.* **79** 075109
[17] Ji M, Brédy R, Chen L, Bernard J, Concina B, Montagne G, Cassimi A and Martin S 2013 *Phys. Scr.* **T156** 014091
[18] Ling Y and Lifshitz C 1998 *J. Phys. Chem. A* **102** 708
[19] Peko B and Stephen T 2000 *Nucl. Instrum. Methods Phys. Res., Sect B* **171** 597
[20] Takahashi N, Adachi Y, Saito M and Haruyama Y 2013 *Nucl. Instrum. Methods Phys. Res., Sect B* **315** 51
[21] Ling Y Y, Gotkis Y and Lifshitz C 1995 *Eur. J. Mass Spectrom.* **1** 41
[22] Ji M, Montagne G, Brédy R, Bernard J, Chen L and Martin S 2013 *Phys. Scr.* **T156** 014092
[23] West B, Useli-Bacchitta F, Sabbah H, Blanchet V, Bodi A, Mayer P M and Joblin C 2014 *J. Phys. Chem. A* **118** 7824

Journal of Physics: Conference Series **583** (2015) 012043 doi:10.1088/1742-6596/583/1/012043

Characterization of a double-sided Si(Li) strip Compton polarimeter

K-H Blumenhagen[1,2,3], A Gumberidze[2,4], R Märtin[1,2], N Schell[5], U Spillmann[2], G Weber[1,2] and Th Stöhlker[1,2,3]

[1] Helmholtz-Institut Jena, Fröbelstieg 3, 07743 Jena, Germany
[2] GSI Helmholtzzentrum für Schwerionenforschung GmbH, Planckstraße 1, 64291 Darmstadt, Germany
[3] Institut für Optik und Quantenelektronik, Friedrich-Schiller-Universität Jena, Max-Wien-Platz 1, 07743 Jena, Germany
[4] ExtreMe Matter Institute EMMI, GSI Helmholtzzentrum für Schwerionenforschung, Planckstraße 1, 64291 Darmstadt, Germany
[5] Helmholtz-Zentrum Geesthacht - Zentrum für Material- und Küstenforschung GmbH, Max-Planck-Straße 1, 21502 Geesthacht, Germany

E-mail: `k.-h.blumenhagen@gsi.de`

Abstract. The response of a double-sided segmented Si(Li) detector system has been investigated. The detector has been irradiated with a collimated, highly linearly polarized beam of 53.2 keV photons from the synchrotron radiation source PETRA III at DESY. The detector was mounted on a platform that could be moved with μm precision thus allowing for a defined beam position on the detector surface. In this paper, the effects of the isolation gaps (gap width = 50 μm) between adjacent segments (strips) were studied, in particular with respect to the effect of charge sharing. The fraction of such charge sharing events increases from about 5% (beam hits center of a strip) to over 50% when the beam is focused just on a gap. The fraction of reconstructed Compton scattering events, which is interesting for Compton polarimetry, amounts to about 3% with the beam impinging at a strip center and 2.8% on average. It can therefore be concluded that events related to charge sharing do not critically degrade the performance of the detector as a Compton polarimter.

32.30.Rj, 29.40.Gx, 07.85.Fv

1. Introduction

Heavy few-electron atomic systems provide excellent properties to investigate fundamental light-matter interactions. At the Experimental Storage Ring (ESR) at GSI, Darmstadt highly charged ions (HCI) up to bare uranium are stored and cooled. Photons emitted in collisions with internal targets provide detailed insight into basic photon-matter interaction and are related to e.g. atomic bremsstrahlung, radiative recombination or characteristic transitions. Standard germanium detectors are used to measure the photon energy and angular distribution. Additional information can be extracted from the polarization state of the photons. Since the energy is typically in the hard X-ray regime (few tens of keV to a few 100 keV), the well established methods for polarization measurements for optical and soft X-ray photons fail. For the hard X-ray regime, a technique that has been applied successfully in several experiments [1, 2, 3, 4, 5, 6, 7], is Compton polarimetry [8] with double-sided strip detectors (DSSDs). There, the 2D-angular distribution of photons Compton-scattered inside the detector is measured to determine the degree and direction of (linear) polarization of the incident photon beam. For this method not only a dedicated detector is required, but also a fast and reliable readout system as well as suitable analysis algorithms. All of these components must be understood in great detail in order to enable precision measurements. In the current report we aim to gain a deeper understanding of the performance of the used 2D Si(Li) polarimeter. In particular, the interaction between the incident photons and matter inside the detector as well as the movement of released charge carriers to the electrodes has to be modeled [9]. By using a well-defined photon beam (energy, position, polarization) available from a synchrotron radiation source, the focus of the current study is to investigate the effect of charge splitting between adjacent detector strips.

2. The 2D Si(Li) polarimeter

The detector has been developed as a dedicated Compton polarimeter within the SPARC collaboration [10, 11]. Its active volume is a lithium-drifted silicon crystal with an area of 80 mm by 80 mm and a thickness of 7 mm. Between the electrodes at the front- and backside a voltage of 800 V is applied. The front and back electrodes are segmented into 32 strips, orthogonal to each other, resulting in a 2D structure of 1024 pseudo-pixels. All 64 strips have a pitch of 2 mm width, resulting in an active area of 64 mm by 64 mm. Adjacent strips are separated by a 50-μm isolation gap. Each strip is read out via its own charge-sensitive preamplifier and an entire chain of subsequent readout electronics. This configuration enables the detection of independent events in different strips, allowing to record their position (2D), energy and time (with the resolution of $\approx$ 50 ns) separately. While the detector crystal has to be kept at 77 K, the preamplifiers are operated at room temperature. This leads to a mediocre energy resolution of $\approx$ 2.1 keV FWHM at 59.54 keV which is mainly determined by the electronic noise of the preamplifiers.

3. Experiment

The experiment was conducted at the High Energy Materials Science Beamline P07 at the synchrotron radiation source PETRA III at DESY, Hamburg [12]. The facility provides a photon beam with tunable energy and a degree of linear polarization of almost 100%. The Si(Li) polarimeter was mounted on a movable platform where it was irradiated directly with the incident beam. The beamline is equipped with remote-controlled collimators and attenuators to adjust the beam size and intensity. Both the detector platform and the collimators can be moved with μm precision. As a first step we determined the calibration that relates the positions of the platform step motors to the beam position on the detector. With this calibration available, any desired beam position on the detector could be set. This was used to perform several scans in the course of the experiment. In this paper, we restrict ourselves to the so-called gap-scan: a horizontal scan over three vertical strips with a stepsize of about 0.05 mm close to the gaps. The scan pattern is depicted in figure 1. For each beam position, events were recorded for about

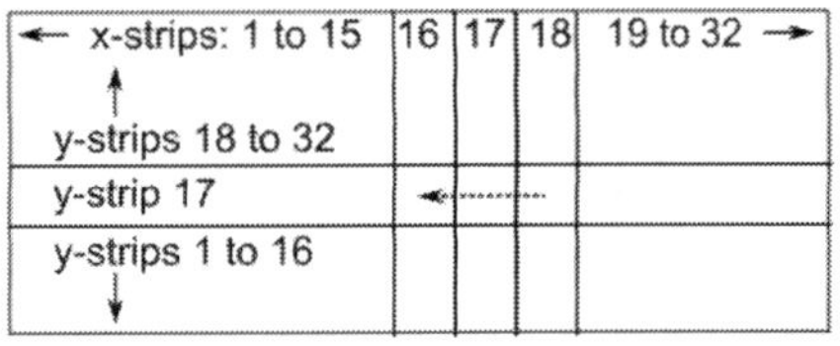

Figure 1. Strip pattern of the detector. The dashed arrow indicates the beam positions for the scan over two vertical gaps.

one minute with an event rate of $\sim$ 1900 Hz. The photon energy was 53.2 keV. A corresponding scan was also performed in the vertical direction. In order to ensure a reliable energy calibration of the detector, several calibration runs were performed during the experiment using standard ^{133}Ba and ^{155}Eu γ-sources.

Journal of Physics: Conference Series **583** (2015) 012043 doi:10.1088/1742-6596/583/1/012043

4. Analysis algorithm

With a DSSD it is possible to reconstruct events of different complexity. The simplest ones are illustrated in figure 2. In figure 2 (a), a single hit is depicted, which corresponds to one energy deposition inside the detector. This event type mainly indicates an absorption of an incident

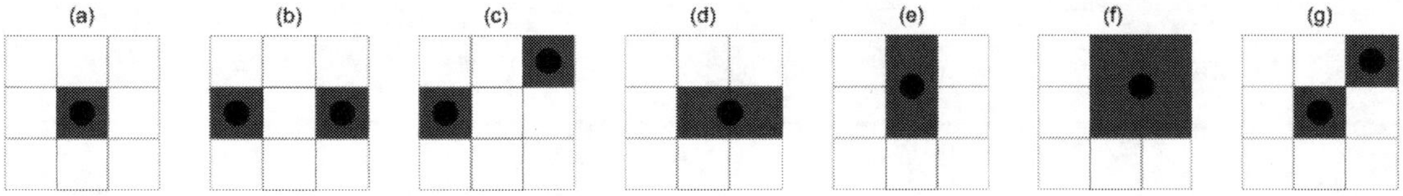

Figure 2. Possible event types in the pseudo-pixel structure of the strip detector: black circles indicate energy depositions, pixels with (without) a signal are grey (white) squares. (a) single hit; (b)+(c) separated double hit; (d)-(f) adjacent double hit due to charge sharing; (g) adjacent double hit from independent energy depositions.

photon. In case of a Compton scattering inside the detector one has two independent energy depositions, one by the recoil electron and one by the scattered photon. These are the events used for polarimetry, but the reconstruction is problematic, if the photon is absorbed very close to the recoil electron, such that signals in neighbouring strips are generated (indicated in figure 2 (g)). First, the direction of the scattered photon has a high uncertainty in this case and second, it cannot be distinguished from the scenario, where a single energy deposition occurred, but charges were collected by more than one strip (figure 2 (d) to (f)). This effect known as charge sharing is the major aspect investigated here. In our analysis, we consider as possible Compton scattering events only separated double hits (figure 2 (b) and (c)), that means on at least one detector side must be at least one strip without signal between the two strips with a signal. The other (adjacent) double hits are attributed to charge sharing events. In order to obtain information about the Compton polarimeter efficiency, for each separated double hit it was checked (by looking at the energy conservation) whether it is related to a Compton scattering event. Therefore the separated double hits shown in the results in section 5 correspond directly to such events that are used for polarization analysis. With this classification into event types, the analysis consists of counting each event type for a given data set which corresponds to one position of the incident beam on the detector. In order to allow a time normalization of the counts, a 10-kHz clock was connected to the data acquisition.

5. Results

Results obtained in this work are event rates and ratios of them as a function of the beam position on the detector. Figures 3 and 4 show plots both for the scan over the vertical strips (detector front side) and the scan over the horizontal strips (detector back side). Position 0 in the plots refers to the center of the central strip. Gaps are located at ± 0.1 cm. Figures 3 (a) and (d) show the total rate of events registered by the detector. For the horizontal strips (d) one can clearly see a decrease in the rate of about 3% at the gap positions. For the vertical strips (a), a decrease of the rate can only be observed at the left gap ($x = -0.1$ cm), and here it amounts to about 1.2%. The missing rate drop at the other gap at $x = 0.1$ cm could not be explained conclusively, one possibility is an instability of the incident beam. In order to cancel this effect, the rate of valid events (valid event: on both detector sides there is at least one strip with a valid signal and total energies on both detector sides do not deviate by more than 6 keV) normalized to the total event rate has been determined (Figures 3 (b) and (e)). For both vertical (b) and horizontal (e) strips the valid events amount to about 97.6% far from the gaps. At the gaps this portion decreases to 96% in case of the vertical strips and 94% for the

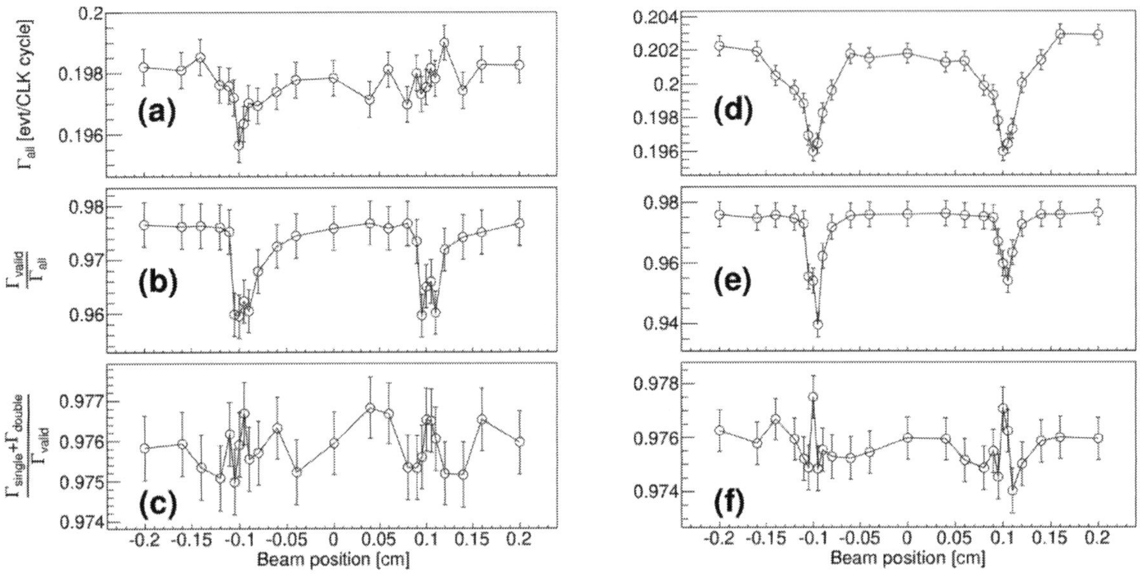

Figure 3. Event rates as a function of the incident beam position on the detector. The scans over the vertical ((a)-(c)) and horizontal ((d)-(f)) strips are shown. Error bars are statistical errors.

horizontal strips. In the next step the composition of the valid events was investigated. Figures 3 (c) and (f) show the sum of single and double (adjacent and separated) hits normalized by the number of valid events. This ratio of about 97.6% varies only by 0.2% with the beam position, which is not statistically significant here. The remaining 2.4% of valid events – which were not analyzed further – correspond to separated double hits that were not classified as Compton events (random coincidences) and events where more than two strips on at least one detector side had a valid signal, which include multiple scatterings. The contributions to the position-independent fraction of valid events, namely the single hits, the separated double hits and the adjacent double hits are shown individually in figure 4. The proportion of valid events that are single hits is shown in figures 4 (a) for the vertical strips and (d) for the horizontal strips. In both cases the ratio decreases from 90% near the strip centers to 40% at the gaps. This decrease is accompanied by a strong increase in the fraction of adjacent double hits (figures 4 (c) and (f)): from 5% near strip centers to 56% at the gaps. On the other hand, the variation in the fraction of separated double hits (figures 4 (b) and (e)) is less pronounced: between 2.4% and 2.9% for the vertical strips (b) and between 1.8% and 3% for the horizontal strips (e). A noticable feature in the curves for single hits and adjacent double hits is the asymmetry of the peaks around the gaps. The origin of this asymmetry could not be determined from the data, but one possibility is a tilt of the detector. Since usually in experiments the exact position of incident particles is not known, average fractions were calculated for the curves in figure 4. The values are as follows: 84.2% (a), 2.8% (b), 10.5% (c), 86.5% (d), 2.8% (e) and 8.2% (f).

6. Conclusion

In this work the effect of the isolation gaps of a double-sided Si(Li) strip detector has been investigated for incident 53.2-keV photons. The average proportion of valid events that are adjacent double hits (charge sharing events) is about 10%. For separated double hits (Compton scattering events) the value is 2.8%, while the corresponding maximum value is about 3%. From that follows that the gaps cause the proportion of valid events that are Compton scattering

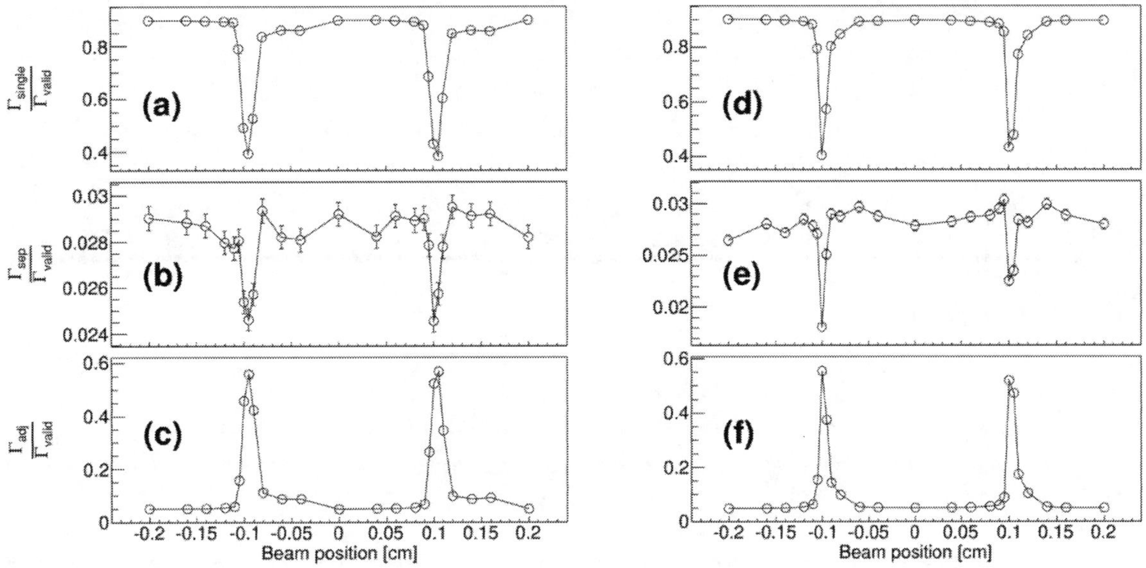

Figure 4. Rates of single and double hits as a function of the incident beam position on the detector. The scans over the vertical ((a)-(c)) and horizontal ((d)-(f)) strips are shown. Error bars are statistical errors.

events to decrease on average by less than 7%. The effect of the gaps is therefore not crucial for Compton polarimetry. We are currently developing a simulation code for the drift of charges inside the detector crystal to model the effects of charge splitting between neighbouring segments for a large range of incident photon energies.

Acknowledgments
We would like to thank Max Schwemlein for his support in planning and conducting the experiment. Parts of this research were carried out at the light source PETRA III at DESY, a member of the Helmholtz Association (HGF). This work was supported by the Alliance Program of the Helmholtz Association (HA216/EMMI).

References

[1] Hess S *et al* 2009 *J. Phys.: Conf. Ser.* **163** 12072
[2] Hess S *et al* 2009 *J. Phys.: Conf. Ser.* **194** 012025
[3] Weber G *et al* 2010 *Phys. Rev. Lett.* **105** 243002
[4] Märtin R *et al* 2012 *Phys. Rev. Lett.* **108** 264801
[5] Märtin R *et al* 2013 *Phys. Scr.* **2013** 014070
[6] Blumenhagen K-H *et al* 2013 *Phys. Scr.* **2013** 014102
[7] Spillmann U *et al* 2013 *Phys. Scr.* **2013** 014103
[8] Lei F *et al* 1997, *Space. Sci. Rev.* **82** 309-388
[9] Weber G *et al* 2011 *Phys. Scr.* **2011** 014034
[10] Weber G *et al* 2010 *JINST* **5** C07010
[11] Protic D *et al* 2005 *IEEE Transactions on Nuclear Science* **52** 3181-85
[12] Schell N *et al* 2014 *Mater. Sci. Forum* **772** 57-61

Journal of Physics: Conference Series **583** (2015) 012044 doi:10.1088/1742-6596/583/1/012044

A miniature EBIT with ion extraction for isolating highly charged ions

S Fogwell Hoogerheide and J N Tan

National Institute of Standards and Technology, Gaithersburg, MD 20899, USA

E-mail: shannon.hoogerheide@nist.gov, joseph.tan@nist.gov

Abstract. A room-temperature miniature electron beam ion trap (EBIT) is being developed for the production of charge states with a relatively low ionization threshold. A unitary Penning trap is modified slightly to provide the magnetic field and electric potential necessary for ion production via electron impact in this compact EBIT. This design allows radial access for in-EBIT spectroscopy as well as extraction of highly-charged ions for isolation at low energy to investigate proposed experiments. A fast micro-channel plate is used as a time-of-flight detector to study the initial production of helium and neon ions. Planned work would also involve the use of a Wien filter to select a single charge state to be isolated in a secondary ion trap for various studies. For instance, fully stripped ions can be captured for recombination experiments to form one-electron ions in high-angular momentum Rydberg states.

1. Introduction

Recent theoretical studies have reported special features in some highly charged ions that could potentially be useful for improved atomic clocks [1, 2], quantum information processing [1], measurements of fundamental constants [3, 1, 4], and tests of quantum electrodynamics [3]. Exploration or realization of these atomic systems requires the production of ions with low ionization threshold, as well as isolation in ultra-high-vacuum (UHV) for manipulation at low energy. We are developing a miniature electron beam ion source/trap (EBIS/T) that could be useful for exploring some of these possibilities. For instance, an ion source that fully strips atoms with $Z < 11$ could facilitate the production of one-electron ions in high-angular-momentum Rydberg states, with optical electric-dipole transitions which are well-suited for an independent measurement of the Rydberg constant [3, 5].

2. Highly charged ions in compact Penning traps

Recent work at the National Institute of Standards and Technology (NIST) has used a unitary Penning trap [6, 7] to capture highly charged ions extracted from an EBIT. An analyzing magnet in the existing ion-extraction beam-line of the NIST EBIT was used to select a single-charge state for isolation in this secondary ion trap, which has been made very compact by integrating NdFeB magnets into the electrode structure [6]. Since the trapping volume is relatively small in a unitary Penning trap, ion capture efficiency relies greatly upon beam focusing, ion pulse-width compression, proper timing of the capture trigger, and energy matching between the EBIT and unitary Penning trap. With careful optimization, roughly 10^3 ions have been captured with an initial energy spread of ≈ 5 eV, which is about 60 times lower than in the ion source [7]. An

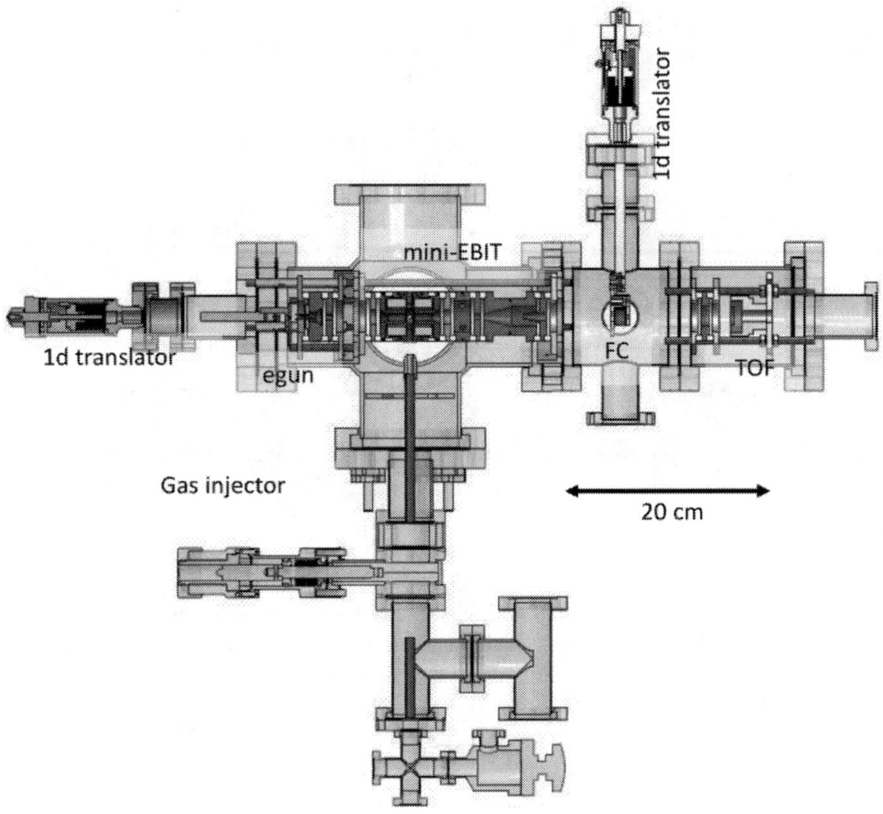

Figure 1. Prototype miniature EBIS/T under study. The electron gun (egun) is mounted on a translation stage. A retractable Faraday cup (FC) or a time-of-flight (TOF) detector can be selected to analyze the extracted ion pulse. A gas injector provides the initial source of atoms for ionization.

interesting feature is that the low energy is obtained within 1 ms of ion capture. This can be useful in experiments for which the time constants of known cooling mechanisms are too long. An illustration is provided in the recent lifetime measurement of the $3d\ ^2D_{5/2}$ metastable spin state in Kr XVIII, a relativistic Rydberg ion [8]. On a longer time scale, lower temperature may be attainable via other mechanisms, such as evaporative cooling [9]. Additional work is underway studying cooling and electron-capture processes of ions isolated in a unitary Penning trap.

3. A miniature electron beam ion source/trap

Compact electron beam ion sources and traps have been constructed using rare-earth permanent magnets in various configurations [10, 11]. The system being developed at NIST uses a two-magnet unitary Penning trap [6], but with its three electrodes shorted together as a single central drift tube that also provides the magnetic field. In this prototype, a simple annular disk is added to each end as end-cap electrodes in order to generate the electrostatic potential required for axial confinement of highly charged ions. The resulting miniature EBIS/T is illustrated in figure 1, located in the main six-way-cross chamber. The electron beam is provided by a magnetically shielded Pierce gun. The electron gun is mounted on a translation stage to assist with focusing the electron beam into the EBIT and optimizing the production of charge states with relatively

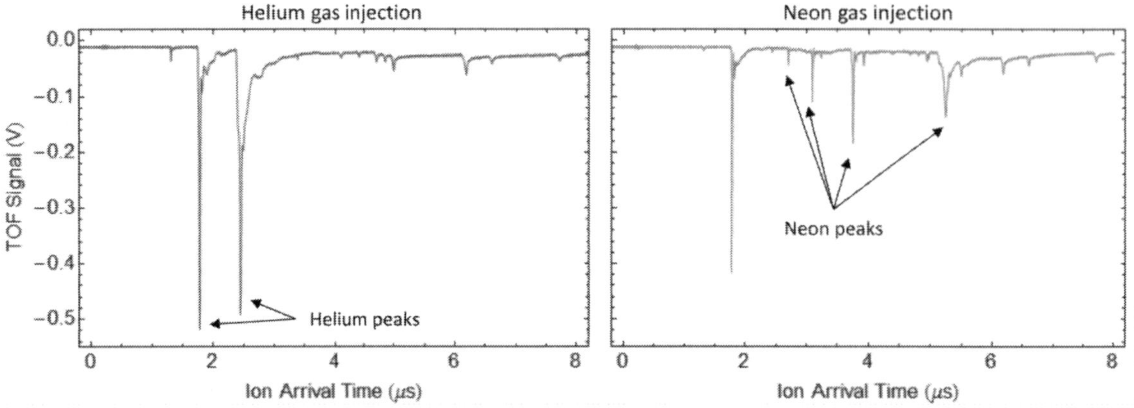

Figure 2. Observation of initial ion extraction signals using the TOF detector, corresponding to the ionization of injected helium gas (left) and injected neon gas (right). The same EBIT operating conditions were used in each case.

low ionization thresholds. A gas injector is mounted below the main chamber, with its nozzle aligned radially into one of the four equidistant holes in the mid-plane of the EBIT to serve as the initial source of atoms selected for electron-impact ionization. The remaining three holes are available for in-EBIT spectroscopy. Highly charged ions can also be extracted in pulses for further study. A retractable Faraday cup or a fast multi-channel plate time-of-flight (TOF) detector, located downstream, can be selected to analyze the extracted ion pulses. A diagram of the full apparatus is shown in figure 1.

Multiple charge states are created by electron-impact ionization inside the EBIT. When ejected to the TOF detector, higher charge states have shorter transit times to the detector. Figure 2 shows some early TOF signals corresponding to the initial extraction of helium ions (left) and neon ions (right), under the same EBIT operating conditions. The production and pulsed extraction of ions in various conditions is under study.

4. Planned work

We are interested in using the miniature EBIS/T for studies involving low ionization energy or isolation of fully stripped ions for experiments to produce one-electron ions in high angular momentum states. A Wien filter and additional ion beam optics will allow for selection of a single-charge state and better control of the extracted ions to optimize capture in a secondary ion trap. Planned charge exchange experiments with extracted ions stored in a secondary ion trap will utilize a rubidium beam oven apparatus [12], currently in development, that provides a beam of Rydberg rubidium atoms via laser excitation.

Some metastable states of interest have very long lifetimes (see, *e.g.*, Ref. [1]). On the other hand, the storage lifetime of ions isolated in a secondary trap is dependent on the background gas pressure. For instance, ion storage lifetimes in a unitary Penning trap are less than 1 s at the pressures typically achievable in room-temperature UHV systems [6]. In order to increase the ion storage lifetime for studying very-long-lived metastable states, we plan to incorporate a Gifford-McMahon-type cryocooler into the apparatus, which would substantially lower the background gas pressures in the ion capture trap.

Journal of Physics: Conference Series **583** (2015) 012044 doi:10.1088/1742-6596/583/1/012044

5. Summary

Theoretical studies have unveiled highly charged ions with special properties that can contribute to significant advances in several areas of precision measurements and atomic physics [1, 2, 3, 4]. Recent work at NIST used a unitary Penning trap to store highly charged ions at relatively low energy for precise measurements [6, 7, 8]. Initial extraction of ions from a miniature EBIS/T has been observed. Work is underway to optimize ion production and to capture the extracted ions in a secondary trap. An interesting application would be to produce one-electron ions in high angular momentum states [3]. Other possible applications include exploratory studies of highly charged ions that are potentially useful for ultra-precise atomic clocks [1, 2], quantum information processing [1], or new measurements of possible time-variation of the fine structure constant [1, 4].

Acknowledgments

We thank N D Guise for his work on the initial design of the miniature EBIT. The work of S Fogwell Hoogerheide was supported by a National Research Council Research Associateship at NIST.

References

[1] Safronova M S, Dzuba V A, Flambaum V V, Safronova U I, Porsev S G and Kozlov M G 2014 *Phys. Rev. Lett.* **113**(3) 030801
[2] Derevianko A, Dzuba V A and Flambaum V V 2012 *Phys. Rev. Lett.* **109**(18) 180801
[3] Jentschura U D, Mohr P J, Tan J N and Wundt B J 2008 *Phys. Rev. Lett.* **100**(16) 160404
[4] Berengut J C, Dzuba V A and Flambaum V V 2010 *Phys. Rev. Lett.* **105**(12) 120801
[5] Pohl R, Gilman R, Miller G A and Pachucki K 2013 *Annu. Rev. Nucl. Part. Sci.* **63** 175–204
[6] Tan J N, Brewer S M and Guise N D 2012 *Rev. Sci. Instrum.* **83** 023103
[7] Brewer S M, Guise N D and Tan J N 2013 *Phys. Rev. A* **88**(6) 063403
[8] Guise N D, Tan J N, Brewer S M, Fischer C F and Jönsson P 2014 *Phys. Rev. A* **89**(4) 040502
[9] Hobein M, Solders A, Suhonen M, Liu Y and Schuch R 2011 *Phys. Rev. Lett.* **106**(1) 013002
[10] Motohashi K, Moriya A, Yamada H and Tsurubuchi S 2000 *Rev. Sci. Instrum.* **71** 890–2
[11] Xiao J, Fei Z, Yang Y, Jin X, Lu D, Shen Y, Liljeby L, Hutton R and Zou Y 2012 *Rev. Sci. Instrum.* **83** 013303
[12] Bell S C, Junker M, Jasperse M, Turner L D, Lin Y J, Spielman I B and Scholten R E 2010 *Rev. Sci. Instrum.* **81** 013105

Journal of Physics: Conference Series **583** (2015) 012045 doi:10.1088/1742-6596/583/1/012045

RF-bunching of relativistic $^{12}C^{3+}$ ion beam for laser cooling experiment at the CSRe

H B Wang [1,2], W Q Wen[1], X Ma[1], Z K Huang[1,2], D C Zhang[1], M Bussmann[3], D F A Winters[4], Y J Yuan[1], X L Zhu[1], D M Zhao[1], R S Mao[1], J Li[1], L J Mao[1], J C Yang[1], H W Zhao[1], H S Xu[1], G Q Xiao[1] and J W Xia[1]

[1]Institute of Modern Physics, Chinese Academy of Sciences, 730000 Lanzhou, China
[2]University of Chinese Academy of Sciences, 100049 Beijing, China
[3]Helmholtz-Zentrum Dresden-Rossendorf, 01328 Dresden, Germany
[4]GSI Helmholtzzentrum für Schwerionenforschung GmbH, D-64291 Darmstadt, Germany

E-mail: x.ma@impcas.ac.cn

Abstract. To prepare the upcoming experiment of laser cooling of relativistic $^{12}C^{3+}$ ion beams at the experimental cooler storage ring (CSRe), a test experiment was performed with $^{12}C^{3+}$ ion beams at an energy of 122 MeV/u on the CSRe, at the Institute of Modern Physics, Lanzhou, China. In this experiment, the main storage ring of CSRm was employed to accumulate and accelerate the ion beam which was injected into the CSRe for the experiments. The number of $^{12}C^{3+}$ ions at the CSRe reached 5×10^8 for every injection, which satisfied the experimental requirement. To fulfil the laser cooling experiment, the $^{12}C^{3+}$ ion beams were bunched by sinusoidal waveforms with fixed and sweeping frequencies, respectively. A resonant Schottky pick-up was employed to record the Schottky spectra of these ion beams. The test experimental results demonstrated that the RF-buncher and diagnostic systems at the CSRe worked well and the CSRe was very stable with $^{12}C^{3+}$ ion beams, hereby the CSRe is suitable for laser cooling experiment.

1. Introduction

Realizing a crystalline ion beam is one of the ultimate goals of laser cooling of heavy ion beams at the storage rings. Laser cooling has been considered to be the only one method to realize the crystalline ion beams at the storage ring because of its extremely narrow and strong cooling force [1, 2]. The other motivation for laser cooling at the experimental cooler storage ring (CSRe) at IMP and the ESR at GSI is to extend this technique to the future large accelerator facilities, such as the Facility for Antiprotons and Ion Research (FAIR) at Darmstadt in Germany [3] and the High Intensity heavy ion Accelerator Facility (HIAF) at Lanzhou in China [4]. The precision nuclear and atomic experiments with highly charged ions will greatly benefit from ion beams with ultra-low momentum spread. Therefore powerful cooling techniques are important for future high-quality relativistic heavy ion beams.

Compared to the normal laser cooling experiment at traps [5, 6] and low-energy storage rings [7], laser cooling of relativistic heavy ions can only rely on laser beams counter-propagating with the ion beams and the laser cooling force is predicted to principally increase with the third power of the ion energy due to the fast transition rates in highly charged ions and relativistic Doppler shift [8]. Fortunately, the method by combining only one laser beam and a RF-buncher to cool the ion beams at

Journal of Physics: Conference Series **583** (2015) 012045 doi:10.1088/1742-6596/583/1/012045

storage rings has already been developed by Hangst et al [9], and was applied in many other storage rings such as the TSR [10] and the ESR [11-13] in Germany, and S-LSR [14, 15]in Japan.

To prepare the upcoming experiment of laser cooling of relativistic $^{12}C^{3+}$ ion beams at the CSRe, a test experiment of $^{12}C^{3+}$ ion beams with the required beam energy for laser cooling experiment was performed. All of the beam diagnostic system and newly installed RF-buncher were employed during the experiment. The experimental results demonstrated that the CSRe is suitable for the laser cooling experiment.

2. Experimental setup

The test experiment was performed with $^{12}C^{3+}$ ion beams on the CSRe, at the Institute of Modern Physics, Lanzhou, China. In the experiment $^{12}C^{3+}$ ions were produced by an Electron Cyclotron Resonance (ECR) ion source. Then the ions were accelerated by a Sector Focused Cyclotron (SFC), and injected into the main Cooler Storage Ring (CSRm). After accumulation and acceleration in the CSRm, the ion beams were extracted and injected into the CSRe at the energy of 122 MeV/u (the ion velocity of 47% of the speed of light). Since the circumference of the CSRe is 128.8 m, this beam energy leads to the revolution frequency of 1.087 MHz. A schematic view of the experimental setup at the CSRe is shown in figure 1 and the parameter settings of the CSRe for this experiment are listed in Table 1.

There are two straight sections both with a length of 30 m in the CSRe. On one section there are diagnostic systems and the internal gas target and the laser cooling experiment will be performed at this straight section as shown in figure 1. The other section installed electron cooler system. In this experiment, the lifetime of ion beam was measured by a DC current transformer (DCCT). As shown in figure 2, the typical multi-injected beam current of the $^{12}C^{3+}$ ion beams is about 300 μA which corresponds to approximately 5.7×10^8 ions circulating in the ring. The beam lifetime is determined to be less than 5 seconds by an exponential fitting. The too short lifetime is because of the bad vacuum condition at the CSRe (about 2×10^{-10} mbar). The resonant Schottky pick-up was used to monitor the revolution frequency and the longitudinal momentum spread of the stored ions. A spectrum analyzer (Tektronix RSA5100A) was used to record the Schottky spectra to get the dynamics and the momentum spread of the beams. Two ultraviolet sensitive photomultiplier tubes (UV PMTs, Eelectron Tube 9403B) and a movable ultraviolet sensitive Channeltron photomultiplier (UV CPM, PHOTONIS CEM4869) were installed to observe the fluorescence produced by the interactions between the ions and the residual gas in the CSRe [16] and the fluorescence from the de-excitation of the projectile ions excited by the laser and the interactions between the ions and the residual gas in the CSRe.

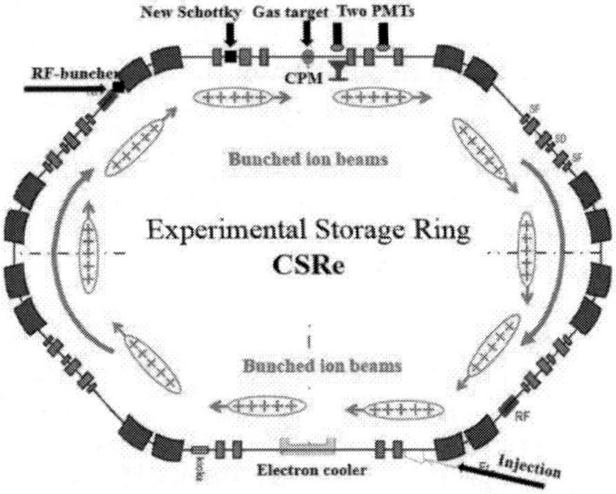

Figure 1 Schematic view of the experimental setup at the CSRe, the locations of the resonant Schottky pick-up, RF-buncher, UV PMTs and the UV CPM are shown. The ion beam will be bunched to ten bunches at the CSRe if the RF-buncher operated at the 10th harmonic of the revolution frequency.

Table 1. Parameters of the CSRe in this experiment.

Parameters	**CSRe**
Circumference	128.80 m
Ion species	$^{12}C^{3+}$
Beam energy	122 MeV/u
Relativistic β,γ	0.47, 1.13
Revolution freq.	1.087 MHz
Transition factor γ_t	2.629
Slip factor η	0.64
Lifetime	~ 5 s
Harmonic number h	10, 20, 35, 50
$\Delta p/p$ inside the bucket	$< 2\times10^{-5}$

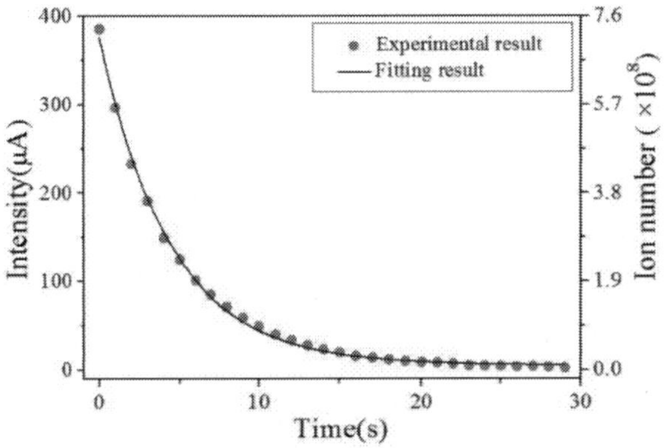

Figure 2 The beam current versus storage time measured by the DCCT, and the beam lifetime is 4.45 ± 0.07s.

In order to produce a counteract-force for the laser resonant scattering force during laser cooling experiment, a RF-buncher was installed at the CSRe, for details see Wen et al. [17], we only give a brief introduction here. The sinusoidal waveforms were produced by signal generator (Tektronic AFG3235). After amplification, these signals were added on the RF-buncher to modulate longitudinally the ion beam. The frequency of the RF-bunching was set as the harmonic of the revolution frequency $f_{bunch} = h\times f_{rev} = h\times v/C$, where v is the velocity of the ion beam, C is the circumference of the storage ring and f_{rev} is the revolution frequency of $^{12}C^{3+}$ ions and the harmonic number h gives the number of ion bunches in the storage ring. In this experiment, the frequency of the RF-buncher was operated from the 10th to 50th harmonic of the revolution frequency and the amplitudes (peak-to-peak value) were set from 100 mV to 500 mV, respectively. Since the beam lifetime (< 5 s) was too short to investigate the longitudinal dynamics systematically, as a result, the beam width and the bunch length were not measured. In order to perform the planned laser cooling experiments at the CSRe, the vacuum has to be improved at least one order of magnitude by baking of the whole storage ring which is under progressing. The relative longitudinal momentum spread of the ions inside the bucket has been determined to be less than 2×10^{-5}. The synchrotron frequency, which is oscillation frequency of the ions in the bucket, could be extracted from the Schottky spectrum [18].

3. Longitudinal RF-bunching of relativistic $^{12}C^{3+}$ ion beam

3.1. RF-bunching by fixed frequency sinusoidal-waveforms

To test the RF-buncher system, the injected ion beams were bunched by radio-frequency sinusoidal waveforms at various harmonic numbers and bunching voltages. The Schottky spectra of RF-bunched $^{12}C^{3+}$ at 35th harmonic number with an amplitude of 300 mV and 50th harmonic with an amplitude of 500 mV are shown in figure 3. The horizontal axis and the vertical axis are frequency and time, respectively. In figure 3 (a) and (b), there are many peaks in the Schottky spectra, and the distance between every adjacent peaks are the synchronous oscillation frequency ω_s of the ions inside the bucket, which can be written as [19]:

$$\omega_s = \frac{\omega_{rev}}{\beta}\sqrt{\frac{qeh\eta U_b}{2\pi\gamma mc^2}} \tag{1}$$

where ω_{rev} is the revolution frequency of the ion beams, β the relativistic factor, q the charge state of the ions, h the harmonic number, η the momentum slip factor, γ the relativistic Lorentz factor, c the velocity of the light, and U_b the effective bunching voltage which can be written as

$$U_b = 2U_0 \sin\frac{\pi hL}{C} \tag{2}$$

where U_0 is the voltage added on the RF-buncher, L the length of the parallel plate of the RF-buncher, and C the circumference of the CSRe.

By using equation (1), the effective bunching voltages can be calculated from the extracted synchrotron frequency of the Schottky spectra. The effective RF-bunching voltage in the figure 3 (a) and (b) are about 38 V and 145 V, respectively. These bunching voltages are enough to produce the counteract-force for the laser cooling experiments.

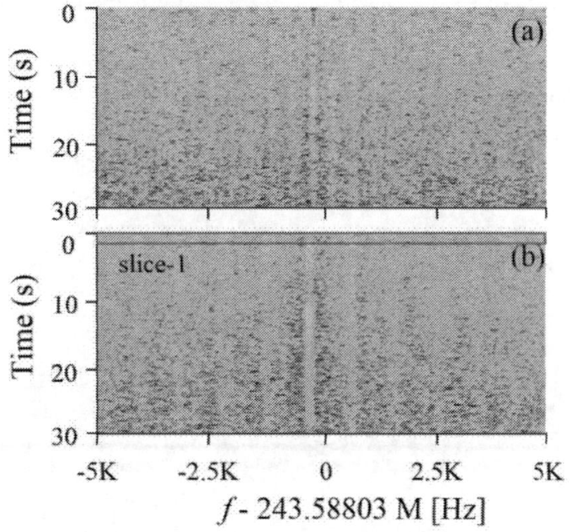

Figure 3 Schottky spectra for RF-bunched $^{12}C^{3+}$ ion beams bunched by sinusoidal waveforms. For figure (a) the harmonic number is 35th and the voltage is 300 mV while for figure (b) the harmonic number is 50th and the voltage is 500 mV.

3.2. Longitudinal momentum spread of $^{12}C^{3+}$ ion beam inside of the bucket

Since laser cooling is only working at the longitudinal direction of the ion beams, the longitudinal momentum spread of the ion beams is the most important index during the laser cooling experiment. The Schottky spectrum shown in figure 4 is the projection of the slice-1 in figure 3 (b), the horizontal axis is frequency and the vertical axis is the Schottky noise power intensity. The longitudinal momentum spread can be obtained by the frequency distribution of the ion beam by the equation

$$\frac{\Delta p}{p} = \frac{1}{\eta}\frac{\Delta f}{f} \tag{3}$$

where $\Delta f / f$ can be derived from the measured Schottky spectrum, and η is the momentum slip factor of the accelerator, $\eta = 1/\gamma^2 - 1/\gamma_t^2$, where γ is the relativity Lorentz factor and γ_t is transition energy of the accelerator. The Gaussian fitting result of the Schottky spectrum is show in figure 4 with a red line, from which we can deduce that the relative momentum spread is about 1.25×10^{-5}. But one should note, the relative momentum spread is the result of the bucket acceptance, not resulted from electron-cooling.

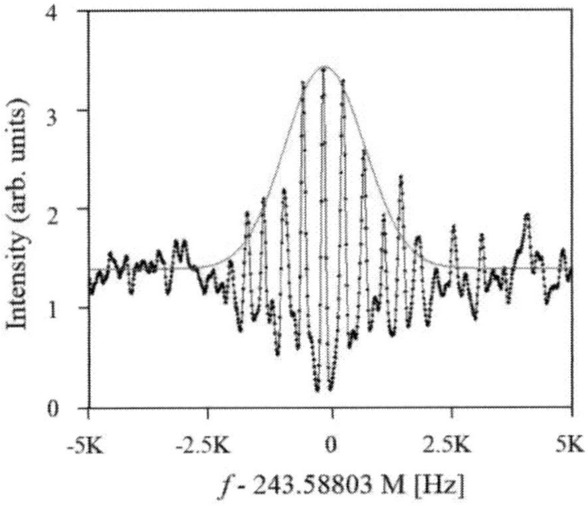

Figure 4 Schottky spectrum of the projection of slice-1 in figure 3 (b), the momentum spread is about 1.25×10^{-5} obtained from Gaussian fitting (a red line) which indicates the acceptance of the bucket.

3.3. RF-bunching of $^{12}C^{3+}$ ion beams by sweeping frequency sinusoidal-waveforms

The observed Schottky spectrum by sweeping the frequency of the RF-buncher is shown in figure 5, the RF-buncher was operated at the 35th harmonic number and the voltage of 500 mV. The sweeping speed was 5 Hz per second. From figure 5, it can be seen that the ions were bunched and scanned by the RF-buncher. The ion beams will be cooled when a resonant laser beam is applied with this RF-bunched and swept ion beams during the next laser cooling experiment.

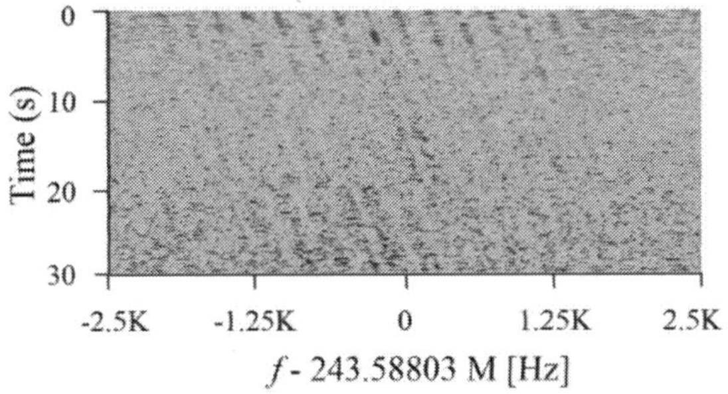

Figure 5 Schottky spectrum for RF-bunched $^{12}C^{3+}$ ions bunched by a sinusoidal waveform of sweeping frequency at the speed of 5 Hz/s at the harmonic number of 35th and the voltage of 500 mV.

3.4. Observation of $^{12}C^{3+}$ and $^{16}O^{4+}$ by Schottky noise signals

The ions of $^{12}C^{3+}$ and $^{16}O^{4+}$ have almost the same charge-mass ratio, therefore, it is not possible to separate them from the ion source and at the CSRm. However, since these two ion species still have a very small difference of charge-mass ratio, the two ion beams could be separated after electron cooling and be observed on the Schottky spectrum. This observation has already been confirmed at the experimental storage ring ESR at GSI. The frequency distribution of the ion beams at the storage ring can be written as [20, 21]:

$$\frac{\Delta f}{f} = -\frac{1}{\gamma_t^2} \times \frac{\Delta(m/q)}{m/q} + \left(1 - \frac{\gamma^2}{\gamma_t^2}\right) \times \frac{\Delta v}{v} \tag{4}$$

Where f is the evolution frequency, m is the mass, q is the amount of charge, γ is the relativistic Lorentz factor, γ_t is the transition energy of the accelerator, and $\Delta v/v$ is the velocity spread. By employing the electron cooling, $\Delta v/v$ can be considered to be negligible small (zero), and therefore equation (4) becomes

$$\frac{\Delta f}{f} = -\frac{1}{\gamma_t^2} \times \frac{\Delta(m/q)}{m/q} \tag{5}$$

By careful calculation, the relative frequency difference between the $^{12}C^{3+}$ and $^{16}O^{4+}$ is 4.6×10^{-5}. Since the Schottky pick-up was operated at 243 MHz, the 224th harmonic of the revolution of the ion beam, the Δf of $^{12}C^{3+}$ and $^{16}O^{4+}$ should be about 11.2 kHz, which can be easily observed in the Schottky spectrum. As mentioned before, after improving the vacuum of the CSRe, the ratio and lifetimes of these two ion species can be measured directly by the Schottky spectrum. In addition, the sympathetic cooling of $^{16}O^{4+}$ by laser cooled $^{12}C^{3+}$ is under consideration. The simulation work has already been started and the detailed discussion of this subject will be given elsewhere.

4. Conclusion

A test experiment of $^{12}C^{3+}$ ion beams was performed at the CSRe to prepare the upcoming laser cooling of relativistic Li-like carbon ion beams. All of the experimental setup and parameters which will be used for the next experiment were tested during this beam time. The test experiment showed that the resonant Schottky pick-up system and the RF-buncher worked well for $^{12}C^{3+}$ ion beams at an energy of 122 MeV/u by electron cooling at the CSRe and the CSRe is very stable for every injection beam lifetime. The problem of vacuum of the CSRe will be solved during summer maintenance time, before the next beam time.

In the forthcoming laser cooling experiment, an ultraviolet sensitive CPM system which has been installed inside of the vacuum will be used to observe fluorescence signals, and a cw laser system [22] and a pulsed laser system [23] will be transported from TU Darmstadt and HZDR to IMP, respectively. The relativistic lithium-like $^{12}C^{3+}$ ion beams will be cooled by combining a scanning cw laser beam and a RF-buncher after the pre-electron-cooling. This is a very promising way to enlarge the momentum acceptance of the laser force by scanning the bunching frequency or the laser wavelength. Both methods will be used to investigate the dynamics of the ultra-cold ion beams and strong-coupling effects at the CSRe. In addition, we will have a possibility for the first time to realize laser cooling experiment by combining a pulsed laser system and a cw laser system. This would be greatly helpful to study laser cooling and make this promising technique ready for routine application before future storage rings become operation, e.g. FAIR in Darmstadt, Germany and HIAF in Lanzhou, China. The precision laser spectroscopy experiments of Li-like and Na-like highly charged ions will also be implemented in future storage ring facilities [24].

Acknowledgments

The experiment was performed within the Laser Cooling Working Group. This work was supported by National Natural Science Foundation of China through grant no. 11221064 and GJHZ1305.

References

[1] Habs D, and Grimm R 1995 *Annu. Rev. Nucl. Part. S.* **45** 391-428.
[2] Schramm U, and Habs D 2004 *Prog. Part. Nucl. Phys.* **35** 583-677.
[3] Henning W F 2004 *Nucl. Instrum. Meth. B* **214** 211–215
[4] Yang J C, Xia J W, Xiao G Q, Xu H S, Zhao H W, Zhou X H, Ma X W, He Y, Ma L Z, Gao D Q, Meng J, Xu Z, Mao R S, Zhang W, Wang Y Y, Sun L T, Yuan Y J, Yuan P, Zhan W L, Shi J,

Chai W P, Yin D Y, Li P, Li J, Mao L J, Zhang J Q, and Sheng L N 2013 *Nucl. Instrum. Meth. B* **317** 263-265.

[5] Birkl G, Kassner S, and Walther H 1992 *Nature* **357** 310-313.

[6] Schätz T, Schramm U, and Habs D 2001 *Nature* **412** 717-720.

[7] Schröder S, Klein R, Boos N, Gerhard M, Grieser R, Huber G, Karafillidis A, Krieg M, Schmidt N, Kühl T, Neumann R, Balykin V, Grieser M, Habs D, Jaeschke E, Krämer D, Kristensen M, Music M, Petrich W, Schwalm D, Sigray P, Steck M, Wanner B, and Wolf A 1990 *Phys. Rev. Lett.* **64** 2901-2904.

[8] Schramm U, Bussmann M, and Habs D 2004 *Nucl. Instrum. Meth. A* **532** 348-356.

[9] Hangst J S, Nielsen J S, Poulsen O, Shi P, and Schiffer J P 1995 *Phys. Rev. Lett.* **74** 4432-4435.

[10] Eisenbarth U, Mudrich M, Eike B, Grieser M, Grimm R, Luger V, Schätz T, Schramm U, Schwalm D, and Weidemüller M 2000 *Hyperfine Interact.* **127** 223-235.

[11] Schramm U, Bussmann M, Habs D, Kuhl T, Beller P, Franzke B, Nolden F, Steck M, Saathoff G, Reinhardt S, and Karpuk S 2006 *AIP Conference Proceedings* **821** 501-509.

[12] Bussmann M, Habs D, Schramm U, Beckert K, Beller P, Franzke B, Kozhuharov C, Kühl T, Nörtershäuser W, Nolden F, Steck M, Karpuk S, Geppert C, Novotny C, Saathoff G, and Reinhardt S 2007 *Proceedings of the Workshop on Beam Cooling and Related Topics, COOL 07* **1** 226-229.

[13] Bussmann M, Schramm U, Habs D, Steck M, Kühl T, Beckert K, Beller P, Franzke B, Nörtershäuser W, Geppert C, Novotny C, Kluge J, Nolden F, Stöhlker T, Kozhuharov C, Reinhardt S, Saathoff G, and Karpuk S 2007 *J. Phys.: Conf. Ser.* **88** 012043.

[14] Tanabe M, Ishikawa T, Nakao M, Souda H, Ikegami M, Shirai T, Tongu H, and Noda A 2008 *Appl. Phys. Express* **1** 028001.

[15] Nakao M, Hiromasa T, Souda H, Tanabe M, Ishikawa T, Tongu H, Noda A, Jimbo K, Shirai T, Grieser M, Okamoto H, and Smirnov A V 2012 *Phys. Rev. ST Accel. Beams* **15** 110102.

[16] Wen W Q, Lochmann M, Ma X, Bussmann M, Winters D F A, Nörtershäuser W, Botermann B, Geppert C, Frömmgen N, Hammen M, Hannen V, Jöhren R, Kühl T, Litvinov Y A, Sánchez R, Stöhlker T, Vollbrecht J, Weinheimer C, Dimopoulou C, Nolden F, and Steck M 2013 *Nucl. Instrum. Meth. A* **711** 90-95.

[17] Wen W Q, Ma X, Bussmann M, Yuan Y J, Zhang D C, Winters D F A, Zhu X L, Li J, Liu H P, Zhao D M, Wang Z S, Mao R S, Zhao T C, Wu J X, Ma X M, Yan T L, Li G H, Yang X D, Liu Y, Yang J C, Xia J W, and Xu H S 2014 *Nucl. Instrum. Meth. A* **736** 75-80

[18] Wen W Q, Ma X W, Zhang D C, Zhu X L, Meng L J, Jie L, Liu H P, Zhao D M, Wang Z S, Mao R S, Zhao T C, Wu J X, Ma X M, Yan T L, Li G H, Yang X D, Liu Y, Yang J C, Yuan Y J, Xia J W, Xu H S, Xiao G Q, and Zhao H W 2013 *Chinese Phys. C* **37** 107005

[19] Boussard D 1986 *CERN* **86-11** 749-782.

[20] Bosch F, Litvinov Y A, and Stöhlker T 2013 *Prog. Part. Nucl. Phys.* **73** 84-140.

[21] Litvinov Y A, and Bosch F 2011 *Rep. Prog. Phys.* **74** 016301.

[21] Beck T, Rein B, and Walther T 2013 *CLEO: 2013, Optical Society of America, San Jose, California* **JTh2A** 66.

[23] Siebold M, Hein J, Wandt C, Klingebiel S, Krausz F and Karsch S 2008 *Opt. Express* **16** 3674-3679.

[24] Backe H 2007 *Hyperfine Interact.* **171** 93-107.

Journal of Physics: Conference Series **583** (2015) 012046 doi:10.1088/1742-6596/583/1/012046

Modeling low energy x-ray interactions with biological material at the CUEBIT

J Klingenberger[1], M Schott[2], T Kimmel[3], D Medlin[3], A Gall[3], M Rusin[4], D Dean[4], E Takacs[3]

[1]School of Computing, Clemson University, 100 McAdams Hall, Clemson, SC 29634, USA
[2]Department of Electrical Engineering and Computer Science, Vanderbilt University, 400 24th Ave S Jacobs Hall, Nashville, TN 37212, USA
[3]Department of Physics and Astronomy, Clemson University, 118 Kinard Laboratory, Clemson, SC 29634, USA
[4]Department of Bioengineering, Clemson University, 301 Rhodes Research Center, Clemson, SC 29634, USA

E-mail: klinge2@clemson.edu

Abstract. Recent developments at Clemson University have established the need to model the production of x-rays using a highly charged ion beam generated by the Clemson University Electron Beam Ion Trap (CUEBIT). A Geant4 modeling environment has been developed on Clemson University's Palmetto2 supercomputing cluster to simulate the interaction of these x-rays with biological material. Preliminary results of the model have been obtained after performing initial simulations on the computing cluster. Future experiments using the CUEBIT as well as refinements to the Geant4 model are discussed.

1. Introduction

Understanding the primary and secondary effects of radiation on biological material is difficult given the small distance and time scales of these interactions. The primary interaction is either a photoionization or a photon scattering effect that takes place somewhere along the path of the incoming x-ray. The secondary processes include electron scattering events leading to a cascade of chemical and biological events. At Clemson University's Electron Beam Ion Trap (CUEBIT), experiments are being performed involving the irradiation of biological material with low energy x-rays. Monte Carlo computer simulations of x-rays interacting with biological material are also being constructed in an effort to better understand the multi-step processes that lead to measurable effects on the cell function. Importantly, the simulations need to be accurate so that any conclusions made are correct and verifiable. A main goal of the CUEBIT simulations is to track secondary interactions inside biological material that occur as a result of chemical or biological processes.

The computer simulations that model x-rays output by the CUEBIT can be verified by performing experiments with the physical machine. Verification can be done by recording the amount of dose deposited by a phantom in the simulated experiment and comparing the results to a phantom placed in an irradiation chamber attached at the end of the real, physical CUEBIT machine.

Journal of Physics: Conference Series **583** (2015) 012046 doi:10.1088/1742-6596/583/1/012046

Recent publications suggest that low energy radiation may cause a cascade of effects on biological material [1]. A main goal of the CUEBIT simulations is to track these secondary interactions inside biological material that occur as a result of biological or chemical processes. Modeling the pathways of these processes using Monte Carlo simulation software will provide further understanding of these effects.

Using computers to perform Monte Carlo simulations of physical processes is not new. The high-energy physics group CERN maintains a freely obtainable particle simulation toolkit called Geant4 [2]. This software provides a programming environment to instantiate particles with arbitrary properties and run a simulation with user-defined particle interactions enabled. While Geant4 was developed in the scope of high-energy physics, the program is now general purpose enough that simulations of low energy interactions are also possible. Additionally, a medical physics simulation package called GATE [3] has been built on top of Geant4, which provides useful features to simplify common tasks in radiation therapy simulations. GATE is an appropriate program for simulating experiments at the CUEBIT due to its Geant4 base and its additional functionality such as dynamic simulations over time.

In this paper, methods for collecting data from GATE simulations will be described. Additionally, simplifications made in regard to this preliminary modeling work will be explained, as well as refinements to be made in the future. Lastly, a technique for speeding up the runtime of simulations through the use of a supercomputing cluster will be detailed.

2. Results and Discussion

Initial results from a GATE simulation of a model of an irradiation port at the end of the CUEBIT have been generated. In this section, the construction of the model and methods for manipulating the model to extract useful data are discussed. The simulation that was developed is a model of a physical experiment that is planned for the CUEBIT. A hollow irradiation chamber at the end of the highly charged ion beam line curves the beam using a quadrupole bender so that the ion beam is directed vertically upward. At the top of the irradiation chamber, a beryllium window is placed so that the ions impact the window. Upon impact, charge exchange occurs between the highly charged ions and the beryllium window, causing a cascade of interactions that ultimately release low energy x-rays. On top of the irradiation chamber, a cell flask is placed on an irradiation port in the path of these x-rays.

2.1. Description of the model geometry

In order to provide useful data, the model used by the simulation needed to be accurate. Particularly, the thickness of certain materials affects the penetration depth of x-rays. Special consideration went into the choice of material for the bottom of the cell flask, since most common flask materials such as silicate glass absorb x-rays in the energies emitted by the beryllium window. In this design, a cell flask bottomed with a thin Mylar membrane is employed, since Mylar is transparent to x-rays in the energy region of interest [4]. Therefore, it is important to include the layer of Mylar between the x-ray source and cell medium in the model with accurate dimensions.

Biological materials are complex subjects to model. While future refinements of the biological material used in the model are planned, the current model approximates the biological material with a volume of pure water. Given that biological material is largely composed of water, the x-ray dose applied to water should not be significantly different than a dose of x-rays applied to biological material.

2.2. Input of the x-ray energy distribution

Particle simulations are computationally intensive programs. Therefore, it is useful to isolate and track interactions in parts of the system that have a direct effect on the experiment of interest. The distribution of x-ray energies generated through charge exchange processes on the beryllium window can be measured using an x-ray spectrometer. The experimental values obtained from this spectrometer can then be input into the model using a command built into GATE. X-rays are then

Journal of Physics: Conference Series **583** (2015) 012046 doi:10.1088/1742-6596/583/1/012046

generated according to this distribution at the x-ray source in the model, which makes the model physically accurate.

However, the x-ray energy distribution generated by the beryllium window at the CUEBIT has not yet been obtained. In absence of this distribution, x-ray energies of exactly 3.0 keV and 3.5 keV from the known ion decay process were used. These values were approximations of the expected values to be generated by the CUEBIT in this initial modeling work. The choice to use an x-ray distribution rather than model the ion decay and decay product generation was intentional. An x-ray energy distribution collected through experiment will be used as input as soon as such data is available.

2.3. Collection of the absorbed dose data

A computer simulation of an experiment would have no purpose in a scientific context if data could not be extracted from the simulation. GATE offers a useful feature called "actors" that allow for functions to be executed at different stages of the simulation. One particular actor that is useful is the DoseActor. With the DoseActor enabled, the energy deposition into a given volume can be output into a file in a voxelized format such as ROOT or MetaImage. In the simulation model of the CUEBIT experiment (Figure 1), the DoseActor is particularly useful since the deposition of energy can be collected on a per volume basis. A more refined model with volumes representing individual cells will enable energy deposition to be collected per cell basis.

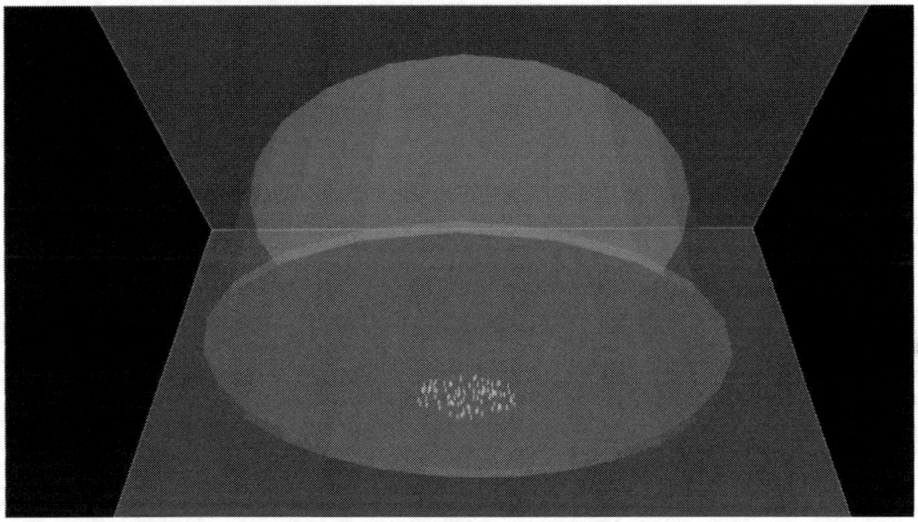

Figure 1. Visualization of the flask irradiation. The gray cylinder is the cell flask, which is filled with a water substrate (bottom of flask). Generated x-rays strike the flask from below.

2.4. Parallelization of the simulation

The complexity of a simulation increases as the number of particles to track increases and as the number of volumes increases, and subsequently the amount of time a simulation takes to complete also increases. Therefore, it is often desirable to simplify a simulation where possible. A parallelization technique was used on the model of the CUEBIT irradiation chamber that enabled multiple particles to be tracked simultaneously. By running the same simulation in multiple processes on the Palmetto2 supercomputing cluster, statistics on energy deposition into the biological material volume could be generated in less time. The speed improvement increases approximately linearly as the number of threads performing the simulation increases. While the current experimental model is not of sufficient complexity to warrant such optimizations, future refinements of the model will contain several thousand volumes of individual cells. Using a supercomputing cluster will provide the critical performance boost needed to compute the simulation with enough particles to guarantee adequate statistical results in a reasonable timeframe.

3. Conclusion

Constructing computer simulations of experiments at the end of the CUEBIT has been shown to be useful in predicting where the majority of the interactions of x-rays with biological material take place. The data collected from models can be compared to results from the irradiation chamber to verify that the models reflect reality. While the current model of the irradiated cell flask contains only a water substitute for biological material, this model will be refined to have cell volumes that contain subvolumes that model organelles. Current Monte Carlo particle tracking packages only model physical interactions, so more work is required to model the chemical and biological effects relevant to the biological material being studied. With the appropriate interactions simulated and a high detail model of the biological material including cellular components, more details of the primary and secondary effects that lead to cell death may be better understood. However, such simulations are computationally expensive on current mid-range hardware, so efficiency optimizations such as parallelization of the simulation are critical to performing the simulations in a reasonable amount of time.

Acknowledgements

NASA EPSCoR Palmetto Academy Program
Clemson University Creative Inquiry (CI) Program
Clemson University Professional Internship & Co-Op (UPIC) Program
Clemson University REU Program in Data-Intensive Computing
Department of Physics and Astronomy at Clemson University

References

Giedzinski E 2013 *Rad. Res.* **164** 540
Allison J *et al* 2006 *IEEE Trans. Nuc. Sci.* **53** 270
Jan S *et al* 2004 *Phys. Med. Biol.* **49** 4543
Medlin D, Heffron W, Siegel A, Wilson R K, Klingenberger A, Gall A, Rusin M, Dean D, Takacs E 2014 *J. Phys.: Conf. Series* (Submitted)

Journal of Physics: Conference Series **583** (2015) 012047 doi:10.1088/1742-6596/583/1/012047

Measuring the stopping power of α particles in compact bone for BNCT

L Provenzano[1,2], L M Rodríguez[1,2], D Fregenal[1,2], G Bernardi[1,2], C Olivares[1], S Altieri[3,4], S Bortolussi[3,4] and S J González[1,2]

[1]Comision Nacional de Energía Atómica (CNEA), Argentina
[2]CONICET, Argentina
[3]Departament of Physics, University of Pavia, Italy
[4]Istituto Nazionale di Fisica Nucleare (INFN), Section of Pavia, Italy

E-mail: srgonzal@cnea.gov.ar

Abstract. The stopping power of α particles in thin films of decalcified sheep femur, in the range of 1.5 to 5.0 MeV incident energy, was measured by transmission of a backscattered beam from a heavy target. Additionally, the film elemental composition was determined by Rutherford Backscattering Spectrometry (RBS). These data will be used to measure boron concentration in thin films of bone using a spectrometry technique developed by the University of Pavia, since the concentration ratio between healthy tissue and tumor is of fundamental importance in Boron Neutron Capture Therapy (BNCT). The present experimental data are compared with numerical simulation results and with tabulated stopping power data of non-decalcified human bone.

1. Introduction

Boron Neutron Capture Therapy (BNCT) is a cancer treatment that was designed to selectively deliver heavy charged particle radiation to tumors at the cellular level. This technique takes advantage of the high Linear Energy Transfer (LET) products of the $^{10}B(n,\alpha)^{7}Li$ reaction and the selective boron uptake by tumor cells. After capturing a thermal neutron, ^{10}B becomes ^{11}B, which immediately disintegrates into ionizing particles that have a range in tissue comparable with cellular dimensions (~12-14 μm). Then, if an adequate amount of boron is selectively concentrated in a tumor and the volume is exposed to thermal neutrons, a higher radiation dose is absorbed by the tumor in comparison to adjacent normal tissue.

The establishment of an appropriate boron concentration ratio between healthy tissue and tumor after administration of a boron compound is one of the fundamental aspects that ensure the clinical effectiveness of BNCT. In order to assess a proper dosimetry, it is thus necessary to know the boron concentration of the tissues irradiated by neutrons. The charged particle spectrometry developed in the University of Pavia [1] allows determining boron concentrations down to 1 ppm in thin soft tissue sections. Extending this technique to analyze hard tissues of interest in osteosarcoma disease requires the knowledge of the stopping power of α particles in these tissues. However, as thin samples are required, the calcium concentration of bone were reduced so that it can be cut with conventional cryostat. Transmission experiments using α particles in thin films of compact bone were performed in an energy range from 1.5 to 5.0 MeV, range wich includes those energies resulting from the neutron capture reaction.

2. Materials and Methods.

2.1. Sheep femur samples.

Compact femur from a sheep was decalcified using Ethylenediaminetetraacetic acid disodium salt (EDTANa) 0.5 M. The decalcifying method consisted in the immersion of small samples of sheep

Journal of Physics: Conference Series **583** (2015) 012047 doi:10.1088/1742-6596/583/1/012047

femur in the EDTANa solution, renewed weekly, during three weeks. After decalcification the samples were sectioned in different nominal thicknesses between 10 μm and 30 μm. The area and weight of each sample were also measured. Pictures of the samples were taken using a stereomicroscope and the program ImagePro™ [2] was employed to measure their area. The weight was measured using a digital scale with a resolution of 0.001 mg.

The elemental composition of the decalcified sheep femur was measured by using Rutherford Backscattering Spectrometry (RBS) in 30 μm thickness sample. In brief, RBS consists in a monoenergetic ion beam (here α particles) backscattered by nuclei of the elements presents in the sample, with an outgoing energy that depends on the nuclei mass. The energy distribution of the scattered α particles extends from a maximum, for nuclei at the sample surface, to lower energies due to the beam energy loss in the incoming and outgoing trajectories for nuclei at the bulk [3]. The measured spectra were analyzed with the SIMNRA software [4] that fits the experimental data by varying elemental concentrations, using a database of scattering cross sections. For the present work, as an initial guess was used the composition reported by ICRU 46 [5] and adding the elements of the decalcifying solution.

2.2. Experimental set-up.

The experimental set-up for the stopping power measurements is shown in figure 1. High doses delivered by the direct α particles beam could damage the sample as well as the detector, and in order to avoid this effect a secondary scattered beam was used instead. The primary beam coming from the accelerator impinges normally on the main target consisting in an oxide film grown on silicon substrate. The thickness of the oxide layer (~50 nm) assures a secondary scattered α particles beam of well-defined energy to the purpose of a transmission stopping experiment. By using two heavy elements oxide, uranium (UOx) and zirconium (ZrOx), secondary beams with different energies near to the maximum of the accelerator capability can be obtained.

A detector positioned 40° apart from the beam line measures the energy distribution of the secondary α particles beam. Calibration spectra were recorded for each ion beam energy and both oxide targets. Afterwards, spectra were measured with different sections of decalcified compact bone interposed in the path of the secondary beam. These films placed in front of the detector move the spectra to lower energies. The shifting of the spectra allows deriving the stopping power [6]. In order to work within the range of interest, the energy of main ion beam was set between 1.5 and 5.0 MeV, in a NEC 5SDH, 1.7 MV Tandem Accelerator at Centro Atómico Bariloche.

The measurements of elemental composition using RBS was performed with a 2.0 MeV α particles beam, and the detector placed 15° apart from the beam line. The energy loss simulations were carried out with SRIM [7] software, using the previously measured composition.

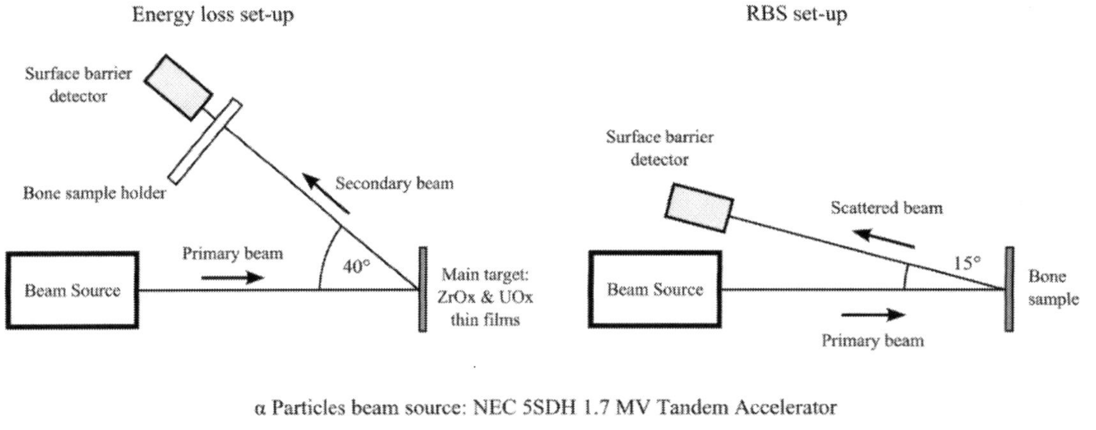

Figure 1. Experimental set-up.

3. Results and Discussion.

Figure 2 shows the experimental RBS spectrum for a decalcified thick bone sample alongside the SIMNRA fitting. The contribution of each chemical element is also depicted. On the right side the resulting elemental composition of the sample in atomic percentage is listed.

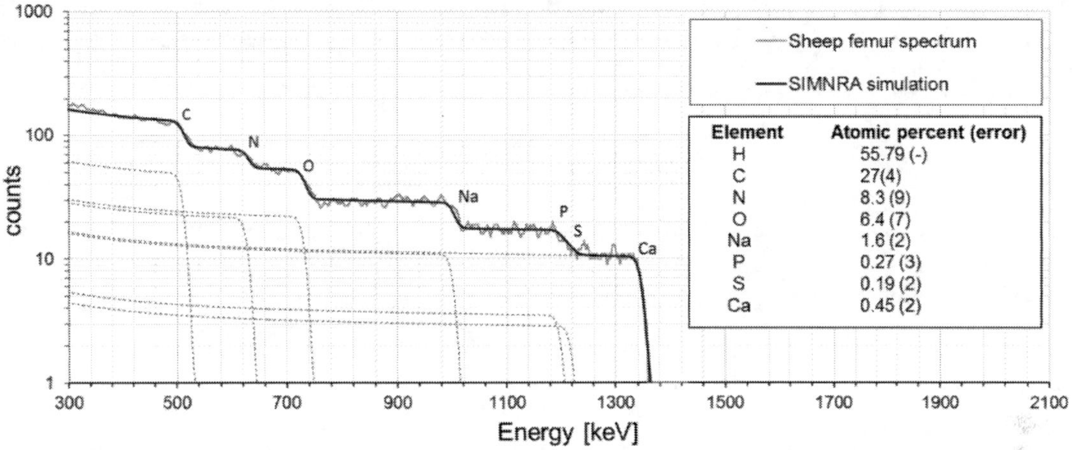

Element	Atomic percent (error)
H	55.79 (-)
C	27(4)
N	8.3 (9)
O	6.4 (7)
Na	1.6 (2)
P	0.27 (3)
S	0.19 (2)
Ca	0.45 (2)

Figure 2. Measured α particles RBS spectrum of a decalcified sheep femur sample with SIMNRA fittings and chemical composition results (inset table).

Figure 3 shows the calibration spectra for uranium and zirconium oxides films (UOx and ZrOx, respectively) measured with 4 MeV beam energy. Note that the low thickness of the film mimic the behaviour of an α particles source with a narrow energy distribution. In figure 4, the UOx calibration spectrum and the one obtained after the secondary beam passed through a 0.905±0.004 mg/cm^2 femur sample are presented. The shift in the energy of the main peak belonging to the U scattering as well as the modification in the plateau corresponding to the substrate (Si) are clearly seen in the plot.

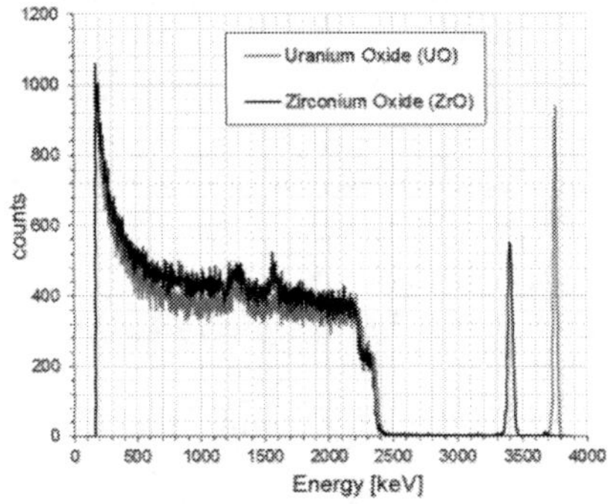

Figure 3. Calibration spectra of UOx and ZrOx targets with 4 MeV α particles main beam.

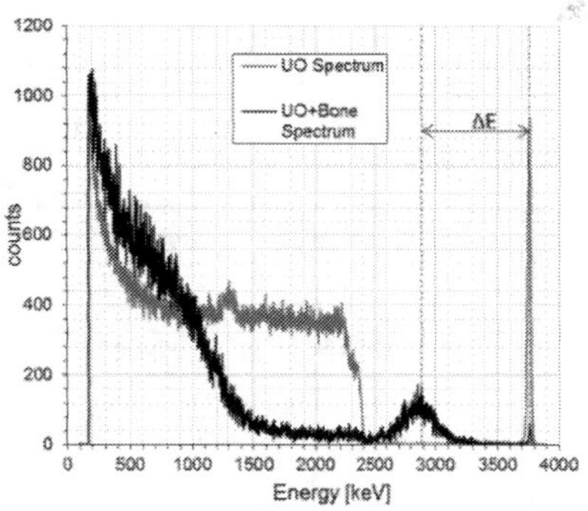

Figure 4. Comparison between the spectra of UOx target and the corresponding one with a 0.905±0.004 mg/cm^2 femur sample interposed.

The results of the measured α particles energy loss as a function of the initial particle energy for two different sample thicknesses are shown in figure 5. From this experimental data, the functional relation between the final energy of α particles and the surface density (mg/cm^2) of the samples was determined. A 9th degree polynomial was used to interpolate the discrete results and also to obtain an easily differentiable smooth curve.

Journal of Physics: Conference Series **583** (2015) 012047 doi:10.1088/1742-6596/583/1/012047

In Figure 6 the α particles stopping power calculated from the differentiation of the polynomial fit (with 95% prediction bounds) is compared with the values reported in ICRU 49[8] for human compact bone, only for reference. SRIM simulation results based on the sheep femur composition measured in this work is also plotted. As can be seen, the simulation overestimates the experimental stopping power. Given that the decalcification process acts from the surface to the center of the bone sample, different thin slices might have different elemental composition. Then, the overestimation could be due to the fact that the energy loss and the elemental composition measurements were performed in different sheep femur slices.

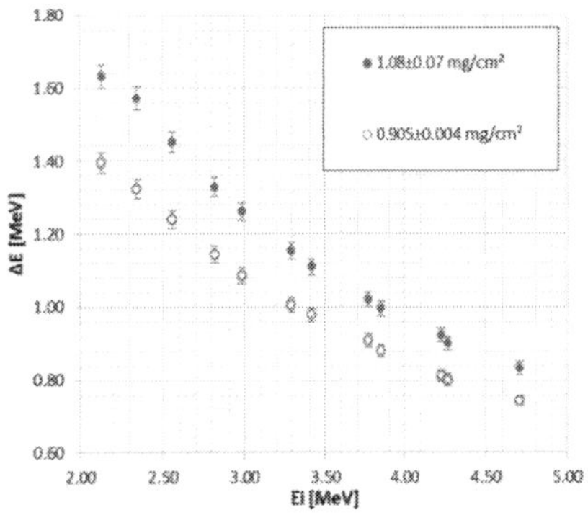

Figure 5. Energy loss of α particles in decalcified compact bone (sheep femur) as a function of the initial particle energy for two sample thicknesses.

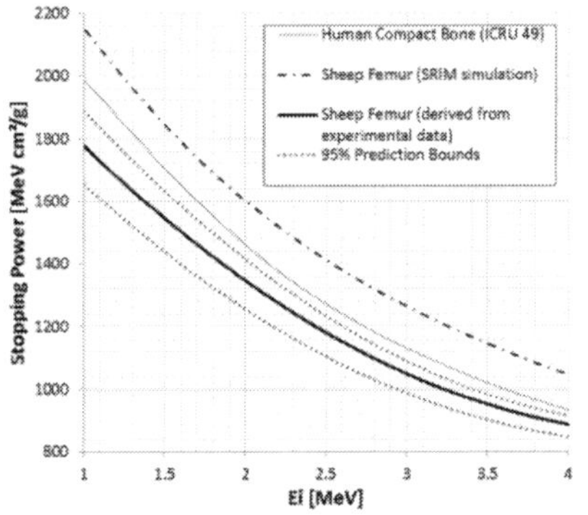

Figure 6. Stopping power of α particles derived from experimental data, compared with SRIM simulation results and data for human compact bone reported in ICRU 49.

4. Conclusions and future work.

Energy loss of α particles as a function of incident energy was measured in sheep femur samples with different thicknesses, from these experimental results the α particle stopping power was derived. Simulations of the stopping power made with SRIM software were also carried out using the chemical composition of compact bone measured with RBS on a thick sample of the same femur.

Further experiments to study the impact in the stopping power of different chemical composition for consecutive slices from the same decalcified bone are in progress.

References

[1] Bortolussi S and Altieri S 2013 *Radiat. Environ. Biophys.* **52** 493

[2] *Image-Pro Plus Reference Guide*, Media Cyberneics Inc., United States, 2009

[3] Alford T, Feldman L and Mayer J 2007 *Fundamentals of Nanoscale Film Analysis* (New york: Springer).

[4] Mayer M 1997 *SIMNRA User's Guide*, Report IPP 9/113, Max-Planck-Institut für Plasmaphysik, Garching, Germany, 1997

[5] ICRU Report 46 - Photon, Electron, Proton and Neutron Interaction Data for Body Tissues 1992

[6] Liguori Neto R, Added N and Coutinho F 2000 *Nuc. Ins. and Met. in Phy. Res. Sect. B* **161–163** 159

[7] Ziegler J, Biersack J and Littmark U 1985 *SRIM:The Stopping and Range of Ions in Matter (New York: Pergamon)*

[8] ICRU Report 49 - Stopping Powers and Ranges for Protons and Alpha Particles 1994

Journal of Physics: Conference Series **583** (2015) 012048 doi:10.1088/1742-6596/583/1/012048

Development of an x-ray irradiation port for biomedical applications at the CUEBIT facility

D Medlin[1], W Heffron[2], A Siegel[3], K Wilson[1], A Klingenberger[1], A Gall[1], M Rusin[4], D Dean[4], E Takacs[1]

[1] Department of Physics and Astronomy, Clemson University, 118 Kinard Laboratory, Clemson, SC 29634, USA

[2] Biology Department, Wofford College, 429 N Church Street, Spartanburg, SC 29303, USA

[3] Department of Physics, Wofford College, 429 N Church Street, Spartanburg, SC 29303, USA

[4] Department of Bioengineering, Clemson University, 301 Rhodes Research Center, Clemson, SC 29634, USA
Email: dmedlin@g.clemson.edu

Abstract. Because of the importance of x-ray interactions in modern medicine, efforts must be made to combine the fields of biology and physics. This paper reviews the development of an x-ray irradiation port that allows us to study the interaction of x-rays generated by highly charged ions with biological material, such as stem and cancer cells. Our goal is to better understand these interactions in order to improve the techniques of x-ray therapy by narrowing and specifically selecting the range of radiation energies applied. Using the Clemson University Electron Beam Ion Trap (CUEBIT), the generation of quasi-monochromatic x-rays from highly charged ions is possible. In order to maintain the integrity of the cells being studied, the cell culture needs to be oriented horizontally during the irradiation process. This poses a problem, as the highly charged ion beam generated at the CUEBIT is also oriented horizontally. Therefore, we have designed a system that employs a quadrupole bender that directs the ion beam vertically, which allows for the production of x-rays directly under the cell culture. The experimental station consists of a vacuum chamber that attaches to the end of the beam line. This chamber houses the quadrupole bender, a beryllium window for generating x-rays, and the interface between the beryllium window and the cell culture. X-rays must transmit through the bottom of a flask before they interact with the cells. Hence, we implement a procedure to replace the bottom of the flask with a thin layer of Mylar, allowing x-rays to penetrate through easily. We will use this system to study the effects of monochromatic x-rays on stem cells, cancer cells, and their associated proteins.

1. Introduction

At the Clemson University Electron Beam Ion Trap (CUEBIT) facility, multiple disciplines have combined their efforts to understand how radiation interacts with biological material with the primary goal of developing better imaging and therapy technologies. The development of a practical and effective experimental apparatus is the first step in the advancement towards this goal.

The application of ionizing radiation to control or kill malignant cells has been studied since 1896, and the mechanisms of action are well understood. Radiation therapy works through the creation of free radicals that induce cell death due to events such as double-stranded DNA breaks. Oxygen acts as a radiosensitizing agent, which increases the effectiveness of radiation therapy by boosting the amount of free radical production. However, cancer cells exist in a state of low oxygen concentration known as hypoxia. Therefore, healthy tissues in normal oxygen-rich states are more susceptible to free radical production. This is a problem because external beam radiation therapy – the most common form of radiotherapy – sends radiation through superficial, oxygen rich, healthy tissue to reach the target area. Currently, radiation technologies irradiate cells with a broad spectrum of x-ray energies in a multi-treatment process called fractionation. Given that many bonds between low-Z elements exist in cancerous and healthy tissue, statistically at least some of the x-rays will damage healthy tissue. Even though fractionation allows the healthy tissue to recover, it is not perfectly preventative of serious healthy tissue damage [1]. The development of new techniques and technologies are paramount for the advancement towards safe and more effective radiation therapies.

It has been shown that radiation can damage many different types of cells through a cascade of physical, chemical and biological processes [2,3]. Since stem cells are an integral part of the regeneration and healing processes in the body, a secondary goal of our research is to understand how ionizing radiation affects the differentiation and proliferation of stem cells.

The two hypotheses of this research are: (1) soft, monochromatic x-rays can damage cancer cells and (2) these x-rays can pause the cell cycle of stem cells. The production of soft, quasi-monochromatic x-rays is possible at the CUEBIT facility. Thus, the effects that quasi-monochromatic x-rays have on cancer and stem cells will be tested here using the apparatus discussed in this paper.

2. Development of the x-ray irradiation port:

The CUEBIT produces highly charged ions by passing a high-energy electron beam through a cloud of noble gases, stripping them of their atoms through impact ionization. These ions are then transported down the beam line where additional experiments can be conducted. In order to study the effects of radiation on cell cultures, an experimental apparatus that attaches to the end of the beam line and produces x-rays directly at the cell culture is imperative.

The beam line at the CUEBIT is oriented horizontally with respect to the floor. For the purposes of this research, a horizontal beam line makes it considerably more difficult to test biological samples. The cell culture dish must be sitting upright to support a cell culture with nutrients for an extended period of time. A quadrupole bender is used to bend the horizontal ion beam 90 degrees upwards. This allows the beam to focus on a beryllium window perpendicularly under the cell culture dish.

X-rays are produced via hollow atom decay during recombination at the beryllium window. An interface between the beryllium window and cell culture dish was made that holds the bottom of the dish approximately 1 mm above the window. Thus, the x-rays will not be significantly absorbed by particles in the atmosphere. A custom cell culture dish was developed in order to allow soft x-rays to transmit through with optimal transmission. A schematic of the x-ray irradiation port and its constituent parts can be seen in Figure 1.

Journal of Physics: Conference Series **583** (2015) 012048 doi:10.1088/1742-6596/583/1/012048

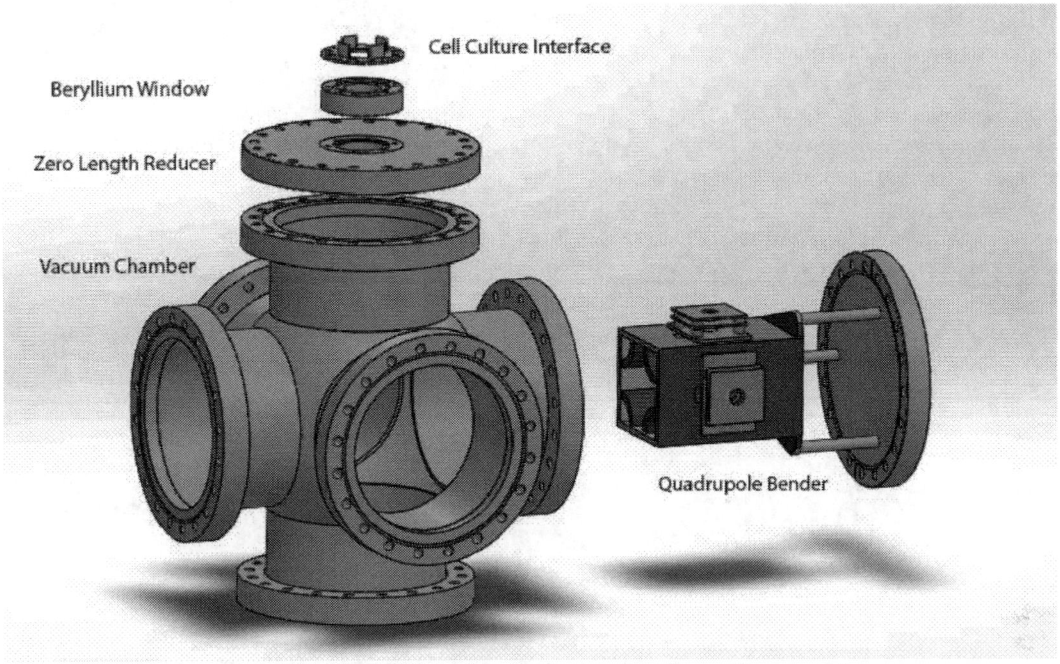

Figure 1. X-ray irradiation port and constituent parts.

2.1 Vacuum Chamber

The vacuum chamber is designed to achieve ultrahigh vacuum pressures on the order of $1.0*10^{-9}$ mbar. This allows the transport of ions through the chamber with minimum recombination. The vacuum chamber houses the quadrupole bender, which must be placed precisely with respect to the ion beam. The chamber has multiple attachment ports that can be used for spectroscopy or beam line imaging.

2.2 Quadrupole Bender

The quadrupole bender used was developed by DREEBIT [4]. It consists of four electrodes of equal potentials and alternating polarities. The incoming highly charged ions in the energy range of about 1*Q keV to 20*Q keV are repelled by the positive electrodes and attracted by the negative electrodes, thus the ions are bent around the first negative electrode by 90 degrees. Q represents the charge of the incoming ions. By changing the voltages applied to the electrodes, different ion beams can be used.

2.3 Beryllium Window

After the beam is bent upward, it is focused onto a 125 μm thick beryllium window, which acts as the barrier between ultrahigh vacuum and the lab's atmosphere. The window is welded to a conflat flange that is attached to the vacuum chamber. Thus, it is grounded and acts as an infinite electron source. As the ions approach the window, they pick up electrons off its surface forming hollow atoms [5]. These electrons will cascade into the lower orbitals late in the decay process, releasing quasi-monochromatic x-rays in all directions. X-rays transmit through the beryllium window and cell culture dish with minimal absorption.

2.4 Cell Culture Interface

The cell culture interface is designed to hold the cell culture close to the beryllium window (approximately 1mm). This allows the x-rays to reach the cell culture with little absorption with atmospheric particles. It attaches to the flange that supports the window and has a hollow cylindrical port that houses the cell culture dish.

2.5 Custom Cell Culture Dish

Several experiments were performed to determine if x-rays would transmit through cell culture dishes. Using a x-ray fluorescence setup, a range of x-ray lines were directed through standard and custom made cell culture dishes. X-ray lines originated from fluorescing a sample holder, covered with multivitamin powder, by a broadband pyroelectric x-ray source [6]. The results of these experiments reveal that soft x-rays are not capable of transmitting through a typical plastic or silicone based cell culture dish. Therefore, a custom six-micrometer Mylar-bottomed cell culture dish was designed and tested. This experiment confirmed that soft x-rays would transmit through with low absorption. Figure 2 shows the x-ray spectrums after transmission through (a) plastic and (b) Mylar bottom dishes.

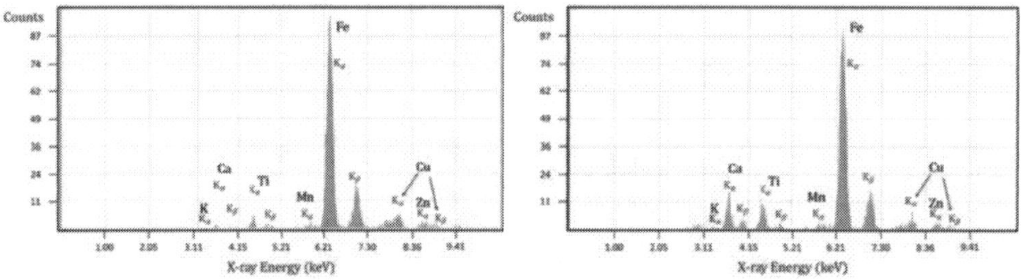

Figure 2. X-ray transmission through plastic (left) and Mylar (right).

3. Conclusion

The development of the x-ray irradiation port at the CUEBIT facility has created much interest in medical and biological physics. Many experiments are being developed in order to better understand how we can use this technology for biological applications. However, preliminary experiments must be performed. The irradiation port will be assembled with an x-ray spectrometer in place of the cell culture to determine the spectra produced with different ion charge states. Additionally, the ion beam collimators must be adjusted to focus the beam onto the beryllium window, which will maximize x-ray production.

Although the production of monochromatic x-rays is possible with simpler technologies, the CUEBIT provides several advantages. The main advantage is that the energy in the admitted x-rays is stored and transported in the potential energy of the highly charged ions. Additionally, charged particles are easy to manipulate, transport, and store. Lastly, at the time of x-ray generation, other radiation – electrons and photons – are generated, which allows for the determination of the exact position of x-ray production.

Computer simulations will also be used to aid with analysis of the results from cell irradiations. With Monte Carlo simulations, the tracks of particles through the biological material can be understood exactly through statistical methods. More developments need to be made in modeling biological material to a high degree of accuracy in current simulation software in order for the results to be meaningful.

Acknowledgements

This work was supported in part by the NASA EPSCoR Palmetto Academy Program, the Clemson University Creative Inquiry (CI) Program, the University Professional Internship & Co-Op (UPIC) Program, the Clemson University REU Program in Data-Intensive Computing, and the Department of Physics and Astronomy at Clemson University.

Journal of Physics: Conference Series **583** (2015) 012048 doi:10.1088/1742-6596/583/1/012048

References

[1] Hanahan D and Weinberg R A 2011 *Cell* **144** 646.
[2] Mestres M, Caballin M R, Barrios L, Ribas M and Barquinero J F 2008 *Radiat. Res.* **170** 93.
[3] Abolfath R M, Carlson D J, Chen Z J and Nath R 2013 *Phys. Med. Biol.* **85** 7143.
[4] Takacs E, Kimmel T D, Brandenburg K H, Wilson R K, Gall A C, Harriss J E and Sosolik C E 2014 Proceedings of the International Symposium on Electron Beam Ion Sources and Traps, EBIST'14, East Lansing, MI, 2014 (In print)
[5] Winter H and Aumayr F 1999 *J. Phys. B: At. Mol. Opt. Phys.* **32** R39.
[6] Brownridge J D and Raboy S 1999 *J. Appl. Phys.* **86** 1.

Journal of Physics: Conference Series **583** (2015) 012049 doi:10.1088/1742-6596/583/1/012049

Proton track structure code in biological matter

M A Quinto[1], J M Monti[2], M E Galassi[2], P F Weck[3], O A Fojón[2], J Hanssen[2], R D Rivarola[2] and C Champion[1]

[1]Université Bordeaux, CNRS/IN2P3, Centre d'Etudes Nucléaires de Bordeaux Gradignan, CENBG, Gradignan, France
[2]Instituto de Física Rosario, CONICET, Universidad Nacional de Rosario, Rosario, Argentina
[3]Sandia National Laboratories, Albuquerque, NM, USA

E-mail: quinto@cenbg.in2p3.fr

Abstract. Several numerical codes for proton and electron transport in water - a commonly used surrogate of the living matter - have been reported in the literature. In the current work, we report on a home-made *step-by-step* Monte Carlo code, called *TILDA-V*, based on a complete set of multiple-differential and total cross sections for describing all the inelastic processes occurring throughout the slowing-down of protons in water and DNA.

1. Introduction

Understanding the radiation-induced effects at the cellular level is of prime importance in predicting the fate of irradiated biological organisms. Thus, whether it is in radiobiology to identify the DNA critical lesions or in medicine to adapt the radio-therapeutic protocols, an accurate knowledge of the numerous interactions induced by charged particles in living matter is required. To do that, Monte-Carlo track-structure codes represent the most suitable and powerful tools, in particular for modeling the full slowing-down of the ionizing particles in biological matter. It is worth noting that those numerical codes are reliable only if the input database used for modeling the charged particle induced interactions is precise and complete enough. In this context, several numerical codes for proton and electron transport in water - a commonly used surrogate of the living matter - have been reported in the literature. Some of these existing codes use theoretical models for describing the physical processes but the majority of them remains nevertheless based on semi-empirical fits to experimental cross sections (for a review, see [1-2] and references therein). In this context, the current work aims at going beyond this artifice with the development of a Monte Carlo code, called *TILDA-V*, based on a complete set of quantum-mechanically calculated multiple differential and total cross sections for describing all the inelastic processes occurring throughout the slowing-down of protons in water and DNA. ***TILDA-V*** (a French acronym for **T**ransport d'**I**ons **L**ourds **D**ans l'**A**qua & le **V**ivo) refers to an extension of the *TILDA* Monte Carlo code previously developed by Champion for modeling heavy ion and secondary electron histories in liquid and gaseous water for impact energies ranging from 10 keV/amu to 100 MeV/amu [1].

Journal of Physics: Conference Series **583** (2015) 012049 doi:10.1088/1742-6596/583/1/012049

2. TILDA-V: a full-differential quantum-mechanically based Monte Carlo code

TILDA-V represents an event-by-event charged particle transport simulation, which includes a variety of theoretical models independently developed within the quantum mechanical framework for describing the single ionizing processes induced by proton and electron impact in water and DNA (see [3-5] and [6-7], respectively). Besides, due to the lack of theoretical support in its present form, *TILDA-V* also includes semi-empirical models for describing the excitation processes both induced by H^+ and H^0 impact as well as the electron loss and the electron capture processes induced by H^0 impact. Some of them are currently under study within the quantum mechanical approach (see Figure 1).

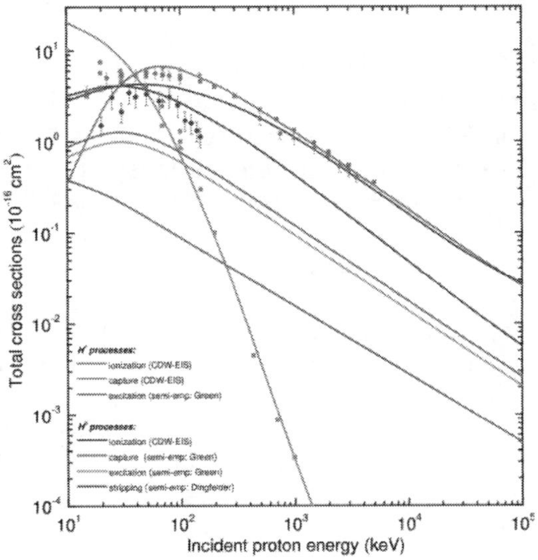

Figure 1. (Color online) Total cross sections for proton/hydrogen induced interactions in water. Theoretical and semi-empirical predictions (solid lines): ionization [3-4], capture [5] and excitation [8] induced by proton impact; ionization [9], capture [8], excitation [8] and stripping [10] induced by hydrogen impact. Experimental data: ionization and capture induced by proton impact (red squares, stars and circles [11-13], and magenta triangles, squares and circles [14-16], respectively) and hydrogen-induced ionization (blue circles [17]).

2.1. *Biological matter modeling: from water to DNA*

Whether it is to describe the biological matter by water or by including some DNA components, the biological medium has been modeled by means of molecular wave functions, all based on quantum mechanical calculations. In the first case, we followed the SCF-LCAO (*self-consistent field - linear combination of atomic orbitals*) approach reported by Moccia [18] who described the water molecule by means of single-center wave functions, all centered at a common origin (the oxygen atom). The latter refer to the equilibrium configurations calculated in the self-consistent field method and agree very well with the experimental geometrical and energetic properties of the water molecule. On the other hand, due the complexity of their multi-center nature, the DNA components were described by using the Gaussian 09 software based on the restricted Hartree-Fock method with geometry optimization a the RHF/3-21G level of theory. Their equilibrium geometries in the gas phase were obtained without symmetry constraints applied and the simulated molecular wave functions were described as linear combinations of atomic orbitals (for more details we refer the reader to our previous work [4]).

2.2. *Particle tracking in TILDA-V: proton, neutral hydrogen and electron transport*

From a general point of view, the charged particle transport simulation comprises series of sampling steps, which determine the distance between two successive interactions as well as the type of interaction occurring at the selected point. In all cases, the interaction type is randomly selected according to the relative magnitude of the individual *total* cross sections of all the processes (ionization, capture, excitation induced by proton and ionization, capture, excitation and electron loss induced by hydrogen atom) while the complete kinematics of the interaction is defined thanks to the *multiple differential* cross sections of the corresponding process.

Thus, if the selected process is ionization or stripping, singly and doubly differential cross sections are successively used to determine the kinetic energy as well as the ejection direction of the secondary electron. Then, the incident particle energy is reduced by the total energy (kinetic + potential) transferred during the collision. Besides, let us note that the charge state of the primary proton particle may also change according to the selected collision. Thus, the electron capture will decrease the initial proton charge (from +1 to 0) whereas the electron loss process (stripping) will increase the hydrogen charge (from 0 to +1). Note that the formation of H^- may also occur, in particular when the hydrogen atom captures an electron. However, due to its short lifetime, the negative ion H^- will rapidly eject its "additional" electron leading then to a final state characterized by a free electron and a hydrogen atom. The secondary electrons also liberated are then followed by means of a complete set of cross sections including triply, doubly and singly differential cross sections needed to characterize the complete kinematics of the electron-induced collisions. All these steps are repeated for all primary and secondary particles until their kinetic energy falls below a predetermined cut-off value. For the primary particles, namely, proton and hydrogen, the cut-off energy is here fixed at 10 keV while for the secondary electrons the latter is fixed at 7.4 eV *i.e.* the excitation threshold of the water molecule. Secondary electrons with kinetic energies lower than this threshold are not followed and assumed as locally absorbed by the medium.

3. Simulation of proton transport in water

TILDA-V is able to provide the coordinates of all the interactions as well as the type of collision together with the energy loss, including the deposited as well as the kinetic energy of the resultant particle(s) potentially created. An accurate dosimetry of any proton-beam irradiation may then be achieved from the macroscopic (cellular) to the microscopic (subcellular) scale. To illustrate this feature, we report in Figures 2 and 3 a comparison between our simulations and available data in terms of CSDA (Continuous Slowing-Down Approximation) range and electronic stopping power for proton energies ranging from 10 keV to 100 MeV. In both cases, a very good agreement is observed.

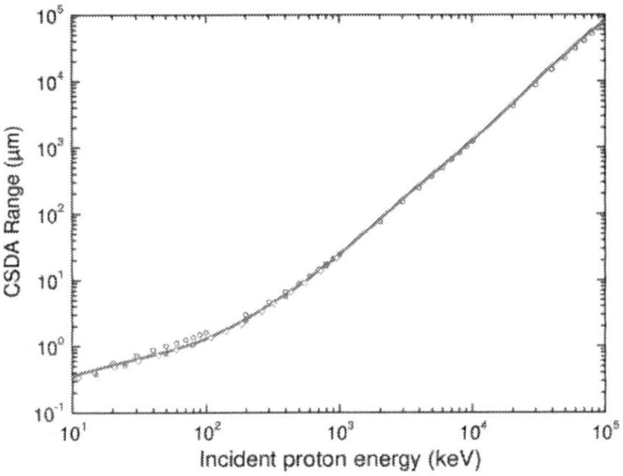

Figure 2. (Color online) Proton range in water: computed prediction (red line) compared with published data (diamonds [19], circles [20] and stars [21]).

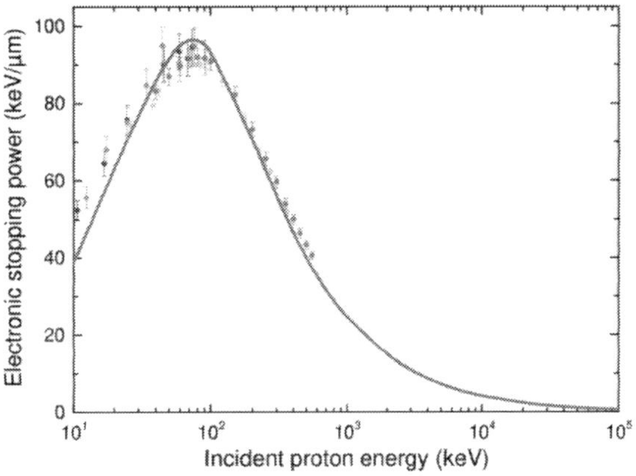

Figure 3. (Color online) Electronic stopping power for H^+/H^0 in water: computed prediction (red line) compared with experimental data (green, blue, orange and magenta circles [22-25]).

Journal of Physics: Conference Series **583** (2015) 012049 doi:10.1088/1742-6596/583/1/012049

4. Conclusions

In the current work, we have briefly reported on the development of a home-made Monte Carlo code called *TILDA-V* able to describe the full proton track structure in water and DNA within the energy range 10 keV-100 MeV. The current version is based on a complete set of quantum-mechanical and semi-empirical multiple differential and total cross sections for modeling all the electron- and proton/hydrogen-induced interactions in water and biological targets (DNA nucleobases and sugar-phosphate backbone). Preliminary comparisons with available data in terms of proton range and stopping power in water have clearly pointed out the performance of our code. Further results will be reported in a forthcoming study focused on proton tracking in a realistic biological medium.

Acknowledgments

This work has been developed as part of the activities planned in the Programme PICS 5921 (THEOS) as well as in the INSERM project PhysiCancer called ''MICRONAUTE''. Sandia National Laboratories is a multi-program laboratory managed and operated by Sandia Corporation, a wholly owned subsidiary of Lockheed Martin Corporation, for the U.S. Department of Energy's National Nuclear Security Administration under contract DE-AC04-94AL85000.

References

[1] Champion C, L'Hoir A, Politis M F, Fainstein P D, Rivarola R D and Chetioui A 2005 *Radiat. Res.* **163** 222
[2] Champion C 2003 *Phys. Med. Biol.* **48** 2147
[3] Champion C, Lekadir H, Galassi M E, Fojón O, Rivarola R D and Hanssen J 2010 *Phys. Med. Biol.* **55** 6053
[4] Galassi M E, Champion C, Weck P F, Rivarola R D, Fojón O and Hanssen J 2012 *Phys. Med. Biol.* **57** 2081
[5] Champion C, Weck P F, Lekadir H, Galassi M E, Fojón O, Abufager P, Rivarola R D and Hanssen J 2012 *Phys. Med. Biol.* **57** 3039
[6] Champion C 2010 *Phys. Med. Biol.* **55** 11
[7] Champion C 2013 *J. Chem. Phys.* **138** 184306
[8] Green A E S and McNeal R J 1971 *J. Geophys. Res.* **76** 133
[9] Champion C, Monti J M, Quinto M A, Weck P F, Rivarola R D, Fojón O A and Hanssen J, private communication
[10] Dingfelder M, Inokuti M and Paretzke H G 2000 *Radiat. Phys. Chem.* **59** 255
[11] Rudd M E, Goffe T V, DuBois R D and Toburen L H 1985 *Phys. Rev. A* **31** 492
[12] Bolorizadeh M A and Rudd M E 1986 *Phys. Rev. A* **33** 888
[13] Luna H, Barros A L F, Wyer J A, Scully S W J, Lecointre J, Garcia P M Y, Sigaud G M, Santos A C F, Senthil V, Shah M B, Latimer C J and Montenegro E C 2007 *Phys. Rev. A* **75** 042711
[14] Dagnat R, Blanc D and Molina D 1970 *J. Phys. B* **3** 1239
[15] Toburen L H 1998 *Radiat. Environ. Biophys.* **37** 221
[16] Gobet F, Farizon B, Farizon M and Gaillard M J 2001 *Phys. Rev. Lett.* **86** 3751
[17] Gobet F, Eden S, Coupier B, Tabet J, Farizon B, Farizon M, Gaillard M J, Ouaskit S, Carre M and Märk T D 2006 *Chem. Phys. Lett.* **421** 68
[18] Moccia R 1964 *J. Chem. Phys.* **40** 2186
[19] Janni J F 1982 Atomic Data Nuclear tables **27** 149
[20] Uehera S, Toburen L H and Nikjoo H 2010 *Int. J. Radiat. Biol.* **77** 139
[21] International Commission on Radiation Units and Measurements 1993 ICRU Report 49 ICRU, Washington DC
[22] Mitterschiffthaler C and Bauer P 1990 *Nucl. Instrum. Methods Phys. Res. B* **48** 58
[23] Phillips J A 1953 *Phys Rev* **90** 532
[24] Baek W Y, Grosswendt B and Willems G 2006 *Radiat. Prot. Dosimetry* **122** 32
[25] Reynolds H K , Dunbar D N F, Wenzel W A and Whaling W 1953 *Phys. Rev.* **92** 742

AUTHOR INDEX

AUTHOR INDEX